T0331858

PHYSICS AND ENGINEERING OF GRADED-INDEX MEDIA

Optical materials with varying refractive indices are called graded-index (GRIN) media and they are widely used within many industries, including telecommunications and medical imaging. Another recent application is space division multiplexing, an enormously improved technique for optical data transmission. The book synthesizes recent research developments in this growing field, presenting both the underlying physical principles behind optical propagation in GRIN media and the most important engineering applications.

The principles of wave optics are employed for solving Maxwell's equations inside a GRIN medium, ensuring that diffractive effects are fully included. The mathematical development builds gradually and a variety of exact and approximate techniques for solving practical problems are included, in addition to coverage of modern topics such as optical vortices, photonic spin-orbit coupling, photonic crystals, and metamaterials. This text will be useful for graduate students and researchers working in optics, photonics, and optical communications.

GOVIND P. AGRAWAL is James C. Wyant Professor of Optics at the University of Rochester. He is a Fellow of Optica, a Life Fellow of the IEEE, and a Distinguished Fellow of the Optical Society of India. He has been awarded the IEEE Photonics Society Quantum Electronics Award, the Riker University Award for Excellence in Graduate Teaching, the Esther Hoffman Beller Medal, the Max Born Award of Optica, and the Quantum Electronics Prize of the European Physical Society. He has also served as Editor-in-Chief for the *Optica Journal Advances in Optics and Photonics*. He is author, with Malin Premaratne, of *Theoretical Foundations of Nanoscale Quantum Devices* (Cambridge University Press, 2021).

PHYSICS AND ENGINEERING
OF GRADED-INDEX MEDIA

GOVIND P. AGRAWAL

University of Rochester

CAMBRIDGE
UNIVERSITY PRESS

Shaftesbury Road, Cambridge CB2 8EA, United Kingdom

One Liberty Plaza, 20th Floor, New York, NY 10006, USA

477 Williamstown Road, Port Melbourne, VIC 3207, Australia

314–321, 3rd Floor, Plot 3, Splendor Forum, Jasola District Centre, New Delhi – 110025, India

103 Penang Road, #05–06/07, Visioncrest Commercial, Singapore 238467

Cambridge University Press is part of Cambridge University Press & Assessment, a department of the University of Cambridge.

We share the University's mission to contribute to society through the pursuit of education, learning and research at the highest international levels of excellence.

www.cambridge.org
Information on this title: www.cambridge.org/9781009282079
DOI: 10.1017/9781009282086

First published 2023

A catalogue record for this publication is available from the British Library.

ISBN 978-1-009-28207-9 Hardback

Dedicated to my grandchildren

Contents

Preface

Maxwell proposed as early as 1854 the concept of a graded-index (GRIN) device, even before he developed his celebrated equations. Similar ideas were used by Wood in 1906 and by Luneberg in 1954 for the imaging applications. By 1970, GRIN glasses were fabricated whose refractive index varied radially in a cylindrical fashion. Such glasses were used either in a rod form as flat lenses or drawn into a fiber form, depending on the application. By the year 1980, GRIN fibers were used for the first generation of optical telecommunication systems. Plastic GRIN fibers were developed during the 1990s and are used routinely for transferring data between computers. More recently, silica GRIN fibers have been used for mode-division multiplexing in telecommunication systems and for observing novel nonlinear phenomena. The GRIN concept has also been extended to photonic crystals and metamaterials.

This book is intended to bring together a large amount of recent research material in a well-organized form such that a reader can develop physical understanding based on the fundamentals and apply it to emerging novel applications. Two earlier books in this area, published in 1978 and 2003, focused on the imaging applications and made use of mostly geometrical optics for describing light propagation inside a GRIN medium. The book employs the techniques of wave optics for solving Maxwell's equations inside GRIN media and ensures that the diffractive effects are fully included. The mathematical development builds up slowly and presents a variety of exact and approximate techniques for solving practical problems.

The primary role of this book is as a graduate-level text suitable for students and scientists working in the areas of optics, photonics, and imaging science. An attempt is made to include as much recent material as possible so that students are exposed to the recent advances in the areas covered by the book. The book can also serve as a reference text for researchers already engaged in or wishing to enter the fields where GRIN media are employed. The reference list at the end of each chapter is more elaborate than what is common for a typical textbook. The listing of recent research papers should be useful for researchers using this book as a reference. At the same time, students can benefit from it if they are assigned problems requiring reading of the original research papers. Although written primarily as a research monograph, portions of this book may be useful for graduate-level courses on the subjects of fiber optics, imaging science, and nonlinear optics.

Many persons have contributed to this book either directly or indirectly. It is impossible to mention all of them by name. I thank my Ph.D. and graduate students who took my courses related to electromagnetic theory, optical waveguides, and fiber-optic communication systems for their insightful questions and comments. I am grateful to my colleagues at the Institute of Optics of University of Rochester for numerous discussions and for providing a cordial and productive atmosphere. Last, but not least, I thank my family for understanding why I needed to spend considerable time on preparing the manuscript for this book instead of spending time with them. If the readers find the book useful, I would consider my time was well spent. For the color version of all figures in this book, please visit the resources on our website www.cambridge.org/9781009282079.

Acronyms

Each scientific field has its own jargon. Although an attempt was made to avoid extensive use of acronyms, many still appear throughout the book. Each acronym is defined the first time it appears in a chapter so that the reader does not have to search the entire text to find its meaning. As a further help, all acronyms are listed here in alphabetical order.

CCD	charge-coupled device
CVD	chemical vapor deposition
CW	continuous wave
DGD	differential group delay
EDFA	erbium-doped fiber amplifier
FFT	fast Fourier transform
FWHM	full width at half maximum
FWM	four-wave mixing
GRIN	graded-index
GVD	group-velocity dispersion
LCP	left-circularly polarized
LED	light-emitting diode
LP	linearly polarized
MMF	multi-mode fiber
NLS	nonlinear Schrödinger
OAM	orbital angular momentum
OSA	optical spectrum analyzer
PBG	photonic bandgap
PCF	photonic crystal fiber
PMD	polarization-mode dispersion
QPM	quasi-phase matching

RCP	right-circularly polarized
RIFS	Raman-induced frequency shift
RMS	root mean square
SAM	spin angular momentum
SBS	stimulated Brillouin scattering
SEM	scanning electron microscope
SHG	second harmonic generation
SLM	spatial light modulator
SMF	single-mode fiber
SNR	signal-to-noise ratio
SOP	state of polarization
SPM	self-phase modulation
SRS	stimulated Raman scattering
SSFM	split-step Fourier method
STML	spatiotemporal mode locking
TOD	third-order dispersion
WDM	wavelength-division multiplexing
WKB	Wentzel–Kramers–Brillouin
XPM	cross-phase modulation
ZDWL	zero-dispersion wavelength

1

Introduction

This chapter provides an introduction to the subject known as gradient-index optics. Section 1.1 provides a historical perspective on this subject before introducing the essential concepts needed in later chapters. Section 1.2 is devoted to various types of refractive-index profiles that are employed for making gradient-index devices, with particular emphasis to the parabolic index profile because of its practical importance. In Section 1.3, we discuss the relevant properties of such devices such as optical losses, chromatic dispersion, and intensity dependence of the refractive index occurring at high power levels. The focus of Section 1.4 is on the materials and the techniques used for fabricating gradient-index devices in the form of a rod or a thin fiber. Section 1.5 provides an overview of how the book is organized for presenting a wide body of research carried out during the last 50 years in the area of gradient-index optics.

1.1 Historical Perspective

Propagation of electromagnetic radiation in any medium is affected by its refractive index, denoted as $n(\mathbf{r}, \omega)$ because of its dependence on the frequency ω of the radiation and on the location $\mathbf{r}$ within the medium. In the case of a homogeneous material with uniform density, the dependence of $n(\mathbf{r}, \omega)$ on $\mathbf{r}$ can be ignored. However, the $\mathbf{r}$ dependence of the refractive index must be considered when density variations occur, either naturally (such as in air) or are introduced artificially by grading the refractive index of a material in some fashion. As an example, the phenomenon of mirage results from an index gradient formed in air on a hot day. Such index gradients change with time because of changes in air's temperature and pressure. When density variations in a medium are static (time independent), the medium is referred to as a graded-index (GRIN) medium. We only consider static density variations in this book.

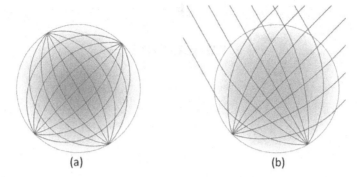

Figure 1.1 Schematic illustration of the GRIN lenses proposed by (a) Maxwell and (b) Luneberg. In both cases, refractive index is the largest at the center and decreases radially toward's the sphere's surface.

Historically, Maxwell proposed more than 160 years ago the concept of a GRIN device, known as the fisheye lens, even before he developed his celebrated equations [1]. The refractive index for such a lens exhibits spherical symmetry and depends on the magnitude of the vector **r**, but not on its direction. Similar ideas were used by Wood [2] in 1906, and by Luneberg in 1954, for imaging applications [3]. Figure 1.1 shows schematically how optical rays bend because of changes in the refractive index inside the GRIN lenses proposed by Maxwell and Luneberg. In both cases, optical rays follow curved paths to come to focus at a point on the sphere's surface.

With advances in glass technology, GRIN glasses could be fabricated by 1970 in which the refractive index varied in a cylindrically symmetric fashion in the plane normal to the direction of propagation. Such GRIN glasses were used either in a rod form [4] or drawn into a fiber form [5], depending on the application. At the same time, planar waveguides were developed in which the refractive index $n(x)$ varied only in one direction normal to the direction of propagation [6–8]. Two books published around 1977 provided a comprehensive account of such GRIN devices [9, 10].

The GRIN fibers were developed during the 1970s and their properties studied extensively in view of their potential applications in the emerging area of optical communications [11]. Indeed, by the year 1980, GRIN fibers were used for the first generation of such systems [12]. Even though telecommunication systems began using single-mode, step-index fibers by 1985, the development of new GRIN materials and devices remained an active area of research. For example, plastic-based GRIN fibers are used routinely for data-transfer applications [13]. One can get a good idea of the intense activity during the 1980s and 1990s by consulting several special issues of the *Applied Optics* journal [14]. Two books also describe the progress realized during this period [15, 16].

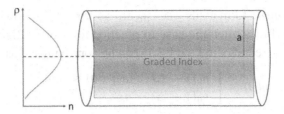

Figure 1.2 Schematic illustration of the refractive-index gradient along a GRIN device.

Starting around 2010, the advent of space-division multiplexing for modern telecommunication systems led to a renewed interest in the use of glass-based GRIN fibers [12, 17, 18]. Since then, the investigation of nonlinear optical phenomena in GRIN fibers has led to major advances. Among these are the topics such as spatiotemporal modulation instability, GRIN solitons, and spatial beam cleanup [19–21]. This book is intended to cover recent research advances and to provide, at the same time, comprehensive coverage of electromagnetic wave propagation inside a GRIN medium.

1.2 Refractive-Index Profiles

The focus of this book is on a GRIN medium whose refractive index varies in a plane normal to the direction of propagation (commonly taken to be the z axis) in a cylindrically symmetric fashion. Figure 1.2 shows schematically how the refractive index varies in such a GRIN rod around its central axis, chosen to be the z axis of the coordinate system. For practical reasons, the refractive index is the largest at the central axis and decreases gradually in all radial directions moving away from the center. In its most general form, the refractive index varies with the radial distance $\rho = \sqrt{x^2 + y^2}$ as [22–24]

$$n^2(\rho) = \begin{cases} n_0^2[1 - 2\Delta f(\rho/a)] & (0 \le \rho \le a) \\ n_0^2(1 - 2\Delta) = n_c^2 & (\rho \ge a), \end{cases} \tag{1.2.1}$$

where n_0 is the maximum value of the refractive index at the center and n_c is its minimum value at $\rho = a$, which is the radius of the cylindrical core enclosing the GRIN region. The function $f(x)$ governs shape of the index profile such that its value is 1 for $x = 1$.

The parameter Δ can be deduced from Eq. (1.2.1) and has the form

$$\Delta = \frac{n_0^2 - n_c^2}{2n_0^2} \approx \frac{n_0 - n_c}{n_0}. \tag{1.2.2}$$

where the approximate form holds when n_c differs from n_0 by at most a few percent so that $\Delta \ll 1$. This is often the case in practice for most GRIN devices. Using $\Delta \ll 1$, the refractive index in Eq. (1.2.1) can be approximated as

$$n(\rho) \approx \begin{cases} n_0[1 - \Delta f(\rho/a)] & (0 \le \rho \le a) \\ n_0(1 - \Delta) = n_c & (\rho > a). \end{cases} \quad (1.2.3)$$

This equation shows that $n(\rho)$ decreases as one moves away from the central axis up to a distance $\rho = a$ in a fashion dictated by the function $f(\rho)$ and takes minimum value n_c in the cladding region $\rho > a$. The GRIN region of radius a constitutes the core of such a GRIN device. The parameter Δ, given in Eq. (1.2.2), represents the fractional decrease in the refractive index across the core and its value is a design parameter for GRIN devices.

The function $f(x)$ governs the shape of the index profile for a GRIN device. This shape depends on the application for which the device is fabricated for and can vary over a wide range. In the case of planar waveguides, even an error function has been used for the shape [8]. In the case of GRIN rods and fibers, it is common to employ a power-law index profile with $f(x) = x^p$, where the exponent p governs the shape of the GRIN region. In this case, the refractive index in the core region varies as [22–24]

$$n^2(\rho) = n_0^2\left[1 - 2\Delta\left(\frac{\rho}{a}\right)^p\right] \qquad (\rho \le a). \quad (1.2.4)$$

Figure 1.3 shows how the refractive-index profile changes when p is varied in the range 1–10 using $n_0 = 1.5$ and $\Delta = 0.06$. The case $p = 2$ corresponds to a parabolic shape of the index profile. Note that the shape becomes closer to a step function for a large value of p such that n remains close to n_0 until one approaches the region near $\rho = a$, where it decreases rapidly and takes the value n_c. A step-index profile, occurring in the limit $p \to \infty$, is used routinely for making step-index fibers. Its use confines light within the core of a step-index fiber through the phenomenon of total internal reflection.

A parabolic index profile, realized for the choice $p = 2$ in Eq. (1.2.4), plays an important role in the literature on GRIN media, and many GRIN devices are designed with such a profile. In this case, we can write Eq. (1.2.4) in the simple form

$$n^2(\rho) = n_0^2(1 - b^2\rho^2), \qquad b = \sqrt{2\Delta}/a. \quad (1.2.5)$$

The parameter b is a measure of the index gradient such that its larger values indicate a faster reduction in the refractive index as ρ is increased. This parameter will play a prominent role in later chapters. As seen from the definition of b in Eq. (1.2.5), its value depends both on the core's radius a and the relative index difference Δ.

Depending on the application, numerical values of the three parameters, a, n_0, and Δ, associated with a GRIN device can vary over a wide range. We classify

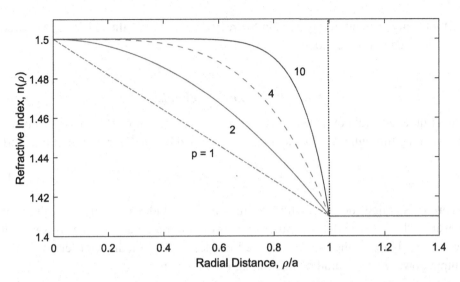

Figure 1.3 Refractive index $n(\rho)$ plotted as a function of the ratio ρ/a for several values of the parameter p. The vertical dotted line at $\rho = a$ separates the core and cladding regions of such GRIN devices.

GRIN devices into two broad groups based on the core's radius a. The value of a exceeds 1 mm for GRIN rods used to make lenses and similar optical elements. In contrast, a is restricted to much smaller values in the range of 10–30 μm for GRIN fibers used for telecommunication applications, among other things. The parameter Δ also varies for these two groups of GRIN devices. Its typical value is around 0.01 for GRIN fibers but can exceed 0.05 for GRIN rods. The value of n_0 depends on the material used for making a GRIN device. In the case of silica glass, n_0 is about 1.45. For plastics, n_0 is closer to 1.5.

We can estimate the value of the parameter b for GRIN rods and fibers from Eq. (1.2.5) by using the values of a and Δ. For GRIN rods, typical values of b are near 0.3 mm^{-1}. In contrast, b is around 5 mm^{-1} for GRIN fibers. Another relevant parameter of a GRIN device is its numerical aperture (NA). As indicated in Section 3.1, it depends on the values of n_0 and Δ as $\mathrm{NA} = n_0\sqrt{2\Delta}$. The NA of a GRIN rod is close to 0.5 when $n_0 = 1.5$ and $\Delta = 0.05$. It is lower for GRIN fibers and has values of 0.2 or less.

1.3 Relevant Optical Processes

All materials affect the electromagnetic radiation propagating through them. The most relevant effects are (i) loss of power with distance owing to absorption and scattering, (ii) chromatic dispersion or a frequency-dependent refractive index, and

(iii) intensity-dependent changes in the refractive index of the material. All three are discussed in this section.

1.3.1 Power-Loss Mechanisms

Under quite general conditions, changes in the average power P of an optical beam propagating through a GRIN medium are governed by the Beer–Lambert law [25]:

$$\frac{dP}{dz} = -\alpha P, \qquad (1.3.1)$$

where α is called the attenuation coefficient. It includes not only absorption of power by the material but also other sources of power attenuation such as Rayleigh scattering. If P_{in} is the power launched inside a GRIN medium of length L, the output power P_{out} is found by integrating Eq. (1.3.1) to be

$$P_{out} = P_{in} \exp(-\alpha L). \qquad (1.3.2)$$

It is customary to express α in the decibel units using the relation [12]

$$\alpha \, (\text{dB/m}) = -\frac{10}{L} \log_{10}\left(\frac{P_{out}}{P_{in}}\right) \approx 4.343\alpha. \qquad (1.3.3)$$

Numerical values of the attenuation coefficient α depend both on the material used to make a GRIN device and the wavelength of light launched into it. Figure 1.4 compares the wavelength dependence of measured loss in silica-glass fibers to losses in two types of plastic fibers [26]. As seen there, plastic fibers exhibit much larger losses (> 10 dB/km) compared to those of silica fibers, whose losses can be reduced to below 0.2 dB/km in the wavelength region near 1550 nm. Absorption by the plastic material is the source of high losses in plastic GRIN fibers.

In the case of optical glasses, absorption by the material of the glass is relatively small in the visible and near-infrared regions. However, even small amounts of impurities can increase this loss considerably. In the case of silica fibers, losses can be reduced to below 1 dB/km by eliminating all impurities. For such fibers, the dominant contribution to α arises from Rayleigh scattering, which is a fundamental loss mechanism arising from local microscopic fluctuations in the density of glass used to make the fiber. Glass molecules move randomly in the molten state and freeze in place during cooling. Resulting density fluctuations produce random fluctuations in the refractive index on a scale smaller than the optical wavelength λ. These fluctuations are the source of Rayleigh scattering, whose cross section varies as λ^{-4} [25]. As seen in Figure 1.4, silica's loss resulting from Rayleigh scattering exceeds 1 dB/km in the visible region but is reduced to below 0.2 dB/km in the infrared region near 1550 nm used for modern optical communication systems.

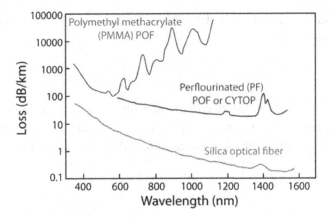

Figure 1.4 Wavelength dependence of the loss in silica fibers and losses in two types of plastic fibers. Note the logarithmic scale in units of dB/km. (After Ref. [26]; ©2014 IOP.)

1.3.2 Chromatic Dispersion

In the case of a GRIN medium, the refractive index $n(\rho, \omega)$ depends both on the spatial location ρ and the frequency ω. Chromatic dispersion has its origin in the frequency dependence of the refractive index. As we shall see in Section 2.1, it is the frequency dependence of the propagation constant, defined as $\beta(\omega) = n(\omega)(\omega/c)$, where c is the speed of light in vacuum, that governs the dispersive properties of any material. When the spectrum of incident light is narrower compared to its central frequency ω_0, we can expand $\beta(\omega)$ in a Taylor series as

$$\beta(\omega) = \beta_0 + \beta_1(\Delta\omega) + \tfrac{1}{2}\beta_2(\Delta\omega)^2 + \ldots, \tag{1.3.4}$$

where $\Delta\omega = \omega - \omega_0$ and $\beta_m = (d^m\beta/d\omega^m)_{\omega=\omega_0}$.

In Eq. (1.3.4), β_1 is related inversely to the group velocity v_g and is responsible for the group delay, $\tau_g = \beta_1 L$, over a length L. The parameter β_2, representing the second derivative of β, is called the group-velocity dispersion (GVD) parameter. This parameter will play an important role in chapters dealing with the propagation of optical pulses inside a GRIN medium. For pulses shorter than 1 ps, it is sometimes necessary to consider the cubic term containing β_3 in the Taylor series in Eq. (1.3.4). This parameter is referred to as the third-order dispersion parameter.

The dispersion parameter β_1 can be calculated for any GRIN medium by taking the frequency derivative of β as

$$\beta_1 = \frac{d\beta}{d\omega} = \frac{n_g}{c}, \qquad n_g = n + \omega\frac{dn}{d\omega}, \tag{1.3.5}$$

where n_g is called the group index. It can be employed to calculate the GVD parameter in the form

$$\beta_2 = \frac{d\beta_1}{d\omega} = \frac{1}{c}\frac{dn_g}{d\omega}. \tag{1.3.6}$$

The sign of β_2 depends on the sign of the derivative $dn_g/d\omega$ and can be positive or negative in different spectral regions for glasses used to make a GRIN device. A related dispersion parameter D is also used for GVD; it is defined as

$$D = \frac{d\beta_1}{d\lambda} = -\frac{\lambda}{c}\frac{d^2n}{d\lambda^2}. \tag{1.3.7}$$

It is easy to show that D is related to β_2 by the relation $D = -(2\pi c/\lambda^2)\beta_2$ and its sign is opposite to that of β_2.

On a fundamental level, the origin of dispersion is related to the atomic resonance frequencies at which a material absorbs electromagnetic radiation. Far from such resonances, the refractive index is well approximated by the Sellmeier equation [27],

$$n^2(\omega) = 1 + \sum_{j=1}^{M} \frac{B_j\omega_j^2}{\omega_j^2 - \omega^2}, \tag{1.3.8}$$

where ω_j is the resonance frequency and B_j is the oscillator strength. The parameters B_j and ω_j are obtained empirically by fitting the measured dispersion curve to Eq. (1.3.8) with $M = 3$. For pure silica glass, these parameters are found to be [27] $B_1 = 0.6961663$, $B_2 = 0.4079426$, $B_3 = 0.8974794$, $\lambda_1 = 0.0684043$ μm, $\lambda_2 = 0.1162414$ μm, and $\lambda_3 = 9.896161$ μm, where $\lambda_j = 2\pi c/\omega_j$ for $j = 1$ to 3.

We can use Eq. (1.3.8) to calculate the frequency dependence of n and n_g for the silica glass without an index gradient. Figure 1.5 shows this dependence in the wavelength range 0.6–1.6 μm. The group-delay parameter is obtained using $\beta_1 = n_g/c$. Even though n decreases monotonically with λ in the entire wavelength range, β_1 exhibits a shallow minimum for silica glass at the specific wavelength, $\lambda = 1.276$ μm, marked by the dotted vertical line in Figure 1.5. This wavelength is called the zero-dispersion wavelength (denoted by λ_{ZD}) because the GVD parameter β_2 vanishes at this wavelength.

Figure 1.6 shows how the dispersion parameters β_2 and D vary with wavelength λ for silica glass (no index gradient) using Eqs. (1.3.6) and (1.3.7). As expected, both β_2 and D vanish at λ_{ZD} near 1.27 μm and change sign for longer wavelengths. It is common to refer to negative values of β_2 as the GVD being anomalous. The curve marked d_{12} shows the differential group delay, $d_{12} = \beta_1(\lambda_1) - \beta_1(\lambda_2)$, using a reference wavelength $\lambda_2 = 0.8$ μm. It shows the relative delay of a pulse as its central wavelength λ_1 is varied.

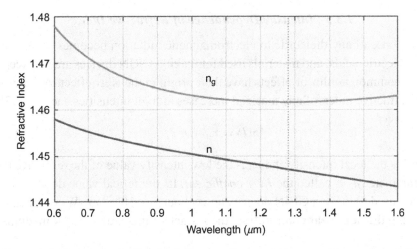

Figure 1.5 Variation of refractive index n and group index n_g with wavelength for fused silica. The dotted line indicates the zero-dispersion wavelength.

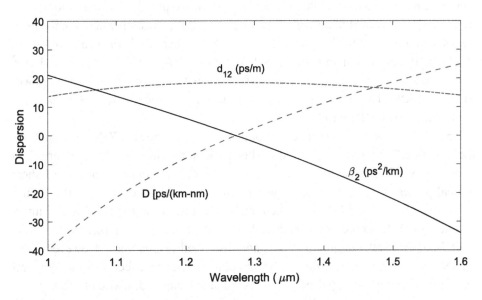

Figure 1.6 Wavelength dependence of β_2, D, and d_{12} for silica glass.

The situation changes considerably when silica glass is used to make a GRIN device. It will be seen in Chapter 2, that the propagation constant β, and hence all dispersion parameters, become mode-dependent for any GRIN medium. In particular, one must consider the intermodal group delay resulting from different values of β_1 for different modes. This topic is covered in Sections 2.4 and 4.1 in the context of optical pulses.

1.3.3 Intensity Dependence of Refractive Index

The response of any dielectric to electromagnetic radiation becomes nonlinear for intense electric fields, and materials used for making GRIN devices are no exception. Several common nonlinear effects have their origin in the Kerr effect. According to it, the refractive index of any material increases at high intensities such that [28]

$$n(I) = n_0 + n_2 I, \tag{1.3.9}$$

where I is the local intensity and n_0 is the low-intensity value of the refractive index. The parameter n_2 is called the *Kerr coefficient*. Its numerical value depends on the material used to make a GRIN device and is about 3×10^{-20} m^2/W for silica glass.

Adding the nonlinear contribution, the refractive index of a GRIN medium has the form

$$n(\mathbf{r}, \omega, I) = n_0(\omega)[1 - \Delta f(\mathbf{r})] + n_2 I(\mathbf{r}), \tag{1.3.10}$$

where the dependence on all three variables is shown explicitly. The maximum value of the nonlinear contribution, $\delta n = n_2 I(\mathbf{r})$, occurs at the location where the intensity peaks. Denoting this peak value with $I_0 = P_0/A_e$, where P_0 is the peak power and A_e is the effective beam area, $\delta n = n_2 P_0/A_e$. As an example, if we use $A_e = 1$ cm^2 and $n_2 = 3 \times 10^{-20}$ m^2/W, $\delta n = 3 \times 10^{-13}$ even at a relatively high peak power of $P_0 = 1$ kW. This value is too small to have any impact when a CW beam is launched inside a GRIN rod.

There are two ways to enhance the nonlinear effects inside a GRIN medium. First, the beam's effective area A_e is reduced considerably when GRIN fibers are used with a core radius close to 10 μm. Second, if a beam containing a train of short optical pulses is used, the peak power P_0 of the pulse can exceed 1 MW. Using $A_e = 10^{-10}$ m^2 for a GRIN fiber, the nonlinear contribution to the refractive index (about 3×10^{-4}) is much smaller than Δ, indicating that it is not likely to affect the GRIN-induced self-imaging phenomenon discussed in Chapter 3. However, if the GRIN fiber is long enough, the nonlinear contribution can affect both the temporal and spectral features of a pulsed beam. As noted in Chapter 5, it also produces novel spatiotemporal features that have been studied extensively in recent years [19–21].

The intensity dependence of the refractive index leads to several nonlinear effects; the two most common ones are known as *self-focusing* and *self-phase modulation* (SPM). The phenomenon of self-focusing is relevant for GRIN media because it can compress an optical beam and compete with the GRIN-induced focusing. Moreover, it leads to a beam's collapse above a certain critical power level [28].

SPM is relevant only for pulsed optical beams. It produces a self-induced phase shift that is different for different parts of the same pulse because of its intensity dependence. Its magnitude can be obtained from Eq. (1.3.10). After a distance L, the nonlinear phase shift is given by

$$\phi_{NL}(\mathbf{r}, t) = (n_2 k_0 L)I(\mathbf{r}, t), \qquad k_0 = 2\pi/\lambda. \tag{1.3.11}$$

The phase shift depends both on the spatial location $\mathbf{r}$ and t because the intensity of a pulsed beam varies with time. This feature is called SPM because a pulse modifies its own phase. The spatial dependence produces a curved wavefront and leads to self-focusing when $n_2 > 0$ and self-defocusing when $n_2 < 0$. The time dependence produces spectral broadening of a pulsed beam [20]. It can also lead to the formation of optical solitons in the anomalous-GVD region of a GRIN fiber [29].

A related nonlinear phenomenon, known as cross-phase modulation (XPM), refers to the nonlinear phase shift of an optical field induced by another field having a different wavelength or state of polarization. Its origin can be understood by noting that the total electric field, in the case of two fields of different wavelengths but the same state of polarization, is given by

$$\mathbf{E}(\mathbf{r}, t) = \hat{\mathbf{x}} \text{Re}[E_1 \exp(-i\omega_1 t) + E_2 \exp(-i\omega_2 t)]. \tag{1.3.12}$$

In this situation, the nonlinear phase shift at the frequency ω_1 is found to be [20]

$$\phi_{NL} = n_2(\omega_1/c)L(|E_1|^2 + 2|E_2|^2). \tag{1.3.13}$$

The first term is the SPM term seen in Eq. (1.3.11). The second term is due to XPM because it represents a phase shift induced by another beam at a different wavelength. The Kerr effect also leads to four-wave mixing when a phase-matching condition is satisfied [20].

1.4 GRIN Materials and Fabrication

Both glasses and plastics are used for making GRIN devices, which are divided into two broad categories based on their transverse dimensions – GRIN rods and GRIN fibers. The radius a of the cylindrical region over which $n(\rho)$ varies exceeds 1 mm for GRIN rods. In contrast, a is restricted to values in the range of 10–30 μm for GRIN fibers. Silica glass is commonly used for making GRIN fibers, but GRIN rods can be made with other materials as well.

In spit of the amorphous nature of glass, its refractive index can be treated as being constant over lengths longer than a wavelength. Thus, glass in its natural state is a homogeneous medium. The important question is how one can modify the refractive index of a glass such that it varies in a prescribed manner with the distance ρ from the central axis. From a fundamental perspective, the refractive index of a material depends on its density and can only be changed by making its density nonuniform. As early as 1980, a review of the GRIN materials listed six

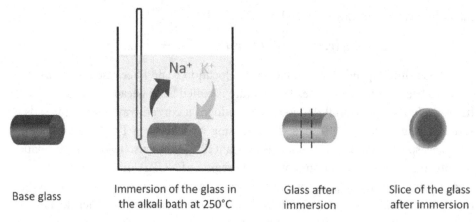

Figure 1.7 Schematic of the ion-exchange process used to make GRIN rods. (After Ref. [33]; ©2021 CC BY.)

different techniques developed for this purpose [30]. In one approach, a boron-rich glass rod is bombarded with neutrons. As neutrons enter the glass, its refractive index decreases because neutrons change the boron's concentration. The decrease is the largest in the outer region of the glass rod and becomes negligible near the rod's center, resulting in a gradient in the glass's refractive index.

Another common method is known as the ion-exchange technique [31–33]. In this approach, shown schematically in Figure 1.7, a base glass with heavier ions (such as K^+) is immersed into a molten salt bath containing lighter alkali ions. At a suitable temperature, lighter ions diffuse into the glass and replace the heavy ions present within the glass [4]. This ionic exchange creates a gradient in the refractive index of the glass because fewer and fewer ions are replaced as one moves from the outer surface toward the central region of the glass. Diffusion of ions into the glass is affected by several factors, such as the strength of bonds that hold the glass ions in their lattice sites and the relative mobilities of various ions inside the glass network. As the mobility of ions depends on temperature, a higher temperature helps in practice. An external electric field is also used to speed up the ion-exchange process and to control the shape of the resulting refractive-index profile [34].

An ion-stuffing technique has also been used for making GRIN devices. In this case, a special glass is chosen whose structure (or phase) changes when heated suitably. The changed part is soluble in a specific acid. When the glass is immersed in this acid, the diffusion of acid into the glass removes less and less material from the inner regions of the glass, resulting in the formation of an index gradient. The main issue with this technique is that very few glasses exist that exhibit phase separation, and the phase separation is not always uniform.

A different technique, known as chemical vapor deposition (CVD), is employed

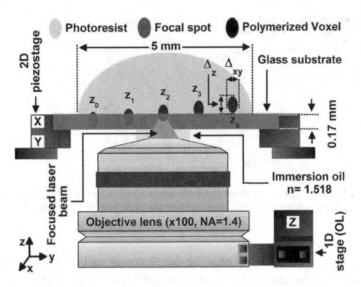

Figure 1.8 Schematic of the direct laser writing scheme used for making GRIN devices. (After Ref. [38]; ©2020 CC BY.)

for making GRIN fibers [35]. The fabrication of such fibers involves two stages. First, a cylindrical rod of about 2 cm diameter is produced with the desired refractive-index profile by depositing silica, layer by layer, inside of a silica tube. An index-increasing dopant material (such as germania) is added to each layer such that the density of dopants varies from one layer to another and is larger in the central region of the tube [36]. The resulting structure is called a preform, which is drawn into the form of a GRIN fiber using a draw tower. Both fabrication stages involve sophisticated technology to ensure the uniformity of the index profile within the fiber's core. A polymerization technique can also be used when a plastic material is used for making GRIN fibers [13]. In this approach, the monomer of an organic material is differentially changed into a polymer by irradiation of ultraviolet radiation.

A technique called *direct laser writing* has been developed in recent years [37–39]. As shown in Figure 1.8, it uses point-by-point exposure of a plastic material to intense laser pulses to change its local refractive index through polymerization. The exposure dose is varied during scanning by varying the laser's power to produce an index gradient. This technique employs a lithographic approach in the sense that an arbitrary refractive-index profile can be created by using a photoresist that is exposed to laser pulses, which modify the refractive index through a multiphoton polymerization process [39].

A nanoscale engineering technique was used in 2018 for making GRIN fibers [40]. In this approach, the core of the preform is made by combining two types of glass rods, made of pure silica and Ge-doped silica, in such a fashion that the density

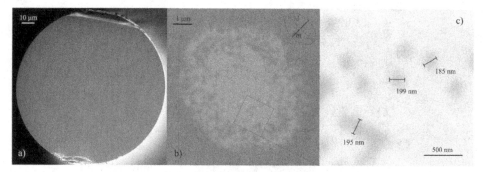

Figure 1.9 SEM images at three magnification levels of a GRIN finer made with the nanoscale engineering technique. (After Ref. [40]; ©2018 CC BY.)

of Ge-doped rods is the largest near the core's center and decreases gradually as the radial distance increases. As a result, the refractive index is effectively graded inside the fiber's core when the preform is drawn into a thin fiber. This technique was used to produce a GRIN fiber with a parabolic index profile. A preform was first made by stacking more than 2100 rods with a core diameter of 0.45 mm inside a tube of silica glass (see Figure 10.8 of Chapter 10). The fiber was drawn using this preform until the diameter of each rod was reduced to nearly 190 nm. Figure 1.9 shows images of this fiber's cross section at three magnification levels that were obtained with a scanning electron microscope (SEM). Part (a) shows the whole fiber, part (b) shows its core, and part (c) shows individual nanosize rods within the core. Such a nanostructure-based technique is not limited to the circular geometry and is flexible enough to produce GRIN fibers with any index profile.

1.5   Overview of Book's Contents

This book provides a comprehensive coverage of the physics of optical phenomena inside a GRIN medium and uses it for understanding the operation of engineered devices such as GRIN lenses, sensors, endoscopes, and tweezers. The first chapter begins with a historical perspective and introduces different refractive-index profiles that can be employed for making a GRIN device. It also focuses on the relevant properties of a GRIN medium, such as power loss, chromatic dispersion, and intensity dependence of the refractive index at high power levels, which may affect the operation of practical devices. The techniques used to fabricate GRIN devices are also discussed in this chapter.

Chapter 2 starts with Maxwell's equations and uses them to obtain the wave equation that must be solved to understand the propagation of optical waves inside a GRIN medium. This equation is solved in Section 2.2 to obtain the modes of a

GRIN medium for a parabolic index profile. Non-parabolic profiles are considered in Section 2.3. When an optical beam is launched into a GRIN medium, it excites different modes with different efficiencies. This issue is discussed in Section 2.4, where we also consider multimode interference and intermodal dispersion. Section 2.5 describes several non-modal techniques that can be used for studying wave propagation in GRIN media.

The focus of Chapter 3 is on the focusing and self-imaging of optical beams inside a GRIN medium. Section 3.1 provides a geometrical-optics perspective and shows why optical rays follow a curved path inside a GRIN medium. The modal expansion is used in Section 3.2 to obtain a propagation kernel that can be used for beams of any spatial shape. It is used to discuss the self-imaging phenomenon. It is also used to study the propagation of Gaussian beams launched on-axis, or offset from the central axis of a GRIN medium. Section 3.3 is devoted to studying how a GRIN rod can be used as a flat lens to focus or to collimate a beam. Imaging characteristics of such a lens are also studied in this section. Several important applications of GRIN devices are discussed in Section 3.4.

Chapter 4 is devoted to the study of the dispersive effects that affect a pulsed beam propagating inside a GRIN medium. In Section 4.1, an equation governing the evolution of optical pulses is obtained. The dispersion parameters appearing in this equation change, depending on which mode is being considered. Section 4.2 focuses on the distortion of optical pulses resulting from differential group delay and group-velocity dispersion. Section 4.3 deals with the effects of linear coupling among the modes occurring because of random variations in the core's shape and size along the fiber's length. A non-modal approach is developed in Section 4.4 for short optical pulses propagating inside a GRIN medium. Section 4.5 describes important applications such as optical communications and biomedical imaging.

The focus of Chapter 5 is on the intensity dependence of the refractive index that leads to a variety of interesting nonlinear effects in the case of GRIN fibers. We consider in Section 5.1 the self-focusing of a CW beam inside a GRIN medium. The pulsed beams are considered in Section 5.2, where we discuss the phenomena of self- and cross-phase modulations. Section 5.3 is devoted to the modulation instability and the formation of multimode solitons. Intermodal nonlinear effects are described in Section 5.4 with emphasis on four-wave mixing and stimulated Raman scattering. Applications such as supercontinuum generation and spatial beam cleanup are discussed in Section 5.5.

Chapter 6 is devoted to the effects related to the presence of loss or gain inside a GRIN medium. Loss is considered in Section 6.1. Except for a reduced power level, its effects are found to be relatively minor. We discuss in Section 6.2 the mechanisms used for providing optical gain inside an active GRIN medium. Section 6.3 is devoted to Raman amplifiers and lasers, built with GRIN fibers and pumped suitably to provide the optical gain. Parametric amplifiers are discussed in Section 6.4, together with the phase matching required for four-wave mixing to occur inside them. The focus of Section 6.5 is on amplifiers and lasers made by doping a GRIN fiber with rare-earth ions. Section 6.6 includes the nonlinear effects and describes how spatial solitons and similaritons can form under suitable conditions inside an active GRIN medium.

Chapter 7 focuses on a nonuniform GRIN fiber, tapered to induce variations in the refractive index along its length, in addition to the transverse radial variations. Section 7.1 describes the ray-optics and wave-optics techniques that can be used for studying propagation of optical beams inside such a medium. Section 7.2 discusses the impact of tapering on the periodic self-imaging for several different tapering profiles. The analogy between a GRIN medium and a harmonic oscillator is exploited in Section 7.3, where we discuss several quantum-mechanical techniques. Section 7.4 is devoted to the case of periodic axial variations, induced by changing the core's radius in a periodic fashion.

Chapter 8 is devoted to the formation of optical vortices inside a GRIN medium. After discussing important polarization concepts in Section 8.1, we use in Section 8.2 specific combinations of the modes that act as vortices with different states of polarizations. Cylindrical vector beams with radial or azimuthal polarization are also discussed. In Section 8.3, we present the techniques used for generating different types of vortex beams. Section 8.4 shows that vortex beams also exhibit the self-imaging property during their propagation inside a GRIN medium. The impact of random mode-coupling is also discussed in this section. Vortex-based applications of GRIN fibers are covered in Section 8.5.

The focus of Chapter 9 is on photonic spin-orbit coupling occurring inside a GRIN medium. Section 9.1 describes two physical mechanisms that can produce changes in the states of polarization of an optical beam. The exact wave equation with the term involving the index gradient is solved in Section 9.2 to introduce the path-dependent geometrical phase to discuss the photonic analog of the spin-Hall effect and conversion between the spin and orbital angular momenta. Section 9.3 considers how the scalar LP_{lm} modes change when the index-gradient term is taken into account. A quantum approach is employed in Section 9.4 to find the exact vector modes of a GRIN medium.

Chapter 10 is devoted to two research areas that have attracted considerable attention in recent years – photonic crystals and metamaterials. Both are artificial structures made by combining two or more materials on a subwavelength scale. They can also be used to make GRIN devices with a spatially varying refractive index. Section 10.1 introduces the basic concepts needed to understand the physics behind the photonic crystal and metamaterials. Section 10.2 is devoted to the engineering of GRIN structures based on photonic crystals. Section 10.3 deals with the metamaterial-based GRIN devices, while the GRIN devices based on metasurfaces are discussed in Section 10.4.

The focus of Chapter 11 is on studying the impact of fluctuations in the amplitude or phase of an optical beam propagating inside a GRIN medium. Such beams are called partially coherent. Section 11.1 introduces the basic concepts related to the topic of partial coherence. Section 11.2 considers how the self-imaging phenomenon is affected by the partial coherence of an incoming beam. The evolution of a partially coherent beam inside a GRIN medium is studied in Section 11.3 using its cross-spectral density. The general result is used to discuss changes in the intensity, optical spectrum, and the degree of spatial coherence. The focus of Section 11.4 is on the polarization effects associated with a partially coherent beam. The concept of polarization matrix is used to discuss periodic changes occurring both in the state of polarization and the degree of polarization.

References

[1] J. C. Maxwell, *The Scientific Papers of James Clerk Maxwell*, vol. 1, ed., W. D. Niven (Cambridge University Press, 1890), 76–79.

[2] R. W. Wood, *Phil. Mag. J. Sci.*, Series 6, **12**, 159 (1906); https://doi.org/10.1080/14786440609463529.

[3] R. K. Luneburg, *Mathematical Theory of Optics* (Brown University, 1944).

[4] A. D. Pearson, W. G. French, and E. G. Rawson, *Appl. Phys. Lett.* **15**, 76 (1969); https://doi.org/10.1063/1.1652909.

[5] T. Uchida, M. Furukawa, I. Kitano, K. Koizumi, and H. Matsumura, *IEEE J. Quantum Electron.* **6**, 606 (1970); https://doi.org/10.1109/JQE.1970.1076326.

[6] T. Izawa and H. Nakagome, *Appl. Phys. Lett.* **21**, 584 (1972); https://doi.org/10.1063/1.1654265.

[7] I. P. Kaminow and J. R. Carruthers, *Appl. Phys. Lett.* **22**, 326 (1973); https://doi.org/10.1063/1.1654657.

[8] H. Smithgall and F. W. Dabby, *IEEE J. Quantum Electron.* **9**, 1023 (1973); https://doi.org/10.1109/JQE.1973.1077405.

[9] M. S. Sodha and A. K. Ghatak, *Inhomogeneous Optical Waveguides* (Plenum Press, 1977).

[10] E. W. Marchand, *Gradient Index Optics* (Academic Press, 1978).

[11] R. Olshansky, *Rev. Mod. Phys.* **51**, 341 (1979); https://doi.org/10.1103/RevModPhys.51.341.

[12] G. P. Agrawal, *Fiber-Optic Communication Systems*, 5th ed. (Wiley, 2021).

[13] Y. Koike, *Fundamentals of Plastic Optical Fibers* (Wiley, 2014).

[14] Special issues of Applied Optics publsihed on April 1, 1980, March 15, 1982, February 1, 1983, June 1, 1984, December 15, 1985, October 1, 1986, February 1, 1988, October 1, 1990, December 1, 1990, and September 1, 1992.

[15] Y. A. Kravtsov and Y. I. Orlov, *Geometrical Optics of Inhomogeneous Media* (Springer, 1990).

[16] C. Gomez-Reino, M. V. Perez, and C. Bao, *Gradient Index Optics: Fundamentals and Applications* (Springer, 2003).

[17] D. J. Richardson, J. M. Fini, and L. E. Nelson, *Nat. Photon.* **7**, 354 (2013); https://doi.org/10.1038/nphoton.2013.94.

[18] T. Mizuno, H. Takara, A. Sano, and Y. Miyamoto, *J. Lightwave Technol.* **34**, 1484 (2016); https://doi.org/10.1109/JLT.2016.2615070.

[19] K. Krupa, V. Couderc, A. Tonello et al., Spatiotemporal Nonlinear Dynamics in Multimode Fibers, in *Nonlinear Guided Wave Optics*, Ed. S. Wabnitz (IOP Science, 2017), 14-1 to 14-23.

[20] G. P. Agrawal, *Nonlinear Fiber Optics*, 6th ed. (Academic Press, 2019), 621-683.

[21] K. Krupa, A. Tonello, A. Barthélémy et al., *APL Photon.* **4**, 110901 (2019); https://doi.org/10.1063/1.5119434.

[22] A. Ghatak and K. Thyagarajan, *An Introduction to Fiber Optics* (Cambridge University Press, 1998).

[23] J. A. Buck, *Fundamentals of Optical Fibers*, 2nd ed. (Wiley, 2004).

[24] K. Okamoto, *Fundamentals of Optical Waveguides*, 2nd ed. (Academic Press, 2010).

[25] M. Born and E. Wolf, *Principles of Optics*, 7th ed. (Cambridge University Press, 2020).

[26] N. Ioannides, E. B. Chunga, A. Bachmatiuk et al., *Mater. Res. Express* **1**, 032002 (2014); https://doi.org/10.1088/2053-1591/1/3/032002.

[27] I. H. Malitson, *J. Opt. Soc. Am.* **55**, 1205 (1965); https://doi.org/10.1364/JOSA.55.001205.

[28] R. W. Boyd, *Nonlinear Optics*, 4th ed. (Academic Press, 2020).

[29] A. S. Ahsan and G. P. Agrawal, *Opt. Lett.* **43**, 3345 (2018); https://doi.org/10.1364/OL.43.003345.

[30] D. T. Moore, *Appl. Opt.* **19**, 1035 (1980); https://doi.org/10.1364/AO.19.001035.

[31] J. R. Hensler, U.S. Patent 3,873,408 (1975); https://patents.google.com/patent/US3873408A/en.

[32] R. V. Ramaswamy and R. Srivastava, *J. Lightwave Technol.* **6**, 984 (1988); https://doi.org/10.1109/50.4090.

[33] C. Fourmentin, X. H. Zhang, E. Lavanant et al., *Sci. Rep.* **11**, 11081 (2021); https://doi.org/10.1038/s41598-021-90626-4.

[34] S. N. Houde-Walter and D. T. Moore, *Appl. Opt.* **24**, 4326 (1985); https://doi.org/10.1364/AO.24.004326.

[35] D. B. Keck and R. Olshansky, U.S. Patent 3,904,268 (1975); https://patents.google.com/patent/US3904268A/en.

[36] T. Izawa and S. Sudo, *Optical Fibers: Materials and Fabrication* (Springer, 1987).

[37] A. Zukauskas, I. Matulaitiene, D. Paipulas et al., *Laser Photon. Rev.* **9**, 1863 (2015); https://doi.org/10.1002/lpor.201500170.

[38] Y. Bougdid and Z. Sekkat, *Sci. Rep.* **10**, 10409 (2020); https://doi.org/10.1038/s41598-020-67184-2.

[39] C. R. Ocier, C. A. Richards, D. A. Bacon-Brown et al., *Light Sci. Appl.* **9**, 196 (2020); https://doi.org/10.1038/s41377-020-00431-3.

[40] A. Anuszkiewicz, R. Kasztelanic, A. Filipkowski et al., *Sci. Rep.* **8**, 1 (2018); https://doi.org/10.1038/s41598-018-30284-1.

2

Wave Propagation

Propagation of electromagnetic waves inside a GRIN medium requires the solution of Maxwell's equations. In this chapter, we obtain the wave equation governing the evolution of optical beams in such a medium and discuss the mathematical techniques used to solve it. Section 2.1 starts with Maxwell's equations and uses them to derive a wave equation in the frequency domain. A mode-based technique is used in Section 2.2 for solving the wave equation for a GRIN device fabricated with a parabolic index profile. The properties of both the Hermite–Gauss and the Laguerre–Gauss modes are discussed. Section 2.3 is devoted to other power-law index profiles and employs the Wentzel–Kramers–Brillouin (WKB) method to discuss the properties of modes supported by them. We discuss in Section 2.4 the relative efficiency with which different modes are excited by an optical beam incident on a GRIN medium. The intermodal dispersive effects that become important for pulsed beams are also covered. Section 2.5 describes several nonmodal techniques that can be used for studying wave propagation in GRIN media.

2.1 Wave Equation for a GRIN Medium

In the international system of units, Maxwell's equations take the form [1]

$$\nabla \times \mathbf{E} = -\frac{\partial \mathbf{B}}{\partial t}, \qquad \nabla \cdot \mathbf{D} = \rho_f, \tag{2.1.1}$$

$$\nabla \times \mathbf{H} = \mathbf{J} + \frac{\partial \mathbf{D}}{\partial t}, \qquad \nabla \cdot \mathbf{B} = 0, \tag{2.1.2}$$

where $\mathbf{E}$ and $\mathbf{H}$ are the electric and magnetic field vectors, respectively, and $\mathbf{D}$ and $\mathbf{B}$ the corresponding electric and magnetic flux densities. The current density $\mathbf{J}$ and the charge density ρ_f represent the sources for generating an electromagnetic field. As only dielectric materials such as glasses and plastics are used to make GRIN devices, we set $\mathbf{J} = 0$ and $\rho_f = 0$ because of the absence of free charges in such materials.

2.1.1 Frequency-Domain Approach

It is often simpler to solve Maxwell's equations in the frequency domain. For this purpose, we employ the Fourier transform of the electric field in the form

$$\tilde{\mathbf{E}}(\mathbf{r}, \omega) = \int_{-\infty}^{\infty} \mathbf{E}(\mathbf{r}, t)\, e^{i\omega t}\, dt. \tag{2.1.3}$$

The inverse Fourier transform then has the form

$$\mathbf{E}(\mathbf{r}, t) = \frac{1}{2\pi} \int_{-\infty}^{\infty} \tilde{\mathbf{E}}(\mathbf{r}, \omega)\, e^{-i\omega t}\, d\omega. \tag{2.1.4}$$

The constitutive relations can be used to relate the flux densities at any point $\mathbf{r}$ to the electric and magnetic fields at the same point as

$$\tilde{\mathbf{D}}(\mathbf{r}, \omega) = \epsilon(\mathbf{r}, \omega)\tilde{\mathbf{E}}(\mathbf{r}, \omega), \qquad \tilde{\mathbf{B}}(\mathbf{r}, \omega) = \mu(\mathbf{r}, \omega)\tilde{\mathbf{H}}(\mathbf{r}, \omega), \tag{2.1.5}$$

where ϵ is the electric permittivity and μ is the magnetic permeability of the medium.

Taking the Fourier transform of all variables in Eqs. (2.1.1) and (2.1.2) and making use of Eq. (2.1.5), Maxwell's equations in the frequency domain take the form

$$\nabla \times \tilde{\mathbf{E}} = i\omega\mu\tilde{\mathbf{H}}, \qquad \nabla \cdot (\epsilon\tilde{\mathbf{E}}) = 0, \tag{2.1.6}$$

$$\nabla \times \tilde{\mathbf{H}} = -i\omega\epsilon\tilde{\mathbf{E}}, \qquad \nabla \cdot (\mu\tilde{\mathbf{H}}) = 0. \tag{2.1.7}$$

Taking the curl of first relation in Eq. (2.1.6) and using Eq. (2.1.7), we obtain a single equation for the electric field:

$$\nabla \times \nabla \times \tilde{\mathbf{E}} = \mu\epsilon\omega^2\tilde{\mathbf{E}}. \tag{2.1.8}$$

We simplify this equation with the vector identity

$$\nabla \times \nabla \times \tilde{\mathbf{E}} = \nabla(\nabla \cdot \tilde{\mathbf{E}}) - \nabla^2\tilde{\mathbf{E}}, \tag{2.1.9}$$

and use the divergence condition,

$$\nabla \cdot (\epsilon\tilde{\mathbf{E}}) = \epsilon(\nabla \cdot \tilde{\mathbf{E}}) + \nabla\epsilon \cdot \tilde{\mathbf{E}} = 0 \tag{2.1.10}$$

to obtain the wave equation

$$\nabla^2\tilde{\mathbf{E}} + \mu\epsilon\omega^2\tilde{\mathbf{E}} + \nabla\left(\frac{1}{\epsilon}\nabla\epsilon \cdot \tilde{\mathbf{E}}\right) = 0. \tag{2.1.11}$$

This equation can be used for any inhomogeneous medium because it allows for spatial variations of both ϵ and μ inside such a medium.

In the case of a nonmagnetic medium such as glass, we can use $\mu \approx \mu_0$, where μ_0 has a fixed value of $4\pi \times 10^{-7}$ H/m. The electric permittivity of a GRIN medium can be related to its refractive index n as $\epsilon = \epsilon_0 n^2$, where ϵ_0 has a value such that

$\mu_0\epsilon_0 = 1/c^2$, c being the speed of light ($c = 2.998 \times 10^8$ m/s). Note that $n(\mathbf{r})$ is in general a complex number whose imaginary part is related to the absorption coefficient of the material. We treat $n(\mathbf{r})$ as a real number in this chapter because losses are relatively small in practice for most GRIN devices at wavelengths far from an atomic resonance; their impact is discussed in Section 6.1. Under such conditions, we can write Eq. (2.1.11) as

$$\nabla^2 \tilde{\mathbf{E}} + n^2(\mathbf{r})k_0^2\tilde{\mathbf{E}} + \nabla\left(\frac{1}{n^2}\nabla n^2 \cdot \tilde{\mathbf{E}}\right) = 0, \tag{2.1.12}$$

where $k_0 = \omega/c = 2\pi/\lambda$ is related to the wavelength λ of radiation.

The last term in Eq. (2.1.12) is finite because the index gradient ∇n is not zero for a GRIN medium. The implications of this term will be explored in Chapter 9, where it is shown that its presence leads to the photonic analog of spin-orbit coupling. Its magnitude is, however, small when the refractive index $n(\mathbf{r})$ varies on a spatial scale much longer than a wavelength. We neglect the last term in Eq. (2.1.12) until Chapter 9 and focus on the simpler Helmholtz equation

$$\nabla^2 \tilde{\mathbf{E}} + n^2(\mathbf{r})k_0^2\tilde{\mathbf{E}} = 0. \tag{2.1.13}$$

The magnetic field vector satisfies an identical equation. In general, the vectorial hybrid modes, denoted by HE$_{mn}$ or EH$_{mn}$ for integer values of m and n, are found by solving these two equations, while ensuring that all boundary conditions are satisfied [2–4].

2.1.2 Weakly Guiding Approximation

Equation (2.1.13) indicates that different components of the electric field do not couple to each other when the last term in Eq. (2.1.12) is negligible. As a result, the state of polarization of the field, related to the orientation of the $\mathbf{E}$ vector, does not change. It thus becomes possible to adopt a scalar approach [5] that is reasonably accurate when the parameter Δ in the expression of $n(\mathbf{r})$ in Eq. (1.2.2) satisfies the condition $\Delta \ll 1$. This condition is known as the weakly guiding approximation and is often satisfied in practice ($\Delta \sim 0.01$). In the scalar approach, the direction of the electric field lies in the x–y plane and does not change during the beam's propagation along the z axis aligned with the central axis of the GRIN medium.

There are two ways of solving Eq. (2.1.13). We can solve it as an initial-value problem, where $\tilde{\mathbf{E}}$ is specified at the input plane located at $z = 0$. It is useful to remove rapid phase variations and introduce the slowly varying amplitude $A(x, y, z)$ of the electric field using

$$\tilde{\mathbf{E}}(\mathbf{r}) = \hat{\mathbf{p}}\, A(x, y, z)\, \exp(ikz), \tag{2.1.14}$$

where $\hat{\mathbf{p}}$ is the unit vector representing the state of polarization and the propagation constant k is chosen to be $k = n_0 k_0$. If we use Eq. (2.1.14) in Eq. (2.1.13) and neglect the second derivative $\partial^2 A/\partial z^2$ in the paraxial approximation, $A(x, y, z)$ is found to satisfy

$$2ik\frac{\partial A}{\partial z} + \nabla_T^2 A + [n^2(\mathbf{r})k_0^2 - k^2]A = 0, \qquad \nabla_T^2 A = \frac{\partial^2 A}{\partial x^2} + \frac{\partial^2 A}{\partial y^2}, \qquad (2.1.15)$$

where ∇_T^2 is the transverse part of the Laplacian operator ∇^2. This equation is solved to find $A(x, y, z)$ in terms of $A(x, y, 0)$.

Alternatively, one can adopt the modal approach in which Eq. (2.1.13) is solved as an eigenvalue problem whose eigenfunctions, known as the modes, do not change with z. The vector $\tilde{\mathbf{E}}$ in Eq. (2.1.13) for an optical mode takes the form

$$\tilde{\mathbf{E}} = \hat{\mathbf{p}}\, U(x, y)e^{i\beta z}, \qquad (2.1.16)$$

where the eigenfunction $U(x, y)$ does not depend on z, and β is the propagation constant (or eigenvalue) that is yet to be determined. Using $n(\mathbf{r})$ from Eq. (1.2.1), Eq. (2.1.13) for a GRIN medium is reduced to

$$\nabla_T^2 U + \{n_0^2 k_0^2[1 - 2\Delta f(\rho)] - \beta^2\}U = 0, \qquad (2.1.17)$$

where the function $f(\rho)$ governs the shape of the refractive-index profile with $\rho^2 = x^2 + y^2$. The solutions $U(x, y)$ of this equation exist for only specific values of β. The modal approach can be used in the nonparaxial region because Eq. (2.1.17) does not require the paraxial approximation.

Before solving Eq. (2.1.17) for a specific form of $f(\rho)$, we discuss how the GRIN problem is analogous to the motion of a particle in quantum mechanics (see Appendix A). This analogy is useful because the techniques developed for quantum mechanics can be applied to solving a classical problem related to wave propagation in a GRIN medium. The quantum dynamics of a particle of mass m_p and energy E_p is governed by the time-independent Schrödinger equation [6]

$$\hat{H}\psi = E_p\psi, \qquad \hat{H} = -\frac{\hbar^2}{2m_p}\nabla^2 + V_0(\mathbf{r}), \qquad (2.1.18)$$

where $\hat{H}$ is the Hamiltonian and $V_0(\mathbf{r})$ is the potential energy. This equation can be written in the form

$$\nabla^2\psi + \frac{2m_p}{\hbar^2}[E_p - V_0(\mathbf{r})]\psi = 0. \qquad (2.1.19)$$

A direct comparison with Eq. (2.1.17) shows that the two equations have the same mathematical form. In particular, U plays the role of the wave function ψ, $n_0^2 k_0^2 - \beta^2$ plays the role of the eigenvalue E_p, and the index profile $f(\rho)$ plays the role of the potential $V_0(\mathbf{r})$.

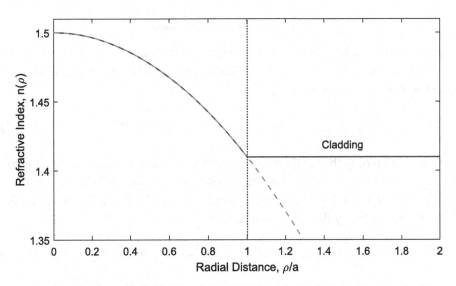

Figure 2.1 Actual index profile of a GRIN fiber (solid line) is compared to its parabolic approximation (dashed line). The dotted vertical line marks the core-cladding boundary.

2.2 Mode-Based Technique

The solutions of Eq. (2.1.17) will clearly depend on the shape of the index profile, governed by the function $f(\rho)$. In fact, this equation cannot be solved in an analytic form for most index profiles, and one must resort to numerical solutions [3], or use an approximate technique such as the WKB method of quantum mechanics [5]. An exception occurs in the case of a parabolic shape of the index profile. The reason is easy to understand if we recall that the potential of a harmonic oscillator is also parabolic in shape. We thus expect a parabolic GRIN profile to have solutions similar to those found for a quantum harmonic oscillator (see Appendix A).

For a parabolic index profile, we use Eq. (1.2.4) with $p = 2$ to obtain

$$f(\rho) = (\rho/a)^2 = (x^2 + y^2)/a^2, \quad 0 \leq \rho \leq a, \tag{2.2.1}$$

with $f(\rho) = 1$ for $\rho > a$. To obtain an analytic solution, it is necessary to assume that Eq. (2.2.1) applies for all values of ρ. Figure 2.1 shows the nature of this approximation by comparing the actual index profile to a parabolic shape. It can be justified only if $U(x, y)$ becomes negligible in the region $\rho \geq a$. Even though the following analysis is not exact, it provides considerable physical insight and its predictions often agree with the experiments.

2.2.1 Hermite–Gauss Modes

We can solve Eq. (2.1.17) using the Cartesian or cylindrical coordinates. To see the analogy between a GRIN medium and a harmonic oscillator, we write Eq. (2.1.17) in the Cartesian coordinates as [5]

$$\frac{\partial^2 U}{\partial x^2} + \frac{\partial^2 U}{\partial y^2} + \{n_0^2 k_0^2 [1 - b^2(x^2 + y^2)] - \beta^2\}U = 0, \qquad (2.2.2)$$

where $b = \sqrt{2\Delta}/a$ is the GRIN parameter introduced in Eq. (1.2.5). This equation can be solved with the method of separation of variables. When we substitute $U(x,y) = F(x)G(y)$ in Eq. (2.2.2) and separate the variables x and y, we obtain the following two ordinary differential equations:

$$\frac{1}{F}\frac{d^2 F}{dx^2} - q^4 x^2 = -C_1, \qquad \frac{1}{G}\frac{d^2 G}{dy^2} - q^4 y^2 = -C_2, \qquad (2.2.3)$$

where the constants C_1 and C_2 and the parameter q satisfy the relations

$$\beta^2 = n_0^2 k_0^2 - C_1 - C_2, \qquad q = \sqrt{n_0 k_0 b}. \qquad (2.2.4)$$

At this point, it is useful to introduce the dimensionless quantities as

$$\xi = qx, \quad \eta = qy, \quad Q_1 = C_1/q^2, \quad Q_2 = C_2/q^2, \qquad (2.2.5)$$

and rewrite Eq. (2.2.3) as

$$\frac{d^2 F}{d\xi^2} + (Q_1 - \xi^2)F = 0, \qquad \frac{d^2 G}{d\eta^2} + (Q_2 - \eta^2)G = 0. \qquad (2.2.6)$$

Each of these equations resembles the time-independent Schrödinger equation obtained for a quantum harmonic oscillator in one dimension (see Appendix A). Their solutions exist only for odd integer values of Q_1 and Q_2 such that

$$Q_1 = 2m + 1, \quad Q_2 = 2n + 1, \quad m, n = 0, 1, 2, \ldots. \qquad (2.2.7)$$

For each value of the integers m and n, the solution involves a Hermite–Gauss function, analogous to a harmonic oscillator. The general solution of Eq. (2.2.2) takes the form [5]

$$U_{mn}(x,y) = \left[N_m H_m(qx)\exp\left(-\tfrac{1}{2}q^2 x^2\right)\right]\left[N_n H_n(qy)\exp\left(-\tfrac{1}{2}q^2 y^2\right)\right], \qquad (2.2.8)$$

where $H_m(qx)$ is a Hermite polynomial of the mth order [7]. A few low-order Hermite polynomials are $H_0(x) = 1$, $H_1(x) = 2x$, $H_2(x) = 4x^2 - 2$, and $H_3(x) = 8x^3 - 12x$. Also, N_m is a normalization constant introduced such that $\iint |U_{mn}(x,y)|^2 dx\,dy = 1$. Its value depends on the integer m as

$$N_m^2 = \frac{q}{\sqrt{\pi}\,2^m m!}. \qquad (2.2.9)$$

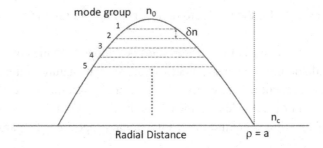

Figure 2.2 Schematic showing the effective mode indices (dashed lines) for several mode groups within parabolic profile of a GRIN medium. The dotted vertical line at $\rho = a$ marks its core-cladding boundary.

Similar to the case of step-index fibers, we can define a dimensional parameter V as

$$V = n_0 k_0 a \sqrt{2\Delta} = n_0 k_0 b a^2 = q^2 a^2. \qquad (2.2.10)$$

Thus, q in Eq. (2.2.8) is related to the V parameter as $q = \sqrt{V}/a$.

The solution $U_{mn}(x, y)$ given in Eq. (2.2.8) represents a specific mode of the GRIN medium, denoted as HG_{mn}. Different modes are obtained for values of the two integers ($m, n = 0, 1, 2, \ldots$). The propagation constant β_{mn} of any mode is found from Eq. (2.2.4). Noting that $C_j = q^2 Q_j$ ($j = 1, 2$) and using Eq. (2.2.7), β_{mn} is found to be

$$\beta_{mn} = [n_0^2 k_0^2 - 2(m + n + 1)q^2]^{1/2} = n_0 k_0 [1 - (m + n + 1)(2b/n_0 k_0)]^{1/2}, \qquad (2.2.11)$$

where we used Eq. (2.2.4). As the ratio b/k_0 is relatively small ($< 10^{-3}$), the second term in Eq. (2.2.11) remains small compared to 1 when $m + n$ is restricted to values well below 1,000. In that situation, we can approximate β_{mn} as

$$\beta_{mn} \approx k_0 [n_0 - (m + n + 1)\delta n], \qquad \delta n = b/k_0 = n_0(2\Delta/V). \qquad (2.2.12)$$

This simplification is equivalent to making the paraxial approximation. The parameter V exceeds 100 for GRIN rods and 10 for multimode GRIN fibers, while Δ is typically ~ 0.01. For this reason, δn remains $< 10^{-3}$ in most cases of practical interest.

It is useful to define the effective mode index as $\bar{n} = \beta/k_0$. If we introduce a single integer as $g = m + n + 1$, we can write $\bar{n}$ in the simple form $\bar{n}_g = n_0 - g\delta n$, where g labels different mode groups. Figure 2.2 shows schematically the ladder-like structure of $\bar{n}_g$ values for several mode groups within the parabolic index profile of a GRIN medium. The first group contains the HG_{00} mode and has the largest value $\bar{n}_1 = n_0 - \delta n$; this mode is called the fundamental mode. As m and n change such that g increases by 1, $\bar{n}_g$ decreases by the same amount δn. The largest value of the integer g is set by the condition that $\bar{n}_g$ must remain larger than the cladding index $n_c = n_0(1 - \Delta)$ for the modes to remain confined within the core of the GRIN medium.

The ladder-like structure of the eigenvalues is analogous to the energy levels of a harmonic oscillator and is a consequence of the parabolic nature of the index profile. It leads to a phenomenon, known as *self-imaging*, which manifests as a periodic revival of any input field incident on the GRIN medium. It will be seen in Chapters 3 and 5 that self-imaging is behind many interesting properties of a parabolic GRIN medium.

The modal profiles in Eq. (2.2.8) take a relatively simple form for low-order Hermite–Gauss modes. For example, the HG$_{00}$ mode has a Gaussian shape governed by

$$U_{00}(x,y) = \frac{q}{\sqrt{\pi}} \exp\left[-\tfrac{1}{2}q^2(x^2 + y^2)\right] = \frac{1}{w_g\sqrt{\pi}} \exp\left(-\frac{x^2 + y^2}{2w_g^2}\right), \quad (2.2.13)$$

where we have introduced the spot size of the fundamental mode of a GRIN device as

$$w_g = 1/q = (n_0 k_0 b)^{-1/2} = a/\sqrt{V}. \quad (2.2.14)$$

The shape of the HG$_{00}$ mode corresponds to a Gaussian beam with the spot size w_g. This parameter plays an important role for all GRIN devices. Its value at a wavelength of 1 μm is about 5 m for standard GRIN fibers (core radius 25 μm) and 15 μm for GRIN rods with a core radius of 1 mm.

The next two modes are degenerate because $\beta_{10} = \beta_{01}$ from Eq. (2.2.11). Their spatial profiles can be written as

$$U_{10}(x,y) = \sqrt{2}\, qx U_{00}(x,y), \qquad U_{01}(x,y) = \sqrt{2}\, qy U_{00}(x,y). \quad (2.2.15)$$

The HG$_{10}$ mode is antisymmetric (or odd) with respect to x, and the HG$_{01}$ mode with respect to y. The amplitudes of both vanish at the central axis of the GRIN medium. Figure 2.3 shows the spatial profiles for a few low-order HG$_{mn}$ modes by plotting $|U_{mn}(x,y)|^2$ on a gray scale in the transverse plane. An important property of all modes is that they are orthogonal to each other and satisfy the relation

$$\iint_{-\infty}^{\infty} U_{m'n'}^*(x,y) U_{mn}(x,y)\, dx\, dy = \delta_{m'm}\delta_{n'n}. \quad (2.2.16)$$

2.2.2 Laguerre–Gauss Modes

In this section we solve Eq. (2.1.17) for a parabolic index profile using the cylindrical coordinates (ρ, ϕ, z):

$$\frac{1}{\rho}\frac{\partial}{\partial \rho}\left(\rho \frac{\partial U}{\partial \rho}\right) + \frac{\partial^2 U}{\partial \phi^2} + [n_0^2 k_0^2(1 - b^2\rho^2) - \beta^2]U = 0. \quad (2.2.17)$$

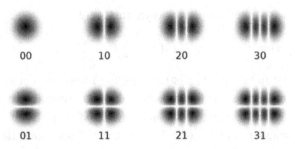

Figure 2.3 Spatial profiles for a few low-order HG$_{mn}$ modes of a GRIN medium with a parabolic index profile.

As before, we use the method of separation of variables and substitute $U(\rho, \phi) = F(\rho)G(\phi)$ in Eq. (2.2.17) and separate the variables ρ and ϕ to obtain the following two equations:

$$\frac{d^2F}{d\rho^2} + \frac{1}{\rho}\frac{dF}{d\rho} + \left[n_0^2 k_0^2(1 - b^2\rho^2) - \beta^2 - \frac{\ell^2}{\rho^2}\right]F = 0 \text{ and} \qquad (2.2.18)$$

$$\frac{d^2G}{d\phi^2} + \ell^2 G = 0, \qquad (2.2.19)$$

where ℓ is a constant. The last equation has solutions of the form $e^{\pm i\ell\phi}$. Because ϕ varies in a periodic fashion such that $G(2\pi) = G(0)$, the constant ℓ can only take integer values ($\ell = 0, 1, 2, \ldots$). It is common to use the even and odd linear combinations of the two solutions and choose $G(\phi) = \cos(\ell\phi)$ or $\sin(\ell\phi)$. Clearly, $\ell = 0$ case is special because $G(\phi) = 1$ for the even mode, and its odd version does not exist. All $\ell = 0$ modes are also radially symmetric because $U(\rho, \phi)$ for them does not depend on the azimuthal angle ϕ. It will be seen in Chapter 8 that the integer ℓ is related to the orbital angular momentum (OAM) associated with a mode or a beam.

For any value of the integer ℓ, the radial Eq. (2.2.18) has multiple solutions in terms of the Laguerre polynomials [7]. Numbering these solutions with another integer $m = 1, 2, \ldots$, the general solution of Eq. (2.2.17) takes the form [5]

$$U_{\ell m}(\rho, \phi) = N_{\ell m}\rho^\ell L_{m-1}^\ell(q^2\rho^2)\exp(-q^2\rho^2/2)\begin{pmatrix}\cos(\ell\phi)\\\sin(\ell\phi)\end{pmatrix}, \qquad (2.2.20)$$

where $q = 1/w_g$, with w_g given in Eq. (2.2.14). The normalization constant $N_{\ell m}$ is found by requiring $\iint |U_{\ell m}|^2\rho d\rho d\phi = 1$:

$$N_{\ell m} = q^{\ell+1}\left[\frac{2(m-1)!}{\pi(\ell + m - 1)!}\right]^{1/2}. \qquad (2.2.21)$$

A few low-order associated Laguerre polynomials are

$$L_0^\ell(x) = 1, \quad L_1^\ell(x) = \ell + 1 - x, \quad L_2^\ell(x) = \tfrac{1}{2}[(\ell + 2)(\ell + 1 - 2x) + x^2]. \quad (2.2.22)$$

In this book, the notation $LP_{\ell m}$ is used for the Laguerre–Gauss modes (rather than $LG_{\ell m}$), where LP is short for linearly polarized. Similar to the HG_{mn} modes, the $LP_{\ell m}$ modes are orthogonal to each other and satisfy Eq. (2.2.16). It is important to stress that the integer m in Eq. (2.2.20) starts at 1. This convention is commonly used in the literature on optical fibers. Sometimes, another integer is introduced as $p = m - 1$ that starts at zero. This choice is made for the LG_{lp} beams propagating in free space [8].

An interesting feature of $LP_{\ell m}$ modes is apparent from Eq. (2.2.20). Except for $\ell = 0$, the odd version of each mode exists with $\sin(\ell\phi)$ in place of $\cos(\ell\phi)$. The propagation constant $\beta_{\ell m}$ for both sets of modes is given by [5]

$$\beta_{\ell m}(\omega) = k_0 n_0 [1 - (\ell + 2m - 1)(2b/n_0 k_0)]^{1/2} \approx k_0 [n_0 - (\ell + 2m - 1)\delta n], \quad (2.2.23)$$

where $\delta n = b/k_0$ such that $b/k_0 \ll 1$. The effective mode index, $\bar{n} = \beta/k_0$, can again be written in the form $\bar{n}_g = n_0 - g\delta n$, with $g = \ell + 2m - 1$, where the integer values of g starting at 1 represent different mode groups. This expression is identical to the one obtained for the Hermite–Gauss modes and results in the same ladder-like structure of the modal refractive indices (see Figure 2.2). This is expected on physical grounds because the eigenvalues of a parabolic GRIN medium should not depend on the coordinate system employed. The modal distributions can become different in different coordinate systems in the presence of degeneracy.

Similar to the HG_{mn} modes, the $LP_{\ell m}$ modes can be arranged in different groups classified by the integer g. The $g = 1$ group has a single member: the LP_{01} mode with $l = 0$ and $m = 1$. The next group with $g = 2$ also has a single member LP_{11}, but it is two-fold degenerate when we include its even and odd versions. The next group with $g = 3$ has two members, LP_{02} and LP_{21}, with three modes. The total group degeneracy increases rapidly as the integer g increases. Moreover, the total number of modes in any group doubles when polarization degeneracy is included. This degeneracy occurs because each mode can exist in two orthogonal polarization states.

The number of modes supported by a GRIN device depends on the index difference δn between two neighboring mode groups. From Eq. (2.2.12), this difference is relatively small for GRIN rods with $\Delta \sim 0.01$ and $V > 100$. For such large values of V, the number of modes may exceed 10,000 because it scales with V as V^2 (see Section 2.3). In contrast, V is about 20 for GRIN fibers fabricated with $a = 25$ μm and can be reduced to below 10 by reducing a further. Such GRIN fibers support only a few mode groups because the effective index, $\bar{n} = n_0 - g\delta n$, should be larger

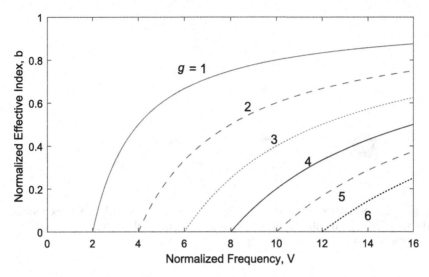

Figure 2.4 Normalized effective index b for several mode groups plotted as a function of the V parameter for GRIN fibers.

than the cladding index, n_c, for any mode confined to the core of a GRIN fiber. If we define a normalized quantity as $b = (\bar{n} - n_c)/(n_0 - n_c)$, a mode can exist as long as $b > 0$. Figure 2.4 shows how b varies with the V parameter for mode groups up to $g = 6$. We see that b approaches zero for different mode groups as V is reduced. Even the modes in the $g = 2$ group cease to exist for $V < 4$. Such a GRIN fiber supports only the LP_{01} mode. We should stress that this mode always exists. Values of b near $V = 2$ in Figure 2.4 are not accurate because the boundary condition at the core-cladding interface has not been applied in this simple analysis.

One may ask how the spatial profiles of LP_{lm} modes are related to those of HG_{mn} modes. From Eq. (2.2.20), the spatial profile $U_{01}(\rho, \phi)$ of the LP_{01} mode is identical to the Gaussian profile in Eq. (2.2.13) for the fundamental HG_{00} mode. The spatial profiles for the even and odd LP_{11} modes are found by using $\ell = m = 1$ in Eq. (2.2.20):

$$U_{11}^{(e,o)}(\rho, \phi) = \sqrt{\frac{2}{\pi}} q^2 \rho \exp\left(-\frac{1}{2} q^2 \rho^2\right) \begin{pmatrix} \cos \phi \\ \sin \phi \end{pmatrix}. \qquad (2.2.24)$$

If we use $x = \rho \cos \phi$ and $y = \rho \sin \phi$, profiles in Eq. (2.2.24) become identical to spatial profiles of the HG_{10} and HG_{01} modes given in Eq. (2.2.15).

It is easy to conclude from Eq. (2.2.20) that all modes for which $\ell = 0$ are radially symmetric modes with the spatial profiles

$$U_{0m}(\rho, \phi) = N_{0m} L_{m-1}(q^2 \rho^2) \exp(-q^2 \rho^2 / 2), \qquad (2.2.25)$$

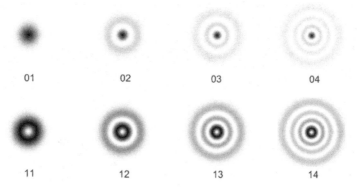

Figure 2.5 Spatial profiles of the LP_{0m} modes (top) and LP_{1m} vortex modes (bottom), identified with the values of l and m. A gray scale is used such that white regions correspond to zero intensity.

where $L_n(x)$ is a regular Laguerre polynomial of order n. Top row of Figure 2.5 shows the spatial distributions for a few low-order LP_{0m} modes by plotting their intensity distribution $|U_{0m}|^2$ in the transverse plane. Their intensity patterns contain one or more concentric rings with a bright spot in the center at $\rho = 0$.

When $\ell \neq 0$, the degenerate nature of the even and odd modes for any m implies that a linear combination of them is also a modal solution of the wave equation. A specific combination, formed with a 90° phase shift between the even and odd versions of the same mode, depends on ϕ as $e^{\pm i\ell\phi}$. Clearly, the intensity of such modes does not depend on ϕ and exhibits a radial symmetry. However, the intensity vanishes in the center at $\rho = 0$ for all such modes, as seen in the bottom row of Figure 2.5. This feature results in a phase singularity at $\rho = 0$ because phase changes by $2\pi\ell$ after a round trip around this point. The singularity represents a vortex with an OAM of $\ell\hbar$ [8]. The integer ℓ is also called the topological charge of a vortex. This topic is discussed in Chapter 8.

2.3 Power-Law Index Profiles

The parabolic index profile considered in Section 2.2 is a special case of the power-law profile given in Eq. (1.2.4). We can treat the entire class of such profiles if we use Eq. (2.1.17) with the function

$$f(\rho) = \left(\frac{\rho}{a}\right)^p, \tag{2.3.1}$$

where p can take any value. Although Eq. (2.1.17) cannot be solved in an analytic form when $p \neq 2$, the eigenvalues of various modes (values of β) can be found approximately with the WKB method that is well known in quantum mechanics.

2.3.1 WKB Approximation

The analysis used in Section 2.2 for finding the modes of a parabolic index profile can also be applied to the power-law profile with $f(\rho)$ in Eq. (2.3.1). Using the cylindrical coordinates, we can still write the solution in the form, $U(\rho, \phi) = F(\rho)\cos(\ell\phi)$, where the radial function $F(\rho)$ satisfies the following equation:

$$\frac{d^2F}{d\rho^2} + \frac{1}{\rho}\frac{dF}{d\rho} + \left[k_0^2 n^2(\rho) - \beta^2 - \frac{\ell^2}{\rho^2} \right] F = 0. \tag{2.3.2}$$

The WKB approximation was first applied to this equation in 1973 in the context of GRIN fibers [11].

The WKB approximation was originally developed for a one-dimensional Schrödinger equation of the form [6]

$$\frac{d^2\psi}{dx^2} + [k_0^2 n^2(x) - \beta^2]\psi = 0. \tag{2.3.3}$$

It shows that the eigenvalue β of different modes satisfies the relation

$$\int_{x_1}^{x_2} [k_0^2 n^2(x) - \beta^2]^{1/2}\, dx = \left(m - \tfrac{1}{2} \right)\pi, \quad m = 1, 2, \ldots, \tag{2.3.4}$$

where x_1 and x_2 are the two turning points at which the integrand vanishes.

Equation (2.3.2) can be converted to the form of Eq. (2.3.3) with the transformation $\rho = \rho_0 e^x$. Using this transformation in Eq. (2.3.2), we obtain

$$\frac{d^2F}{dx^2} + \left[(k_0^2 n^2(x) - \beta^2)\rho^2 - \ell^2 \right] F = 0. \tag{2.3.5}$$

When applied to this equation, the WKB approximation provides the relation [9]

$$\int_{\rho_1}^{\rho_2} [k_0^2 n^2(\rho) - \beta^2 - \ell^2/\rho^2]^{1/2}\, d\rho = \left(m - \tfrac{1}{2} \right)\pi, \quad m = 1, 2, \ldots, \tag{2.3.6}$$

where β now depends on two integers ℓ and m. The main difference compared to Eq. (2.3.4) is that both ρ_1 and ρ_2 are positive because ρ varies from 0 to ∞. Figure 2.6 shows schematically their locations at the intersections of the two curves: $f(\rho) = k_0^2 n^2(\rho) - \beta^2$ and $g(\rho) = \ell^2/\rho^2$.

As an example of the usefulness of Eq. (2.3.6), we apply it to a parabolic index profile by choosing $p = 2$ in Eq. (2.3.1). Making the substitution $u = \rho^2$, the integral can be done in a closed form [10]. As the integrand vanishes at the upper and lower limits, the final result is given by

$$\left[\frac{1}{2q^2}(k_0^2 n_0^2 - \beta^2) - \ell \right]\pi = (2m - 1)\pi. \tag{2.3.7}$$

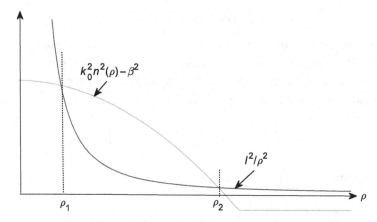

Figure 2.6 Location of the turning points ρ_1 and ρ_2 at the intersections of the two curves shown.

This provides us with the following expression for the modal propagation constants:

$$\beta_{lm} = [k_0^2 n_0^2 - 2(\ell + 2m - 1)q^2]^{1/2} = k_0 n_0 [1 - (\ell + 2m - 1)(2b/n_0 k_0]^{1/2}. \quad (2.3.8)$$

A comparison with Eq. (2.2.23) shows that the WKB approximation leads to the exact expression of β_{lm} derived in Section 2.2.2.

2.3.2 Number of Modes and Propagation Constants

Equation (2.3.6) can be used to find the modal propagation constants approximately when $p \neq 2$ in Eq. (2.3.1), provided $V \gg 1$ and the GRIN device supports a large number of modes. In this case, it is possible to estimate the total number of modes supported by a GRIN device for any value of p. As details are available in several books [2, 3], only the main steps are described here. The propagation constant β depends on two integers ℓ and m. We can count the number of modes up to a certain β by summing over m for values of ℓ ranging from 0 to $\ell(\beta)$. In the limit $V \gg 1$, we replace the sum with the integral [9]

$$N(\beta) = 4 \int_0^{\ell(\beta)} m \, d\ell, \quad (2.3.9)$$

where the factor of 4 accounts for the four-fold degeneracy for all modes with $\ell > 0$. Even though modes for $\ell = 0$ are only two-fold degenerate, this does not lead to a significant error when the total number of modes is large.

We replace m with $m - \frac{1}{2}$ in Eq. (2.3.9) and use its value from Eq. (2.3.6) to obtain

$$N(\beta) = \frac{4}{\pi} \int_0^{\ell(\beta)} \int_{\rho_1}^{\rho_2} [k_0^2 n^2(\rho) - \beta^2 - \ell^2/\rho^2]^{1/2} \, d\rho \, dl. \quad (2.3.10)$$

Because the integral over l is much easier to perform, we change the order of integration. The integral over l extends from 0 to a maximum value, $\ell_0 = \rho[k_0^2 n^2(\rho) - \beta^2]^{1/2}$, at which the integrand vanishes. The ρ integral begins at $\rho = 0$ and extends to a distance R such that $n(R)k_0 = \beta$. Using these results, we can write Eq. (2.3.10) as

$$N(\beta) = \frac{4}{\pi} \int_0^R g(\ell_0) \frac{1}{\rho} \, d\rho, \tag{2.3.11}$$

where the integral over ℓ has a simple form

$$g(\ell_0) = \int_0^{\ell_0} \sqrt{\ell_0^2 - \ell^2} \, d\ell. \tag{2.3.12}$$

Carrying out the integration, we obtain $g(\ell_0) = (\pi/4)\ell_0^2$. Using this result in Eq. (2.3.11) and the expression for l_0, we obtain

$$N(\beta) = \int_0^R [k_0^2 n^2(\rho) - \beta^2]\rho \, d\rho. \tag{2.3.13}$$

Integration over ρ can be performed after using $n^2(\rho)$ from Eq. (1.2.4). The final result is given as

$$N(\beta) = \frac{pV^2}{2(p+2)} \left(\frac{k_0^2 n_0^2 - \beta^2}{2\Delta k_0^2 n_0^2} \right)^{(p+2)/p}. \tag{2.3.14}$$

The preceding equation provides the number of modes whose eigenvalues range from the maximum value $n_0 k_0$ to a specific value β. It can be used to write β in the form

$$\beta = n_0 k_0 \left[1 - 2\Delta \left(\frac{N(\beta)}{N_t} \right)^{p/(p+2)} \right]^{1/2}, \tag{2.3.15}$$

where we have defined N_t as

$$N_t = \frac{p}{p+2} \frac{V^2}{2}. \tag{2.3.16}$$

It is important to realize that we have not found β because $N(\beta)$ itself depends on β. What we have found is an implicit relation for β. However, Eq. (2.3.15) can be used to show that N_t is the total number of modes supported by the GRIN device. This can be seen by noting that, when $N(\beta) = N_t$, the propagation constant is reduced to $\beta = n_c k_0$, which is the lowest possible value of β. For a parabolic index profile ($p = 2$), $N_t = V^2/4$, and this number increases to $V^2/2$ for a step-index profile with $p \gg 1$. As V can exceed 100 for GRIN rods with $a > 1$ mm, such rods support thousands of modes. The number is smaller for GRIN fibers but still exceeds 100 when $a = 25$ μm.

If we compare Eq. (2.3.15) with Eq. (2.3.8), we conclude that for a parabolic index profile with $p = 2$,

$$N(\beta) = (\ell + 2m - 1)^2. \tag{2.3.17}$$

When p does not deviate much from 2, we expect this relationship to hold approximately. Its use allows us to obtain the following useful formula for β_{lm}:

$$\beta_{lm} = n_0 k_0 \left[1 - 2\Delta \left(\frac{2(\ell + 2m - 1)}{V\sqrt{2p/(p+2)}} \right)^{2p/(p+2)} \right]^{1/2}. \tag{2.3.18}$$

This result can be used to analyze GRIN devices whose parabolic profile is not perfect but can be fitted with a power law with p values close to two.

2.4 Modal Excitation and Interference

As we saw in Section 2.3, a GRIN rod or fiber supports a large number of modes that depend on its core size. The important question from a practical standpoint is how many of these modes are excited when an optical beam is launched into such a device. In a typical experiment, the size, shape, and polarization of the input beam are controlled using lenses and other optical elements. Assuming that the beam is polarized linearly in the x-y plane, the electric field at the input plane located at $z = 0$ has the form $\tilde{\mathbf{E}}(x, y, 0) = \hat{\mathbf{p}} A(x, y, 0)$, where $\hat{\mathbf{p}}$ is the unit vector representing the state of polarization and $A(x, y, 0)$ governs the spatial distribution of the incoming beam.

2.4.1 Modal Coupling Efficiency

To find which modes are excited by a beam incident on a GRIN device, we expand $A(x, y, 0)$ in terms of its modes as

$$A(x, y, 0) = \sum_{l=0}^{\infty} \sum_{m=1}^{\infty} C_{lm} U_{lm}(x, y), \tag{2.4.1}$$

where the double sum covers all LP_{lm} modes of the device. The expansion coefficients are found by multiplying this equation with $U_{l'm'}^*(x, y)$ and integrating over the transverse plane. Using the mode-orthogonality relation in Eq. (2.2.16), the result is given by

$$C_{lm} = \iint_{-\infty}^{\infty} U_{lm}^*(x, y) A(x, y, 0) \, dx \, dy. \tag{2.4.2}$$

It shows that the extent of spatial overlap of each mode with the input beam determines how strongly that mode will be excited. Depending on the size and shape of

the input beam, some of these coefficients may vanish. The quantity $|C_{lm}|^2$ provides
a measure of the coupling efficiency.

It is easy to find the amplitude $A(x,y,z)$ from Eq. (2.4.1) after the beam has
propagated inside the GRIN medium to arrive at the location z. As the spatial profile
of each mode does not change during propagation, and it only acquires a phase shift,
$A(x,y,z)$ is given by

$$A(x,y,z) = \sum_{l=0}^{\infty} \sum_{m=1}^{\infty} C_{lm} U_{lm}(x,y) \exp(i\beta_{lm}z), \qquad (2.4.3)$$

where β_{lm} is given in Eq. (2.2.23) in the case of a parabolic index profile. Once C_{lm}
are known from Eq. (2.4.2), this series can be summed to find how the optical beam
evolves inside a GRIN medium.

Although the modal approach provides a solution to the beam-propagation problem in principle, it is not practical for GRIN rods or fibers supporting hundreds or
thousands of modes. We will see in Section 3.2 that the double sum in Eq. (2.4.3)
can be carried out analytically to obtain a propagation kernel that has proven quite
useful in practice. In some GRIN fibers, the core radius is made so small that the
fiber supports only a few modes. The modal approach is useful for such fibers.

As an example, consider the case of a collimated Gaussian beam launched into
a GRIN fiber such that its intensity peaks precisely at the core's center. In this
situation, the input field is given by

$$A(x,y,0) = A_0 \exp(-\rho^2/2w_0^2), \qquad (2.4.4)$$

where A_0 is the peak amplitude of the beam and w_0 is its spot size at $z = 0$. As this
beam is radially symmetric (no ϕ dependence), it is easier to perform the integration
in Eq. (2.4.2) using the cylindrical coordinates:

$$C_{lm} = A_0 \int_0^{\infty} \int_0^{2\pi} U_{lm}^*(\rho, \phi) \exp\left(-\frac{\rho^2}{2w_0^2}\right) \rho \, d\rho \, d\phi. \qquad (2.4.5)$$

When we use U_{lm} given in Eq. (2.2.20) and carry out the integration over ϕ, we find
that C_{lm} vanishes for all values of $l > 0$. As a result, such a Gaussian beam excites
only the radially symmetric LP_{0m} modes. The integrals in Eq. (2.4.5) can be done
analytically for $l = 0$ using the known result [10]

$$\int_0^{\infty} e^{-bx} L_n(x) \, dx = (b-1)^n b^{-(n+1)}. \qquad (2.4.6)$$

The resulting expression for C_{0m} is:

$$C_{0m} = A_0 \frac{\sqrt{\pi}}{q} \left(\frac{1 - q^2 w_0^2}{1 + q^2 w_0^2}\right)^{m-1} \left(\frac{2q^2 w_0^2}{1 + q^2 w_0^2}\right). \qquad (2.4.7)$$

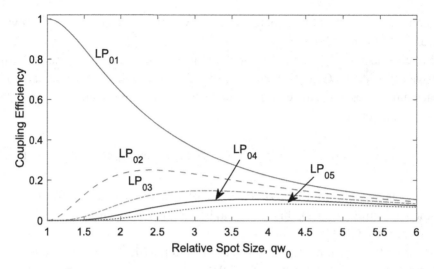

Figure 2.7 Coupling efficiency η_{0m} plotted as a function of qw_0 for five LP_{0m} modes when a Gaussian beam with the spot size w_0 is launched into a GRIN fiber.

In practice, one wants to know the coupling efficiency η, defined as the fraction of input power coupled into a specific mode. Using $P_0 = \int\int |A(x, y, 0)|^2 dx\,dy$ for the input power, the coupling efficiency is given by

$$\eta_{lm} = |C_{lm}|^2/P_0, \tag{2.4.8}$$

In the case of a Gaussian beam with the amplitude given in Eq. (2.4.4), $\eta_{lm} = 0$ for all $l > 0$. Using C_{0m} from Eq. (2.4.7) with $P_0 = (\pi w_0^2)A_0^2$, η_{0m} is found to be

$$\eta_{0m} = \left(\frac{1 - q^2 w_0^2}{1 + q^2 w_0^2}\right)^{2m-2} \left(\frac{2qw_0}{1 + q^2 w_0^2}\right)^2. \tag{2.4.9}$$

This expression depends on a single parameter qw_0 that can be written as the ratio w_0/w_g, where w_g is the spot size of the fundamental mode given in Eq. (2.2.14). As one may expect, the coupling efficiency depends on the spot size w_0 of the input beam and is the largest for the LP_{01} mode, whose shape resembles that of a Gaussian beam. Indeed, η_{01} becomes 1 (coupling efficiency 100%) when the input spot size is matched to that of the LP_{01} mode so that $qw_0 = 1$. Only the fundamental mode of the GRIN fiber is excited in that situation.

When the input spot size w_0 is larger than w_g ($qw_0 > 1$), several LP_{0m} modes of the GRIN fiber are excited simultaneously, but the largest coupling still occurs for the LP_{01} mode. Figure 2.7 shows η_{0m} as a function of qw_0 for the five LP_{0m} modes ($m = 1$–5) for a Gaussian beam with the input spot size w_0. For $qw_0 = 2$, η_{01} is reduced to 64%, about 24% of the input power is coupled in to the LP_{02} mode, and only 12% goes to other higher-order modes. However, when the input beam is so wide that qw_0 exceeds 6, all five modes are almost equally excited, each containing about 10% of the input power. As seen from Eq. (2.4.9), η_{0m} becomes nearly constant in the limit $qw_0 \gg 1$ and takes a relatively small value scaling as $(qw_0)^{-2}$.

Modes with $l \neq 0$ can also be excited if radial symmetry of the coupling configuration is broken. A simple way to realize this in practice is to shift the center of the input Gaussian beam from the core's axis. When the peak of the Gaussian beam is shifted by a distance s along the x axis, the input field in Eq. (2.4.4) is replaced with

$$A(x, y, 0) = A_0 \exp\left(-[(x-s)^2 + y^2]/2w_0^2\right). \tag{2.4.10}$$

Using this form and converting it to the cylindrical coordinates, Eq. (2.4.4) becomes

$$C_{lm} = A_0 \int_0^\infty \int_0^{2\pi} U_{lm}^*(\rho, \phi) \exp\left(-\frac{\rho^2 + s^2}{2w_0^2} + \frac{2s\rho \cos\phi}{2w_0^2}\right) \rho \, d\rho \, d\phi. \tag{2.4.11}$$

The exponential term containing $\cos\phi$ breaks the radial symmetry and ensures that C_{lm} is finite for the modes with $l \neq 0$.

2.4.2 Multimode Interference

When an input beam excites several modes of a GRIN fiber, the intensity at a distance z, defined as $I = |A|^2$, can be written from Eq. (2.4.2) as

$$I(x, y, z) = \left| \sum_{l=0} \sum_{m=1} C_{lm} U_{lm}(x, y) \exp(i\beta_{lm} z) \right|^2, \tag{2.4.12}$$

where the sum is limited to only those values of l and m that correspond to the modes excited by the input beam. As the values of β_{lm} differ for different mode groups, the intensity pattern changes considerably with z because of interference among different modes. Indeed, the output of GRIN fibers exhibits what is referred to as a speckle pattern.

We can write Eq. (2.4.12) in a simpler form using the concept of mode groups identified with the integer $g = l + 2m - 1$. Using Eq. (2.2.23) for β_{lm}, the intensity takes the form

$$I(x, y, z) = \left| \sum_{g=1}^M C_{lm} U_{lm}(x, y) e^{-igbz} \right|^2, \tag{2.4.13}$$

where M is the number of excited mode groups. If only the first two modes are excited, the intensity is given by

$$I(x, y, z) = |C_{01} U_{01}(x, y)|^2 + |C_{11} U_{11}(x, y)|^2 + 2|C_{01} C_{11} U_{01} U_{11}| \cos(bz + \delta\phi), \tag{2.4.14}$$

$\xi = 1$ $\xi = 2$ $\xi = 3$ $\xi = 4$

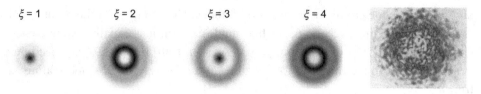

Figure 2.8 Simulated intensity patterns at few distances when an on-axis Gaussian beam is launched into a GRIN fiber. Last image is an actual speckle pattern for a GRIN fiber of 50 μm core diameter.

where $\delta\phi$ is the relative phase difference. When $\delta\phi = 0$, the interference term vanishes at distances such that $bz = (j + \frac{1}{2})\pi$, where j is an integer. At these distances, Eq. (2.4.13) takes the form

$$I(x,y,z) = \sum_{g=1}^{M} |C_{lm}U_{lm}(x,y)|^2 . \tag{2.4.15}$$

At other distances, the interference terms must be included.

As an example, Figure 2.8 shows the output intensity patterns at four distances ($\xi = bz$) when an on-axis Gaussian beam is launched into a GRIN fiber with a large spot size such that multiple LP_{0m} modes are excited with nearly the same coupling efficiency. Even though the radial symmetry is preserved, the spatial pattern is far from being Gaussian and takes distinctly different shapes as the beam propagates inside the GRIN fiber. The pattern loses its radial symmetry when a beam is launched off-center, or when images of objects are transmitted through a GRIN fiber. Moreover, it takes the form of a speckle pattern when random phase variations and mode coupling that invariably occur in any fiber are considered. The last image in Figure 2.8 shows an example of such a pattern at the output of a GRIN fiber when an initially Gaussian beam from a laser is transmitted through it. In recent years, machine-learning techniques have been developed to recover images from such speckle patterns [12–14].

2.4.3 Excitation of a Specific Mode

For some applications, it is desirable to selectively excite a specific mode of a GRIN fiber [15]. Selective excitation of a mode was considered as early as 1978. In one scheme, the peak of a narrow Gaussian beam, whose width was chosen judiciously, was moved away from the core's center to excite different modes of a GRIN fiber [16]. In another, the beam was launched from one side of the fiber through a prism [17]. Such a side-launch technique requires removal of the cladding and is not used in practice. Spatial filtering of a laser beam through an amplitude or phase mask has been found to be a better approach for selective mode excitation. Indeed, mode-division multiplexing was first proposed in 1982 based on the use of such masks [18].

It is easy to understand why an amplitude mask works. If the mask blocks the central part of a Gaussian beam, spatial overlap of the resulting ring-shape beam with the fundamental LP_{01} mode of the fiber may be reduced so much that almost no power is coupled into this mode. As a result, if the beam is still radially symmetric, it will couple most of its power into the LP_{02} mode. The idea can be generalized to shape the input beam such that it overlaps most strongly with a specific LP_{lm} mode.

To understand the operation of a phase mask, we should ask how the phase of a LP_{lm} mode varies in the *x*-*y* plane. It is easy to see from Eq. (2.2.20) that the phase is uniform across this plane for all modes with $\ell = 0$. However, it varies considerably for modes with $\ell \neq 0$ because of the presence of a ϕ-dependent factor. For example, the phase of the $LP_{11}^{(e)}$ mode jumps by π for negative values of *x* such that ϕ is in the range $\pi/2 < \phi < 3\pi/2$. The same phase jump occurs for the $LP_{11}^{(o)}$ mode for negative values of *y*. Multiple phase jumps occur for other modes with $\ell > 1$. Phase masks exploit such phase variations for their operation. If a mask divides a Gaussian beam into two semicircular parts with a π phase jump across them, the beam will couple most of its power into the LP_{11} mode.

After 1985, many techniques were developed that made use of amplitude and phase masks for exciting a specific mode of a GRIN fiber [19–21]. Among these, spatial light modulators provide the most flexible solution. Such devices provide full control of the phase over the entire cross section of an input beam and can be digitally programmed to excite one or more specific modes of a GRIN fiber [22, 23]. With the advent of mode-division multiplexing, the use of volume holographic gratings has also become common [24–26]. Special devices, known as spatial multiplexers, have been developed that allow one to transmit different data streams over different modes of the same GRIN fiber [4, 15].

The even and odd LP_{lm} modes of a GRIN fiber with $\ell \neq 0$ can be used to create a vortex mode that carries some OAM, whose magnitude depends on the integer ℓ [27]. In recent years, several techniques have been developed for generating and propagating such vortex beams over considerable lengths of a GRIN fiber [28–31]. Chapter 8 is devoted to such optical beams.

2.4.4 Intermodal Dispersion

As we saw in Section 2.2, the propagation constant β of each mode varies with the frequency or wavelength of the incident beam. In the case of a CW laser beam, the spectral bandwidth $\delta\omega$ of the beam is typically small enough that we can treat the incident radiation as being monochromatic. In this case, only the transit time of a beam across a GRIN device depends on its central frequency ω_0 because of chromatic dispersion.

The situation becomes more dramatic for a pulsed beam containing a train of short optical pulses. In this case, the spectral width of the beam depends on the duration of pulses. It becomes wide enough for short pulses that the effects of chromatic dispersion may need to be included for pulses shorter than 10 ps. Using $j = l + 2m - 1$ for the mode group j, its propagation constant from Eq. (2.2.23) can be written as

$$\beta_j(\omega) = \frac{\omega}{c}[n_0(\omega) - j\delta n(\omega)], \qquad \delta n = b/(n_0 k_0). \tag{2.4.16}$$

Both n_0 and δn depend on the frequency ω. The contribution of $n_0(\omega)$, called material dispersion, has been discussed in Section 1.3.2. The contribution of $\delta n(\omega)$ is called waveguide dispersion because its magnitude can be controlled by modifying the design of the GRIN device acting as a waveguide.

As was done in Section 1.3.2, we expand $\beta_j(\omega)$ in a Taylor series around ω_0 as

$$\beta_j(\omega) = \beta_{0j} + \beta_{1j}(\omega - \omega_0) + \frac{\beta_{2j}}{2}(\omega - \omega_0)^2 + \cdots, \quad \beta_{mj} = (d^m \beta_j/d\omega^m)_{\omega = \omega_0}. \tag{2.4.17}$$

The parameter β_{1j} is related to the group delay of the jth mode and β_{2j} is the GVD parameter of that mode. These parameters are different for different modes because each mode travels with a different speed inside a GRIN medium. The difference in the group delay for any two modes provides a measure of the differential group delay (DGD). Its presence leads to distortion of an optical pulse when its energy is carried by several modes that arrive at different times at the output end of the fiber. DGD is relatively large for step-index fibers but becomes acceptable for GRIN fibers with a parabolic index profile [11, 32].

We can calculate β_{1j} for a GRIN medium by taking the frequency derivative of $\beta_j(\omega)$ in Eq. (2.4.16) and setting $\omega = \omega_0$:

$$\beta_{1j} = \left[\frac{n_g}{c} - \frac{j}{c}\delta n_g\right]_{\omega = \omega_0}, \tag{2.4.18}$$

where n_g is the group index and δn_g is defined the same way as n_g. The first term provides a constant delay for all modes, but the second term is different for different mode groups. We can use this term to calculate the DGD for any mode, defined as $\tau_{1j} = \beta_{1j} - \beta_{11}$, where we used the fundamental mode ($j = 1$) as the reference mode. Using Eq. (2.4.18), DGD depends on δn as

$$\tau_{1j} = (1 - j)[\delta n + \omega(d\delta n)/d\omega)]. \tag{2.4.19}$$

Taking the frequency derivative of δn, the DGD is found to be related to $d\Delta/d\omega$ as

$$\tau_{1j} = (1 - j)(a\sqrt{2\Delta})^{-1}(d\Delta/d\omega). \tag{2.4.20}$$

Recalling that $\Delta = 1 - n_c/n_0$, we expect this derivative to be small. Typical values of DGD are ~ 0.1 ps/m for $j = 2$ at wavelengths near 1.55 μm, and they increase linearly for higher values of j.

DGD is not desirable for optical communication systems because it leads to distortion and broadening of optical pulses used for data transmission. However, compared to step-index fibers, GRIN fibers with a nearly parabolic index profile exhibit considerably less DGD. This was the reason why such fibers were developed during the 1970s for applications in optical communication systems [4]. More recently, GRIN fibers have been used for space-division multiplexing in such systems.

The DGD expression in Eq. (2.4.20) is derived assuming a parabolic shape for the index profile of a GRIN fiber. One should consider how the DGD depends on the exponent p when the power-law profile given in Eq. (2.3.1) is employed [2, 3]. For this purpose, we can use β in Eq. (2.3.15) for calculating the DGD. Treating N as a constant while taking the frequency derivative and assuming $\Delta \ll 1$, the following result is obtained to the second order in Δ [9]:

$$\beta_1 = \frac{n_0}{c}\left[1 + \Delta\frac{p-2}{p+2}\left(\frac{N}{N_t}\right)^{\frac{p}{p+2}} + \Delta^2\frac{3p-2}{2(p+2)}\left(\frac{N}{N_t}\right)^{\frac{2p}{p+2}}\right]. \tag{2.4.21}$$

If we neglect the Δ^2 term, the group delay becomes the same for all modes for a parabolic index profile with $p = 2$. This result agrees with Eq. (2.4.18). When we include the quadratic term, β_1 becomes mode independent if we choose $p = 2(1 - \Delta)$.

What matters for optical communication systems is the pulse broadening induced by DGD inside a GRIN fiber. It turns out that this broadening can be calculated for a power-law index profile in the WKB approximation [11, 32]. The results show that minimum broadening occurs for an optimum value of p that differs from 2 by a small amount, which depends on the wavelength of pulses, among other things [3]. In the specific case in which an incoming pulse excites all modes of a GRIN fiber with the same coupling efficiency, minimum pulse broadening occurs for $p = 2(1 - 1.2\Delta)$ [32].

The GVD parameter β_{2j} can also be different for different modes. Its value is found by taking the frequency derivative of $\beta_{1j}(\omega)$ given in Eq. (2.4.18) and setting $\omega = \omega_0$:

$$\beta_{2j} = \frac{1}{c}\left[\frac{dn_g}{d\omega} - j\frac{d\delta n_g}{d\omega}\right]_{\omega=\omega_0}. \tag{2.4.22}$$

As before, the first term represents the material's contribution to the GVD (see Figure 1.6), while the second term provides the waveguide contribution that is different for different mode groups. The material's contribution vanishes at the zero-dispersion wavelength of the fiber and is about -25 ps^2/km near 1.55 μm. The second term in Eq. (2.4.21) changes this value by an amount that is different for different modes, but such changes are relatively small at any specific wavelength.

2.5 Optical Propagation in GRIN Media

When an optical beam excites a large number of modes inside a GRIN medium, the use of modal approach becomes impractical. In a non-modal approach, Eq. (2.1.13) is solved as an initial-value problem, where $\tilde{E}$ is specified at the entrance of a GRIN medium located at $z = 0$. The slowly varying amplitude of the electric field is introduced as indicated in Eq. (2.1.14) to obtain Eq. (2.1.15). It is common to choose $k = n_0 k_0$. Using the parabolic index profile given in Eq. (1.2.5) with this choice, Eq. (2.1.15) takes the form

$$2ik\frac{\partial A}{\partial z} + \frac{\partial^2 A}{\partial x^2} + \frac{\partial^2 A}{\partial y^2} - k^2 b^2 (x^2 + y^2)A = 0. \tag{2.5.1}$$

where b is the index gradient defined in Eq. (1.2.5). This equation should be solved to find $A(x, y, z)$ at a distance z in terms of $A(x, y, 0)$. It is useful to introduce normalized variables as $q_1 = x/w_g$, $q_2 = y/w_g$ and $\xi = bz$, where $w_g = (kb)^{-1/2}$. Equation (2.5.1) then becomes

$$2i\frac{\partial A}{\partial \xi} + \frac{\partial^2 A}{\partial q_1^2} + \frac{\partial^2 A}{\partial q_2^2} - (q_1^2 + q_2^2)A = 0. \tag{2.5.2}$$

The problem can be simplified further with the method of separation of variables. As Eq. (2.5.2) is linear, we assume that its solution is separable in the variables q_1 and q_2 and can be factored as $A(q_1, q_2, z) = U(q_1, z)V(q_2, z)$. Using this form in Eq. (2.5.2), it is easy to verify that both $U(q, z)$ and $V(q, z)$ satisfy the same equation:

$$2i\frac{\partial U}{\partial \xi} + \frac{\partial^2 U}{\partial q^2} - q^2 U = 0. \tag{2.5.3}$$

We discuss several techniques that can be used to solve Eq. (2.5.2) or (2.5.3).

2.5.1 Quantum-Mechanical Approach

As discussed in Appendix A, our classical GRIN problem is related to a quantum harmonic oscillator. This can be seen by writing Eq. (2.5.2) in the form of the Schrödinger equation of quantum mechanics (with $\hbar = 1$)

$$i\frac{\partial A}{\partial \xi} = \hat{H}A, \tag{2.5.4}$$

where the Hamiltonian has the form

$$\hat{H} = \frac{1}{2}\left(\hat{p}_1^2 + \hat{p}_2^2\right) + \frac{1}{2}\left(\hat{q}_1^2 + \hat{q}_2^2\right). \tag{2.5.5}$$

Here, $\hat{q}_1$ and $\hat{q}_2$ are the position operators, and the corresponding momentum operators are defined as $\hat{p}_j = -i(d/dq_j)$ for $j = 1, 2$. They satisfy the usual commutation relations:

$$[\hat{q}_i, \hat{q}_j] = 0, \quad [\hat{p}_i, \hat{p}_j] = 0, \quad [\hat{q}_i, \hat{p}_j] = i\delta_{ij}. \quad (2.5.6)$$

The Hamiltonian in Eq. (2.5.5) is identical to that of a two-dimensional harmonic oscillator. Note that ξ plays the role of time for the GRIN problem.

As the Hamiltonian is separable in the q_1 and q_2 dimensions, the wave function can be factored as $A(q_1, q_2, \xi) = \Psi(q_1, \xi)\Psi(q_2, \xi)$. This feature allows us to split the two-dimensional problem into two one-dimensional problems. If we adopt the Dirac's notation to denote a quantum state, we need to solve

$$i\frac{d}{d\xi}|\Psi\rangle = \hat{H}|\Psi\rangle, \quad \hat{H} = \frac{1}{2}(\hat{p}^2 + \hat{q}^2). \quad (2.5.7)$$

This equation is solved in Appendix A by introducing the annihilation and creation operators [33] and writing the Hamiltonian in Eq. (2.5.7) in the form

$$\hat{H} = \hat{a}^\dagger \hat{a} + \tfrac{1}{2}. \quad (2.5.8)$$

The Fock states, which are the eigenstates of this Hamiltonian with the eigenvalues $E_n = n + \tfrac{1}{2}$, correspond to the Hermite–Gauss modes found in Section 2.2. The corresponding eigenvalues lead to the modal propagation constant β_{mn} given in Eq. (2.2.11) when both spatial dimensions are taken onto account. The integers m and n start as 0 and represent different modes of the GRIN medium.

In the Schrödinger picture, operators do not evolve with propagation, but the wave function changes with ξ. Because $\hat{H}$ does not depend on ξ, we can solve Eq. (2.5.7) formally to obtain

$$|\Psi(\xi)\rangle = \hat{U}(\xi)|\Psi(0)\rangle = \exp(-i\xi\hat{H})|\Psi(0)\rangle, \quad (2.5.9)$$

where $\hat{U}(\xi) = \exp(-i\xi\hat{H})$ is a unitary evolution operator. The form of the Hamiltonian in Eq. (2.5.8) allows one to apply many results obtained for harmonic oscillators to a GRIN medium [35–37]. For example, it is possible to use the evolution operator $\hat{U}(\xi)$ in Eq. (2.5.9) to obtain the amplitude $A(x, y, z)$ as a surface integral involving $A(x, y, 0)$ [38]. This integral is used in Section 3.2 to discuss beam propagation inside a GRIN medium.

2.5.2 Moment-Based Quantum Technique

We use in this section the Heisenberg picture to solve Eq. (2.5.4) and make useful predictions that apply to any input beam [34]. In the Heisenberg picture, operators evolve with ξ, but the wave function remains in its initial state at $\xi = 0$. The average of any operator $\hat{O}(\xi)$ is calculated using the relation

$$\langle O(\xi) \rangle = \int A^*(q_1, q_2, 0) \, \hat{O}(\xi) A(q_1, q_2, 0) \, dS, \tag{2.5.10}$$

where the surface integration is over the transverse plane at $\xi = 0$. We can use Heisenberg's equation of motion to find how this average changes inside a GRIN medium:

$$i \frac{d\langle O \rangle}{d\xi} = \langle [\hat{O}, \hat{H}] \rangle, \tag{2.5.11}$$

where we have assumed that $\hat{O}$ does not depend on ξ explicitly. As an example, when $\hat{O} = 1$, the average in Eq. (2.5.10) provides the beam's power at a distance ξ. It follows from Eq. (2.5.11), that this average does not change with ξ, and the initial power of the beam is conserved as it propagates inside the GRIN medium. This is expected on physical grounds, as all losses were ignored in Eq. (2.5.1).

The moments of $\hat{q}_j$ and $\hat{p}_j$ ($j = 1, 2$) can be calculated using

$$\langle q_j^m \rangle = \int q_j^m |A(q_1, q_2)|^2 \, dS, \quad \langle p_j^m \rangle = \int A^*(q_1, q_2) \left(-i \frac{d}{dq_j} \right)^m A(q_1, q_2) \, dS, \tag{2.5.12}$$

where m is an integer. Changes in $\langle q_j^m \rangle$ with ξ are calculated using $\hat{O} = \hat{q}_j^m$ in Eq. (2.5.11). As a simple example, consider the first moment ($m = 1$), representing the location of the intensity peak. It is easy to show that

$$\frac{d\langle q_j \rangle}{d\xi} = -\langle p_j \rangle, \qquad \frac{d\langle p_j \rangle}{d\xi} = \langle q_j \rangle. \tag{2.5.13}$$

These equations can be combined to obtain a harmonic oscillator equation with the general solution:

$$\langle q_j(\xi) \rangle = C_1 \cos \xi + C_2 \sin \xi, \tag{2.5.14}$$

where the constants C_1 and C_2 are found from the initial conditions.

If the input beam is launched such that its peak is shifted from the axis of the GRIN medium and is located at the point (x_0, y_0), the initial conditions are $\langle x \rangle = x_0$ and $\langle y \rangle = y_0$ at $z = 0$. We also need to know the initial values of $\langle p_1 \rangle$ and $\langle p_2 \rangle$. Both are zero when the input beam is collimated with a planar wave front at $z = 0$. Using these values and converting Eq. (2.5.14) to the original variables, we obtain

$$\langle x(z) \rangle = x_0 \cos(bz), \qquad \langle y(z) \rangle = y_0 \cos(bz). \tag{2.5.15}$$

Thus, the location of the beam's center oscillates in a sinusoidal manner with the period $L_p = 2\pi / b$. It follows from Eq. (2.5.15) that the beam's center does not move when it coincides initially with the axis of the GRIN medium. These conclusions hold for any spatial shape of the optical beam.

One can also find the z dependence of a beam's width as it propagates inside a GRIN medium. This width is related to spreading of the wave packet. When the beam is aligned with the core's center, the variance is equal to the second moment

$\langle \rho^2 \rangle$, where $\rho^2 = x^2 + y^2$. In the normalized variables, this variance is calculated using $\sigma^2 = \langle q_1^2 \rangle + \langle q_2^2 \rangle$. As before, we use Eq. (2.5.11) to obtain [34]:

$$\frac{d}{d\xi} \langle q_j^2 \rangle = -\langle q_j p_j + p_j q_j \rangle, \qquad \frac{d}{d\xi} \langle p_j^2 \rangle = \langle q_j p_j + p_j q_j \rangle, \qquad (2.5.16)$$

where $j = 1, 2$. Using again Eq. (2.5.11), we find

$$\frac{d}{d\xi} \langle q_j p_j + p_j q_j \rangle = 2 \langle q_j^2 \rangle - 2 \langle p_j^2 \rangle. \qquad (2.5.17)$$

The preceding results can be combined to obtain the following equation for σ^2:

$$\frac{d^2 \sigma^2}{d\xi^2} = 4 \langle H \rangle - 4 \sigma^2, \qquad (2.5.18)$$

where $\langle H \rangle$ does not change with ξ. This equation has the solution:

$$\sigma^2(\xi) = \langle H \rangle + a_r \cos(2\xi + \phi_r), \qquad (2.5.19)$$

where a_r is the amplitude and ϕ_r is the phase of sinusoidal oscillations. These constants are set by the initial conditions at $\xi = 0$. Using $\sigma = \sigma_0$ and $d\sigma/d\xi = 0$ at $\xi = 0$, we find $a_r = \sigma_0^2 - \langle H \rangle$ and $\phi_r = 0$. Using these values, we obtain

$$\sigma^2(\xi) = \langle H \rangle + [\sigma_0^2 - \langle H \rangle] \cos(2\xi), \qquad (2.5.20)$$

where $\langle H \rangle$ depends on the shape of input beam. Recall $\xi = bz$ and $\sigma = w_r/w_g$, where w_r is the root-mean-square (RMS) width of the beam. Equation (2.5.20) shows that the width of any input beam oscillates inside a grin medium with a period $\xi = \pi$. During each period, the width takes its minimum value at the midpoint $\xi = \pi/2$, and this value depends on the spot size w_g of the fundamental mode. An advantage of the moment method is that it allows us to make specific predictions that apply to all input beams.

2.5.3 Variational Method

Equation (2.5.1), describing evolution of a CW beam inside a GRIN medium, can also be solved with the variational method. This method was applied in 1992 to a nonlinear GRIN medium, assuming a Gaussian shape for the input beam [39]. It allows one to calculate how the beam's parameters (such as its amplitude, width, and phase) vary as the beam propagates inside the medium.

The variational method is well known from classical mechanics and is used in many different contexts [40]. Mathematically, it makes use of the Lagrangian $\mathcal{L}$ defined as

$$\mathcal{L} = \int_{-\infty}^{\infty} \mathcal{L}_d(u, u^*) \, dx \, dy, \qquad (2.5.21)$$

where the Lagrangian density $\mathscr{L}_d$ is a function of the generalized coordinate u and u^*, both of which evolve with z. Minimization of the action, $\mathscr{A} = \int \mathscr{L} \, dz$, requires that $\mathscr{L}_d$ satisfy the Euler–Lagrange equation

$$\frac{\partial}{\partial x} \left(\frac{\partial \mathscr{L}_d}{\partial u_x} \right) + \frac{\partial}{\partial y} \left(\frac{\partial \mathscr{L}_d}{\partial u_y} \right) - \frac{\partial \mathscr{L}_d}{\partial u} = 0, \tag{2.5.22}$$

where a subscript on u denotes a derivative of u with respect to that variable. One can show that Eq. (2.5.1) is obtained using the Lagrangian density

$$\mathscr{L}_d = 2ik \left(A \frac{\partial A^*}{\partial z} - A^* \frac{\partial A}{\partial z} \right) + \left| \frac{\partial A}{\partial x} \right|^2 + \left| \frac{\partial A}{\partial y} \right|^2 + k^2 b^2 (x^2 + y^2) |A|^2, \tag{2.5.23}$$

when A^* acts as the generalized coordinate u in Eq. (2.5.22).

To make further progress, we assume that the input beam has a Gaussian shape and evolves inside the GRIN medium in a self-similar manner. In other words, it retains its Gaussian shape even though its width, amplitude, and wave-front curvature change with z. A suitable form of the amplitude $A(x, y, z)$ is:

$$A(x, y, z) = A_p \exp \left(-\frac{x^2 + y^2}{2w^2} + \frac{ikh}{2} (x^2 + y^2) + i\psi \right). \tag{2.5.24}$$

The four parameters in this equation (A_p, w, h, and ψ) vary with z as the beam propagates. Total power of the beam, $P_0 = \iint |A|^2 \, dx \, dy$, is related to A_p and w as $P_0 = A_p^2(\pi w^2)$.

We substitute $\mathscr{L}_d$ from Eq. (2.5.23) into Eq. (2.5.21), use the form of A from Eq. (2.5.24), and integrate over the transverse coordinates. The resulting Lagrangian is found to be

$$\mathscr{L} = P_0 \left[2k \frac{d\psi}{dz} + k^2 w^2 \frac{dh}{dz} + w^2 \left(\frac{1}{w^4} + k^2 h^2 \right) + k^2 b^2 w^2 \right]. \tag{2.5.25}$$

It is used to obtain the differential equations for the beam's parameters (P_0, w, h, and ψ) by applying the reduced Euler–Lagrange equation,

$$\frac{\partial}{\partial z} \left(\frac{\partial \mathscr{L}}{\partial u_z} \right) - \frac{\partial \mathscr{L}}{\partial u} = 0, \tag{2.5.26}$$

where u plays the role of a parameter. The resulting equations take the form [39]:

$$\frac{dP_0}{dz} = 0, \quad \frac{dw}{dz} = hw, \quad \frac{d\psi}{dz} = -\frac{1}{kw^2}, \tag{2.5.27}$$

$$\frac{dh}{dz} + \left(h^2 + b^2 - \frac{1}{k^2 w^4} \right) = 0. \tag{2.5.28}$$

We combine the w and h equations to obtain the following equation:

$$\frac{d^2w}{dz^2} + b^2w - \frac{1}{k^2w^3} = 0. \tag{2.5.29}$$

If we solve Eq. (2.5.29) to obtain $w(z)$, Eq. (2.5.27) can be used to find h, and ψ. Thus, the problem has been reduced to solving a single second-order differential equation for the beam's width. Physically, the second and third terms in this equation represent the effects of index gradient and diffraction, respectively.

To solve Eq. (2.5.29), we introduce $s = (kb)^{1/2}w = w/w_g$ and $\xi = bz$ to obtain

$$\frac{d^2s}{d\xi^2} + s - \frac{1}{s^3} = 0. \tag{2.5.30}$$

Multiplying Eq. (2.5.30) with $2(ds/d\xi)$ and integrating, s^2 is found to satisfy

$$\frac{ds^2}{d\xi} = 2\sqrt{Cs^2 - (1 + s^4)}. \tag{2.5.31}$$

Using the initial conditions, $s = s_0$ and $ds/d\xi = 0$ when $\xi = 0$ (a planar wave front at $z = 0$), the constant C is found to be $C = s_0^2 + s_0^{-2}$. Noting that $C > 2$ for any value of s_0, the solution of Eq. (2.5.31) takes the form

$$s^2(\xi) = s_0^2 \cos^2 \xi + s_0^{-2} \sin^2 \xi. \tag{2.5.32}$$

In terms of the original variables, the preceding solution becomes

$$w(z) = w_0[\cos^2(2\pi z/L_p) + C_f^2 \sin^2(2\pi z/L_p)]^{1/2}, \tag{2.5.33}$$

where $L_p = 2\pi/b$ and $C_f = (w_g/w_0)^2$ with $w_g = (kb)^{-1/2}$. These two parameters depend on the GRIN parameter b and play an important role in later chapters.

The solution for $w(z)$ shows that the width of the Gaussian beam follows a periodic pattern with the period L_p. The length L_p is also known as the pitch of a GRIN device. Figure 2.9 shows the oscillatory behavior of the beam's width by plotting w/w_0 as a function of distance for several values of the parameter C_f. For $C_f > 1$, the width increases initially because diffraction dominates, takes its maximum value $C_f w_0$ at $z = L_p/4$, recovers its original value w_0 at $z = L_p/2$, and the whole process repeats. The same thing happens For $C_f < 1$ except that the width decreases initially, takes its minimum value $C_f w_0$ at $z = L_p/4$, and then recovers the original value w_0 at $z = L_p/2$.

The case $C_f = 1$ is the only situation in which the beam width does not change and remains fixed to its input value w_0. This is possible if only a single mode of the GRIN medium is excited by the input beam. This is indeed the case because $C_f = 1$

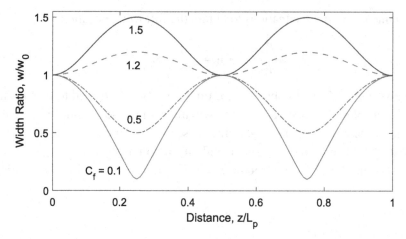

Figure 2.9 Periodic variations of the width of a Gaussian beam inside a GRIN medium. The ratio $w(z)/w_0$ is plotted as a function of the normalized distance z/L_p for several values of the parameter C_f.

is realized for $w_0 = w_g$. It is easy to show that the only the fundamental LP_{01} mode is excited in this situation. For this purpose, we write C_f in the form

$$C_f = \frac{w_g^2}{w_0^2} = \frac{L_p}{2\pi k w_0^2}. \tag{2.5.34}$$

It follows from the discussion in Section 2.4.1, that such a beam couples 100% of its power into the LP_{01} mode, and the size of the beam should not change during its propagation inside the GRIN medium. In contrast, multiple modes are excited when $C_f \neq 1$, and the width follows a periodic pattern as indicated in Eq. (2.5.33).

We use $w(z)$ from Eq. (2.5.33) into Eq. (2.5.27) to calculate A_p, b, and ψ. Clearly, $A_p(z) = A_p(0)w_0/w(z)$ and h depends on the derivative dw/dz. The phase ψ requires an integration and is found to be

$$\psi(\xi) = \int_0^\xi \frac{C_f\, du}{\cos^2 u + C_f^2 \sin^2 u} = \tan^{-1}(C_f \tan \xi), \tag{2.5.35}$$

It will be seen in Section 3.2 that the variational solution in Eq. (2.5.24) agrees with the exact solution obtained by summing over all modes of the GRIN medium. Even though the preceding analysis assumed a Gaussian shape for the input beam, it has been shown that other beam shapes follow a self-similar periodic evolution inside a GRIN medium [41]. This is expected from Eq. (2.5.19) based on the quantum approach, which shows that the periodic evolution of a beam's width is a general feature of a GRIN medium the parabolic index profile. It will be seen in Section 3.2

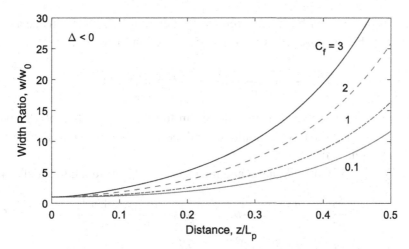

Figure 2.10 Variations of the width of a Gaussian beam inside a GRIN medium whose refractive index increases radially ($\Delta < 0$) from its value at the core's center. The ratio $w(z)/w_0$ is plotted as a function of the normalized distance z/L_p for several values of the parameter C_f.

that this behavior is related to the self-imaging phenomenon occurring inside a GRIN medium for any input beam.

We can apply the variational solution to a GRIN medium in which the refractive index increases radially from its value at core's center, rather than decreasing. This situation amounts to changing the sign of Δ in Eq. (2.3.1) or changing the sign of q^4 in Eq. (2.5.2). Such a change makes q^2, C_f, and L_p imaginary quantities. Using the well-known relations, $\cos(ix) = \cosh(x)$ and $\sin(ix) = i \sinh(x)$, in Eq. (2.5.33), the width of a Gaussian beam is found to evolve with z as

$$w(z) = w_0[1 + (1 + |C_f|^2) \sinh^2(2\pi z/|L_p|)]^{1/2}. \tag{2.5.36}$$

This equation shows that no self-imaging occurs, and the beam's width increases with z without any bound for all values of C_f. Figure 2.10 shows the width increase for values of C_f in the range of 0 to 3. The absence of self-imaging is not surprising because a GRIN medium with $\Delta < 0$ acts as a defocusing lens. Moreover, the GRIN-induced defocusing enhances the diffractive spreading of the Gaussian beam expected in a homogeneous medium. Clearly, such a GRIN medium is not of much practical use.

2.5.4 Eikonal-Based Technique

The variational method required us to choose a specific Gaussian shape for the beam. The question remains open what happens for other beam shapes. It is possible to answer this question for beams that remain cylindrically symmetric inside a GRIN medium. For this purpose, we write Eq. (2.5.1) in cylindrical coordinates as

$$2ik\frac{\partial A}{\partial z} + \frac{\partial^2 A}{\partial \rho^2} + \frac{1}{\rho}\frac{\partial A}{\partial \rho} - k^2 b^2 \rho^2 A = 0. \tag{2.5.37}$$

To solve Eq. (2.5.37), we seek its solution in the form $A = A_0 \exp(ikS)$, where $A_0(\rho, z)$ is related to the beam's amplitude and the eikonal $S(\rho, z)$ governs the shape of the beam's phase front.

From Eq. (2.5.37), the eikonal S and intensity $I = A_0^2$ are found to satisfy [42]:

$$2\frac{\partial S}{\partial z} + \left(\frac{\partial S}{\partial \rho}\right)^2 = -b^2\rho^2 + \frac{1}{k^2 A_0}\left(\frac{\partial^2 A_0}{\partial \rho^2} + \frac{1}{\rho}\frac{\partial A_0}{\partial \rho}\right) \tag{2.5.38}$$

$$\frac{\partial I}{\partial z} + \frac{\partial I}{\partial \rho}\frac{\partial S}{\partial \rho} + \left(\frac{\partial^2 S}{\partial \rho^2} + \frac{1}{\rho}\frac{\partial S}{\partial \rho}\right)I = 0. \tag{2.5.39}$$

The two terms on the right side of Eq. (2.5.38) represent the effects of index gradient and diffraction, respectively. In their absence, we can find S using the method of separation of variables. The solution is $S = \frac{1}{2}\rho^2/(z + R_c)$, where R_c is the radius of curvature of the phase front at $z = 0$. When all terms are included, the general solution takes the form [43]

$$S(\rho, z) = \frac{1}{2}\rho^2 h(z) + g(z), \tag{2.5.40}$$

where h is related to the curvature of the phase front.

By using S from Eq. (2.5.40) into Eq. (2.5.39), the intensity equation becomes

$$\frac{\partial I}{\partial z} + h(z)\rho\frac{\partial I}{\partial \rho} + 2h(z)I = 0. \tag{2.5.41}$$

Its general solution can be written in the form [43]

$$I(\rho, z) = \frac{I_0}{f^2(z)}F\left(\frac{\rho}{w_0 f(z)}\right), \tag{2.5.42}$$

where $F(\cdot)$ is an arbitrary function governing the beam's shape, w_0 is the initial width of the beam at $z = 0$, and $f(z)$ satisfies the differential equation

$$\frac{df}{dz} = h(z)f(z), \qquad f(0) = 1. \tag{2.5.43}$$

The self-similar solution in Eq. (2.5.42) applies to any cylindrically symmetric beam. It indicates that the beam's width scales at a distance z as $w(z) = w_0 f(z)$, and its intensity changes accordingly to ensure the conservation of total power. The scaling factor $f(z)$ depends on the beam's shape and can be found once a shape is specified.

As an example, consider a Gaussian beam with the intensity $I(\rho,0) = I_0 \exp(-\rho^2/w_0^2)$ at $z = 0$. It follows from Eq. (2.5.42) that the intensity at a distance z is given by

$$I(\rho,z) = \frac{I_0 w_0^2}{w^2(z)} \exp\left(-\frac{\rho^2}{w^2(z)}\right), \tag{2.5.44}$$

with $w(z) = w_0 f(z)$. Using $I(\rho,z)$ in Eq. (2.5.38), the eikonal S satisfies

$$2\frac{\partial S}{\partial z} + \left(\frac{\partial S}{\partial \rho}\right)^2 = -b^2\rho^2 + \frac{1}{k^2 w^2}\left(\frac{\rho^2}{w^2} - 2\right). \tag{2.5.45}$$

If we use the form of S given in Eq. (2.5.40) and equate the terms with and without ρ^2 on both sides of Eq. (2.5.45), we obtain the following equations for $h(z)$ and $g(z)$:

$$\frac{dh}{dz} + h^2 + b^2 = \frac{1}{k^2 w^4}, \qquad \frac{dg}{dz} = -\frac{1}{k^2 w^2}. \tag{2.5.46}$$

We can convert the equation for h to an equation for w using Eq. (2.5.43) with $f(z) = w(z)/w_0$. The result is

$$h(z) = \frac{1}{w}\frac{dw}{dz}. \tag{2.5.47}$$

The resulting equation for the width $w(z)$ becomes

$$\frac{d^2 w}{dz^2} + b^2 w - \frac{1}{k^2 w^3} = 0. \tag{2.5.48}$$

This equation is identical to Eq. (2.5.29) obtained with the variational method. The curvature parameter in Eq. (2.5.47) also agrees with the variational result in Eq. (2.5.27).

The b^2 term in Eq. (2.5.48) has its origin in the index gradient, but the last term results from diffraction. This term can be neglected in the geometrical-optics approximation because it scales with k as $1/k^2$. If we do so, Eq. (2.5.48) is reduced to a harmonic-oscillator equation. Using the initial conditions, $w(0) = w_0$ and $dw/dz = 0$ at $z = 0$ for an input beam with a planar wavefront, its solution is simply $w(z) = w_0 \cos(bz)$. This solution becomes invalid near $bz = \pi/2$ because $w(z)$ becomes so small that the diffraction term does not remain negligible. When the diffraction term in Eq. (2.5.48) is included, its solution takes the form given in Eq. (2.5.33).

One may ask how this solution changes when the input beam is not collimated and has a curved wavefront such that $h(0) = 1/R_c$. To solve Eq. (2.5.48) under such conditions, we write it in a normalized form given in Eq. (2.5.30) and integrate it once to obtain Eq. (2.5.31). In terms of the variable $u = s^2$, we need to integrate

$$\frac{du}{d\xi} = 2\sqrt{Cu - (1 + u^2)}, \tag{2.5.49}$$

where $\xi = bz$ and $u = (w/w_g)^2$, and $w_g = (kb)^{-1/2}$ is the spot size of the fundamental mode of the GRIN device. The solution of Eq. (2.5.49) can be written as

$$\int \frac{du}{\sqrt{Cu - (1 + u^2)}} = 2\xi + \phi_0, \tag{2.5.50}$$

where ϕ_0 is a constant of integration. The integral can be done in a closed form [10], and the final result is given by

$$u(\xi) = \tfrac{1}{2}[C + \sqrt{C^2 - 4}\sin(2\xi + \phi_0)]. \tag{2.5.51}$$

We use the initial conditions to find the two constants. Using $u = u_0$ at $\xi = 0$, we obtain

$$\sin\phi_0 = (2u_0 - C)/\sqrt{C^2 - 4}. \tag{2.5.52}$$

The constant C is found from Eq. (2.5.47) and noting that $dw/dz = w_0/R_c$ at $z = 0$, where R_c is the initial radius of curvature of the input beam. Using $du/d\xi = 2u_0/(bR_c)$ at $\xi = 0$, the constant C is found to be

$$C = u_0[1 + (bR_c)^{-2}] + 1/u_0. \tag{2.5.53}$$

Recalling $u = (w/w_g)^2$, the solution in Eq. (2.5.51) becomes

$$w(z) = \frac{w_g}{2}[C + \sqrt{C^2 - 4}\sin(2bz + \phi_0)]^{1/2}. \tag{2.5.54}$$

Even though the solution appears to be complicated for Gaussian beams launched with a curved wavefront, it reduces to the form given in Eq. (2.5.33) when the beam is collimated ($R_c = \infty$).

References

[1] D. K. Kalluri, *Principles of Electromagnetic Waves and Materials* (CRC Press, 2017), chap. 1.
[2] J. A. Buck, *Fundamentals of Optical Fibers*, 2nd ed. (Wiley, 2004).
[3] K. Okamoto, *Fundamentals of Optical Waveguides*, 2nd ed. (Academic Press, 2010).
[4] G. P. Agrawal, *Fiber-Optic Communication Systems*, 5th ed. (Wiley, 2021).
[5] A. Ghatak and K. Thyagarajan, *An Introduction to Fiber Optics* (Cambridge University Press, 1998).
[6] D. J. Griffiths and D. F. Schroeter, *Introduction to Quantum Mechanics*, 3rd ed. (Cambridge University Press, 2018).
[7] M. Abramowitz and I. A. Stegun, Eds., *Handbook of Mathematical Functions* (Dover, 1983).
[8] L. Allen, M. W. Beijersbergen, R. J. C. Spreeuw, and J. P. Woerdman, *Phys. Rev. A* **45**, 8185 (1992); https://doi.org/10.1103/physreva.45.8185.
[9] D. Marcuse, *Light Transmission Optics*, 2nd ed. (Van Nostrand Reinhold, 1982).

[10] I. S. Gradshteyn and I. M. Ryzhik, Eds., *Table of Integrals, Series, and Products*, 6th ed. (Academic Press, 2000).

[11] D. Gloge and E. A. J. Marcatili, *Bell Syst. Tech. J.* **52**, 1563 (1973); https://doi.org/10.1002/j.1538-7305.1973.tb02033.x.

[12] N. Borhani, E. Kakkava, C. Moser, and C. Psaltis, *Optica* **5**, 960 (2018); https://doi.org/10.1364/OPTICA.5.000960.

[13] P. Fan, T. Zhao, and L. Su, *Opt. Express* **27**, 20241 (2019); https://doi.org/10.1364/OE.27.020241.

[14] T. Zhao, S. Ourselin, T. Vercauteren, and W. Xia, *Opt. Express* **28**, 20978 (2020); https://doi.org/10.1364/OE.396734.

[15] Y. Fazea and V. Mezhuyev, *Opt. Fiber Technol.* **45**, 280 (2018); https://doi.org/10.1016/j.yofte.2018.08.004.

[16] L. Jeunhomme and J. P. Pocholle, *Appl. Opt.* **17**, 463 (1978); https://doi.org/10.1364/AO.17.000463.

[17] P. S. Szczepanek and J. W. Berthold, *Appl. Opt.* **17**, 3245 (1978); https://doi.org/10.1364/AO.17.003245.

[18] S. Berdagué and P. Facq, *Appl. Opt.* **21**, 1950 (1982); https://doi.org/10.1364/AO.21.001950.

[19] C.-L. Chen, *Appl. Opt.* **27**, 2353 (1988); https://doi.org/10.1364/AO.27.002353.

[20] F. Dubois, Ph. Emplit, and O. Hugon, *Opt. Lett.* **19**, 433 (1994); https://doi.org/10.1364/OL.19.000433.

[21] W. Q. Thornburg, B. J. Corrado, and X. D. Zhu, *Opt. Lett.* **19**, 454 (1994); https://doi.org/10.1364/OL.19.000453.

[22] G. Stepniak, L. Maksymiuk, and J. Siuzdak, *J. Lightwave Technol.* **29**, 1980 (2011); https://doi.org/10.1109/JLT.2012.2189756.

[23] J. Carpenter and T. D. Wilkinson, *J. Lightwave Technol.* **29**, 1386 (2012); https://doi.org/10.1109/JLT.2011.2155621.

[24] K. Aoki, A. Okamoto, Y. Wakayama, A. Tomita, and S. Honma, *Opt. Lett.* **38**, 769 (2013); https://doi.org/10.1364/OL.38.000769.

[25] Y. Wakayama, A. Okamoto, K. Kawabata, A. Tomita, and K. Sato, *Opt. Express* **21**, 12920 (2013); https://doi.org/10.1364/OE.21.012920.

[26] S. Vyas, P.-H. Wang, and Y. Luo, *Opt. Express* **25**, 23726 (2017); https://doi.org/10.1364/OE.25.023726.

[27] A. M. Yao and M. J. Padgett, *Adv. Opt. Photon.* **3**, 161 (2011); https://doi.org/10.1364/AOP.3.000161.

[28] J. Demas, L. Rishøj, and S. Ramachandran, *Opt. Express* **23**, 28531 (2015); https://doi.org/10.1364/OE.23.028531.

[29] S. Chen and J. Wang, *Sci. Rep.* **7**, 3990 (2017); https://doi.org/10.1038/s41598-017-04380-7.

[30] A. Wang, L. Zhu, L. Wang et al., *Opt. Express* **26**, 10038 (2018); https://doi.org/10.1364/oe.26.010038.

[31] J. Wang, S. Chen, and J. Liu, *APL Photon.* **6**, 060804 (2021); https://doi.org/10.1063/5.0049022.

[32] R. Olshansky and D. B. Keck, *Appl. Opt.* **15**, 483 (1976); https://doi.org/10.1364/AO.15.000483.

[33] J. J. Sakurai and J. Napolitano, *Modern Quantum Mechanics*, 3rd ed. (Cambridge University Press, 2020).

[34] T. Hansson, A. Tonello, T. Mansuryan et al., *Opt. Express* **28**, 24005 (2020); https://doi.org/10.1364/OE.398531.

[35] G. Dattoli, S. Solimeno, and A. Torre, *Phys. Rev. A* **34**, 2646 (1986); https://doi.org/10.1103/PhysRevA.34.2646.

[36] G. Nienhuis and L. Allen, *Phys. Rev. A* **48**, 656 (1993); https://doi.org/10.1103/PhysRevA.48.656.

[37] M. A. M. Marte and S. Stenholm, *Phys. Rev. A* **56**, 2940 (1997); https://doi.org/10.1103/PhysRevA.56.2940.

[38] K. Zhu, X. Sun, X. Wang, H. Tang, and Y. Peng, *Opt. Commun.* **221**, 1 (2003); https://doi.org/10.1016/S0030-4018(03)01402-0.

[39] M. Karlsson, D. Anderson, and M. Desaix, *Opt. Lett.* **17**, 22 (1992); https://doi.org/10.1364/OL.17.000022.

[40] D. Cline, *Variational Principles in Classical Mechanics*, 2nd ed. (University of Rochester Press, 2019).

[41] S. Longhi and D. Janner, *J. Opt. B: Quantum Semiclass. Opt.* **6**, S303 (2004); https://doi.org/10.1088/1464-4266/6/5/019.

[42] M. S. Sodha, A. K. Ghatak, and D. P. S. Malik, *J. Opt. Soc. Am.* **61**, 1492 (1971); https://doi.org/10.1364/OSA.61.001492.

[43] M. S. Sodha and A. K. Ghatak, *Inhomogeneous Optical Waveguides* (Plenum Press, 1977).

3

Focusing and Self-Imaging

We saw in Section 2.3 that the number of modes supported by a GRIN device scales as $V^2/4$, where V depends on the GRIN parameters a and Δ, as seen in Eq. (2.2.10). This parameter is relatively large for GRIN rods ($V > 100$) designed with $a > 1$ mm. The number of modes is so large for such rods that a mode-based theory ceases to be useful. This chapter shows that it is possible to "sum over all modes" and obtain a kernel that governs the propagation of continuous-wave (CW) beams inside GRIN rods. We use this kernel to discuss two phenomena, periodic focusing and self-imaging, that are important for applications of GRIN devices. The chapter is organized as follows. Section 3.1 provides a geometrical-optics perspective and shows why optical rays follow a curved path inside a GRIN medium. The modes obtained in Section 2.2 are used in Section 3.2 to find the propagation kernel and to discuss the phenomenon of self-imaging inside such a medium. Section 3.3 is devoted to studying how a GRIN rod can be used as a flat lens to focus an incoming optical beam. Imaging characteristics of such a lens are also considered in this section. Several important applications of GRIN devices are discussed in Section 3.4.

3.1 Geometrical-Optics Approximation

We focus on a GRIN medium whose refractive index decreases in a cylindrically symmetric fashion from its central axis up to a distance $\rho = a$ such that

$$n(\mathbf{r}) = n_0[1 - \Delta f(\rho)], \quad 0 \le \rho \le a, \tag{3.1.1}$$

where ρ is the radial distance. The minimum value of the refractive index, $n_c = n_0(1 - \Delta)$, occurs at $\rho = a$ and it persists for $\rho > a$.

It is easy to understand from Figure 3.1 why optical rays do not follow a straight path inside a GRIN medium. For any functional form of $f(\rho/a)$, we can replace

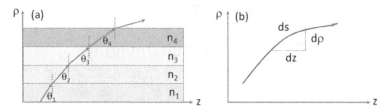

Figure 3.1 (a) Schematic of the multi-layer structure used for a GRIN medium in the ray-optics description. (b) Trajectory of a ray and the notation employed.

the GRIN medium with a multi-layer structure such that each successive layer has a slightly smaller refractive index [1]. Consider a ray traveling at an θ_1 in the first layer (measured from the normal to the interface). When it arrives at the second layer, its direction changes such that $\theta_2 > \theta_1$, as dictated by Snell's law. This angle keeps getting larger until the ray turns around. This is the reason why optical rays inside a GRIN medium remain confined to the vicinity of its core region of radius a.

3.1.1   The Ray Equation and Its Solution

To find the actual path of a ray inside a GRIN medium, one approach makes use of the geometrical-optics approximation that holds well when wavelength λ_0 of light is small relative to other dimensions of interest. In the limit of large k_0, the solution of Eq. (2.1.13) is sought in the form $\tilde{\mathbf{E}} = \mathbf{E_0} \exp(ik_0 S)$. To the leading order in k_0, $\mathbf{E_0}$ remains constant, and $S(\mathbf{r})$ satisfies the eikonal equation [2]:

$$(\nabla S)^2 = n^2(\mathbf{r}). \tag{3.1.2}$$

The surface, $S(\mathbf{r}) = \text{constant}$, represents the phase front of the wave. The unit vector normal to this phase front, $\mathbf{u} = (\nabla S)/n$, points in the direction of the path taken by a geometrical ray. Using $\mathbf{u} = d\mathbf{r}/ds$ for this ray's direction, its path is governed by

$$n\frac{d\mathbf{r}}{ds} = \nabla S, \tag{3.1.3}$$

where ds is the length of a segment along the ray's path (see Figure 3.1). We can eliminate S from this equation by taking its derivative again to obtain the ray equation [3]:

$$\frac{d}{ds}\left(n\frac{d\mathbf{r}}{ds}\right) = \nabla n, \tag{3.1.4}$$

where we used $dS/ds = \mathbf{u} \cdot \nabla S = |\nabla S| = n$.

It is useful to describe a ray's path using the transverse vector $\mathbf{r}_t = (x, y)$, where both x and y are functions of z. We can convert Eq. (3.1.4) to this form by using

$$\frac{d}{dz} = \frac{ds}{dz}\frac{d}{ds} = \sqrt{1 + x'^2 + y'^2}\frac{d}{ds}, \tag{3.1.5}$$

where $x' = dx/dz$ and $y' = dy/dz$ are related to the slope of the ray at the point z, and we used $ds^2 = dx^2 + dy^2 + dz^2$. This relation allows us to write Eq. (3.1.4) in the form

$$h\frac{d}{dz}\left(hn\frac{d\mathbf{r}_t}{dz}\right) = \nabla n \tag{3.1.6}$$

where $h = (1 + x'^2 + y'^2)^{-1/2}$. This equation is simplified considerably in the paraxial approximation. As both x' and y' are relatively small in this approximation, we can use $h \approx 1$ to the first order in x' and y'.

Let us apply Eq. (3.1.6) to a GRIN fiber with a parabolic index profile. Using $f(\rho) = (\rho/a)^2$ in Eq. (3.1.1), the refractive index varies with x and y as

$$n(x, y) = n_0[1 - \tfrac{1}{2}b^2(x^2 + y^2)], \qquad b = \sqrt{2\Delta}/a. \tag{3.1.7}$$

Using $h = 1$, the x component of Eq. (3.1.6) is found to satisfy

$$n\frac{d^2x}{dz^2} = \frac{\partial n}{\partial x} = -n_0 b^2 x. \tag{3.1.8}$$

A similar equation is obtained for the y component. Replacing n with n_0 on the left side of Eq. (3.1.8) (justified for $\Delta \ll 1$), the path of a geometrical ray is governed by

$$\frac{d^2x}{dz^2} + b^2 x = 0, \qquad \frac{d^2y}{dz^2} + b^2 y = 0, \tag{3.1.9}$$

where b plays the role of a spatial frequency. These equations show that both x and y vary with z in a periodic fashion (analogous to a harmonic oscillator). When a ray enters a GRIN medium at the position (x_0, y_0), with the initial slope (x_0', y_0'), its path is governed by

$$x(z) = x_0 \cos(bz) + (x_0'/b)\sin(bz), \qquad y(z) = y_0 \cos(bz) + (y_0'/b)\sin(bz). \tag{3.1.10}$$

3.1.2 Meridional and Skew Rays

Equation (3.1.10) shows that any ray recovers its initial position and direction periodically at distances such that $bz = 2m\pi$, where m is an integer. The period, $L_p = 2\pi/b$, represents the pitch of a GRIN medium. Based on Eq. (3.1.10), rays can be classified into two types, known as meridional and skew rays. Meridional rays remain confined to a plane and their paths always cross the z axis. An example of a meridional ray is provided by the ray for which both y_0 and y_0' vanish initially.

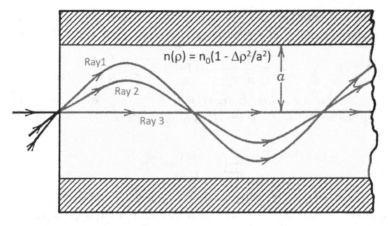

Figure 3.2 Trajectories of three meridional rays propagating at different angles inside a GRIN medium.

As seen from Eq. (3.1.10), the path of such rays lies in the $y = 0$ plane because $y(z)$ remains zero for all values of z. For a meridional ray entering parallel to the z axis with $x_0' = 0$ (called a field ray), the path takes the simple form $x(z) = x_0 \cos(bz)$.

For an axial ray entering with $x_0 = 0$, the path depends on the ray's initial slope x_0'. Figure 3.2 shows schematically the paths of three axial rays inside a GRIN rod fabricated with the parabolic index profile. As expected, all rays come together at the distance L_p. In practice, the spatial shape of any incoming beam is reproduced at that distance. This is the origin of self-imaging in a parabolic GRIN medium.

In contrast to meridional rays, skew rays never cross the central axis of a GRIN medium and follow a complicated path as they move around this axis. Among such rays, helical rays are of special interest because they move along a helical path, maintaining the same distance from the central axis. Consider a skew ray entering with $y_0 = 0$ and $x_0' = 0$. In this case, the solution in Eq. (3.1.10) is reduced to

$$x(z) = x_0 \cos(bz), \qquad y(z) = (y_0'/b) \sin(bz). \tag{3.1.11}$$

If the initial slope satisfies the relation $y_0' = bx_0$, the radial distance of the ray, $\rho = \sqrt{x^2 + y^2}$, remains fixed at its initial value x_0 at all z, and the ray follows a helical path. Using $x = \rho \cos \phi$ and $y = \rho \sin \phi$, we deduce that the azimuthal angle of this helical ray varies linearly with z as $\phi(z) = bz$.

We can use Eq. (3.1.10) to find the numerical aperture (NA) of a GRIN device. Consider a ray entering the front end of a device at $z = 0$ at an initial angle θ_i such that $x_0 = y_0 = 0$. From Snell's law, its angle inside the GRIN device will satisfy $n_0 \sin \theta_0 = n_a \sin \theta_i$, where $n_a \approx 1$ is the air's refractive index. From Eq. (3.1.10), the maximum height of the ray will be x_0'/b at the location $bz = \pi/2$. For this ray to remain confined within the GRIN device, the largest θ_0 should be such that the maximum height equals the radius a of the core region of the device, resulting in the condition $x_0' = ba$. The initial slope of the ray is related to the angle θ_0 as $x_0' = \tan \theta_0$. Using $b = \sqrt{2\Delta}/a$, we find

$$\tan \theta_0 \approx \sin \theta_0 = ba = \sqrt{2\Delta}. \tag{3.1.12}$$

Using this result, the numerical aperture of a GRIN device is found to be

$$NA = n_0 \sin \theta_0 = n_0 \sqrt{2\Delta}. \tag{3.1.13}$$

We can also use Eq. (3.1.10) to find the focal length of a GRIN lens designed to focus a collimated beam incident at its front end. All rays move parallel to the z axis ($x_0' = y_0' = 0$) for such a beam with a planar phase front. Consider a specific ray with an initial position $(x_0, 0)$ at $z = 0$. Its position at a distance z is given by $x(z) = x_0 \cos(bz)$. Clearly, $x(z)$ decreases as z increases and becomes 0 at a distance $z = L_p/4$ where $bz = \pi/2$. As this happens to all rays at the same time, we can say the entire beam focuses down to a point at that distance. Of course, the ray-optics description fails before this point is reached. We saw in Section 2.5 that wave optics predicts a minimum width of the beam at the focal point. Nonetheless, a general conclusion is that any incoming beam undergoes focusing inside a GRIN medium. The focal length of such a GRIN lens is

$$L_f = \frac{1}{4}L_p = \frac{\pi}{2b} = \frac{\pi a}{2\sqrt{2\Delta}}. \tag{3.1.14}$$

The sinusoidal nature of the solution in Eq. (3.1.10) implies that the focused beam will begin to spread soon after the focal point is crossed.

3.1.3 ABCD Matrix and Ray Tracing

We can use Eq. (3.1.10) for tracing optical rays inside a GRIN medium as follows. This equation provides a ray's location (x, y) at a distance z in terms of its initial position and slope at $z = 0$. We calculate the ray's slope at z and obtain

$$x'(z) = dx/dz = -bx_0 \sin(bz) + x_0' \cos(bz). \tag{3.1.15}$$

We can write this equation and Eq. (3.1.10) in a matrix form as

$$\begin{pmatrix} x \\ x' \end{pmatrix} = \begin{pmatrix} A & B \\ C & D \end{pmatrix} \begin{pmatrix} x_0 \\ x_0' \end{pmatrix}, \tag{3.1.16}$$

where the transfer matrix, called an ABCD matrix, is given by [4, 5]

$$\begin{pmatrix} A & B \\ C & D \end{pmatrix} = \begin{pmatrix} \cos(bz) & \sin(bz)/b \\ -b\sin(bz) & \cos(bz) \end{pmatrix}. \tag{3.1.17}$$

The same matrix relates y and y' to their initial values.

As expected on physical grounds, the ABCD matrix is unitary. It can be used to calculate a ray's position and direction at any distance z, if these quantities are known for that ray at $z = 0$. The four elements of the ABCD matrix oscillate in a periodic fashion such that this matrix reduces to a unit matrix at distances that are multiples of the pitch L_p introduced earlier. At a distance $L_f = L_p/4$, the diagonal elements of the ABCD matrix vanish. This distance corresponds to the focal plane of a GRIN lens because all incoming parallel rays meet at its central axis. Clearly, geometrical optics is able to predict both the focusing and self-imaging phenomena. Its major failure is that it cannot provide the minimum size of the focused beam at the focal point.

3.2 Wave-Optics Description

The modes obtained in Section 2.2 are used in this section to study the propagation of a CW beam inside a GRIN medium. Working in the frequency domain at a fixed frequency ω, we drop the tilde on the electric field in Eq. (2.1.13) to simplify the notation. We also assume that the direction of $\mathbf{E}$ remains fixed and treat it as a scalar quantity.

3.2.1 Propagation Kernel for a GRIN Medium

Following Section 2.4.1, an optical beam with the input field $E(x, y, 0)$ excites multiple modes of a GRIN medium such that

$$E(x, y, 0) = \sum_m \sum_n c_{mn} U_{mn}(x, y), \tag{3.2.1}$$

where the sum extends over the whole range of two integers ($m, n = 0$ to ∞). We have chosen to expand E in terms of the Hermite–Gauss modes normalized such that

$$\iint_{-\infty}^{\infty} U_{m'n'}^*(x, y) U_{mn}(x, y) \, dx \, dy = \delta_{m'm} \delta_{n'n}. \tag{3.2.2}$$

The expansion coefficients c_{mn} are found to be [see Eq. (2.4.2)]

$$c_{mn} = \iint_{-\infty}^{\infty} U_{mn}^*(x, y) E(x, y, 0) \, dx \, dy. \tag{3.2.3}$$

Our objective is to find the electric field $E(x, y, z)$ after the beam has arrived at a distance z. As all modes propagate without change in their spatial profiles and only acquire a z-dependent phase shift inside the GRIN medium, $E(x,y,z)$ is obtained by multiplying each mode in Eq. (3.2.1) with a phase factor such that

$$E(x,y,z) = \sum_m \sum_n c_{mn} U_{mn}(x,y) \exp(i\beta_{mn}z). \tag{3.2.4}$$

Substituting c_{mn} from Eq. (3.2.3), we can write the result in the form

$$E(x,y,z) = \iint_{-\infty}^{\infty} K(x,x';y,y')E(x',y',0)\,dx'\,dy', \tag{3.2.5}$$

where the propagation kernel is given by

$$K(x,x';y,y') = \sum_m \sum_n U_{mn}(x,y)U_{mn}^*(x',y')e^{i\beta_{mn}z}. \tag{3.2.6}$$

The form of Eq. (3.2.5) is similar to that found for optical beams propagating inside a homogeneous medium of constant refractive index.

If the double sum in Eq. (3.2.6) can be evaluated in a closed form, the resulting kernel will include all excited modes of a GRIN fiber, without any explicit reference to them. It turns out that the double sum can be done for a parabolic index profile because of a ladder-like structure of the propagation constants (see Figure 2.2) in such a medium [9]. Using Eq. (2.2.8) for $U_{mn}(x,y)$ together with the expression in Eq. (2.2.12) for β_{mn}, the kernel in Eq. (3.2.6) takes the form

$$K(x,x';y,y') = \frac{q^2}{\pi} \exp\left[i(n_0k_0 - b)z - \frac{q^2}{2}(x^2 + y^2 + x'^2 + y'^2)\right]S(x,x')S(y,y'), \tag{3.2.7}$$

where the function $S(x,x')$ is defined as

$$S(x,x') = \sum_{m=0}^{\infty} \frac{1}{m!} H_m(qx)H_m(qx')\left(\tfrac{1}{2}e^{-ibz}\right)^m. \tag{3.2.8}$$

The single sum in this equation can be carried out using Mehler's formula [12]:

$$\sum_{m=0}^{\infty} \frac{1}{m!} H_m(x)H_m(x')(\tfrac{1}{2}s)^m = \frac{1}{\sqrt{1-s^2}} \exp\left[\frac{2sxx' - s^2(x^2 + x'^2)}{(1-s^2)}\right]. \tag{3.2.9}$$

Using the preceding result, the propagation kernel for a parabolic GRIN medium is found to be [9]

$$K(x,x';y,y') = \frac{kbe^{i\psi}}{2\pi i \sin(bz)} \exp\left[\frac{ikb}{2\sin(bz)}[\cos(bz)(x'^2 + y'^2) - 2(xx' + yy')]\right], \tag{3.2.10}$$

where $k = n_0k_0$ and the phase is given by

$$\psi(\mathbf{r}) = kz + \frac{1}{2}kb \cot(bz)(x^2 + y^2). \tag{3.2.11}$$

As one may expect from the discussion in Appendix A, a similar expression is found for the propagator of a quantum harmonic oscillator [13]. The kernel in Eq. (3.2.10) is also related to the fractional Fourier transform in two dimensions [14]; see Appendix B. This kernel was used in 1976 to study the imaging properties of a GRIN lens [15].

In the case of radially symmetric fields, it is common to make use of the cylindrical coordinates (ρ, ϕ, z). In terms of these coordinates, Eq. (3.2.5) becomes

$$E(\rho, \phi, z) = \iint_{-\infty}^{\infty} K(\rho, \rho'; \phi, \phi')E(\rho', \phi', 0)\, \rho' d\rho'\, d\phi', \qquad (3.2.12)$$

and the propagation kernel given in Eq. (3.2.10) takes the form

$$K(\rho, \rho'; \phi, \phi') = \frac{kb e^{ikz}}{2\pi i \sin(bz)} \exp\left[\frac{i}{2}kb\cot(bz)(\rho^2 + \rho'^2) - \frac{ikb\rho\rho'}{\sin(bz)}\cos(\phi - \phi')\right].$$

$$(3.2.13)$$

As the input field does not depend on ϕ' for radially symmetric fields, the integration over ϕ' in Eq. (3.2.12) can be carried out using the known result [16]

$$\int_0^{2\pi} \exp[\pm ip \cos(\phi - \phi')]\, d\phi' = 2\pi J_0(p). \qquad (3.2.14)$$

This result does not depend on ϕ, indicating that the radial symmetry of the input beam is maintained during its propagation inside a GRIN medium. From Eq. (3.2.12), the field $E(\rho, z)$ at a distance z becomes

$$E(\rho, z) = \frac{kb e^{ikz}}{i \sin(bz)} \int_0^{\infty} E(\rho', 0) J_0\left(\frac{kb\rho\rho'}{\sin(bz)}\right) \exp\left[\frac{i}{2}kb\cot(bz)(\rho^2 + \rho'^2)\right] \rho'd\rho'.$$

$$(3.2.15)$$

3.2.2 Self-Imaging in Parabolic GRIN Media

The propagation kernel in Eq. (3.2.10) has sinusoidal functions that are periodic in z. In fact, it can be written in terms of the elements of the ABCD matrix given in Eq. (3.1.17) as $K(x, x'; y, y') = F(x, x')F(y, y')$, where $F(x, x')$ is defined as

$$F(x, x') = \left(\frac{k e^{ikz}}{2\pi iB}\right)^{1/2} \exp\left(\frac{ik}{2B}[Ax'^2 - 2xx' + Dx^2]\right), \qquad (3.2.16)$$

with $A = D = \cos(bz)$ and $B = \sin(bz)/b$. This form of the kernel has been known for a while and has proved useful in several different contexts [6–8].

Let us take the limit $b \to 0$, which amounts to setting $n = n_0$ in Eq. (3.1.1). A GRIN medium is reduced to a homogeneous medium of constant refractive index

n_0 in this limit. Noting that $A = D = 1$ and $B = z$ in the limit $b \to 0$, it is easy to deduce that the kernel in Eq. (3.2.10) is reduced to

$$K_h(x, x'; y, y') = \frac{k e^{ikz}}{2\pi i z} \exp\left(\frac{ik}{2z}[(x - x')^2 + (y - y')^2]\right). \tag{3.2.17}$$

This form of the kernel is well known for a homogeneous medium of constant refractive index [17]. The appearance of $\sin(bz)$ in Eq. (3.2.10) in place of z indicates that an optical beam propagating inside a GRIN medium is likely to exhibit periodic evolution along its length.

To reveal the nature of this periodicity, we consider what happens to the propagation kernel at a distance $z = L_p = 2\pi/b$, where $\sin(bz)$ vanishes. Noting that $\cos(bz) = 1$ at this distance, the kernel in Eq. (3.2.10) can be written in the form $K = f(x - x')f(y - y')e^{ikz}$, where the function $f(x)$ is defined as

$$f(x) = \sqrt{\frac{d}{\pi}} e^{-dx^2}, \qquad d = \frac{kb}{2i \sin(bz)}. \tag{3.2.18}$$

It is easy to show that $\int_{-\infty}^{\infty} f(x)\, dx = 1$. At a distance such that $bz = 2\pi$, d becomes infinitely large, and $f(x)$ is reduced to the delta function $\delta(x)$. As a result, the Kernel in Eq. (3.2.10) takes the form

$$K(x, x'; y, y') = \delta(x - x')\delta(y - y')e^{ikL_p} \tag{3.2.19}$$

The use of this result in Eq. (3.2.5) leads to $E(x, y, L_p) = E(x, y, 0)e^{ikL_p}$. Up to a constant phase factor, the optical field at the distance L_p becomes identical to the input field, resulting in self-imaging of any object located at $z = 0$. Because of the periodic nature of the kernel, this evolution pattern is repeated in all sections of length L_p. The same periodicity was found in Section 3.1 using the ray equation of geometrical optics. The pitch L_p of a GRIN rod is related to the device's parameters as

$$L_p = \frac{2\pi}{b} = \frac{2\pi a}{\sqrt{2\Delta}}. \tag{3.2.20}$$

The numerical value of L_p is about 2 cm for a GRIN rod, if we use $a = 1$ mm and $\Delta = 0.05$ as typical values; L_p is close to 1 mm for GRIN fibers.

Let us ask what happens at the locations $z_m = (m + \frac{1}{2})L_p$ that correspond to the middle of each self-imaging period. At these locations, the two delta functions in Eq. (3.2.19) are replaced with $\delta(x + x')$ and $\delta(y + y')$. As a result, an inverted image of the object at $z = 0$ is formed that is flipped in both transverse directions. If the input at $z = 0$ is radially symmetric, the sign changes have no impact, and the input field at $z = 0$ is reproduced (self-imaged) periodically at distances that are multiples of $\frac{1}{2}L_p$. This period doubles to L_p in the absence of radial symmetry.

Consider what happens at a distance $z = L_f = L_p/4$, which is just one quarter of the pitch. At this distance, $A = D = 0$ and $B = 1/b$ in Eq. (3.2.16) because $bz = \pi/2$. As a result, the kernel takes the form

$$K(x, x'; y, y') = \frac{kbe^{ikz}}{2\pi i} \exp[ikb(xx' + yy')], \tag{3.2.21}$$

It follows from Eq. (3.2.5) that the optical field at $z = L_f$ is just the Fourier transform of the input field (up to a constant). This is a well-known property of any lens [17], and it suggests that a GRIN rod of length L_f acts as a lens and should focus any input beam at $z = L_f$. The focusing of a wide collimated input beam to a small size can be understood by noting that the Fourier transform of a plane wave with constant amplitude in the xy plane is just $\delta(x)\delta(y)$. We have seen in Section 2.5 that even a Gaussian beam gets focused to a small spot size at that distance. The focal length of a GRIN rod with $a = 1$ mm and $\Delta = 0.05$ is about 5 mm, resulting in a compact flat lens that is useful for a variety of imaging applications.

3.2.3 Gaussian-Beam Propagation

As a simple example, we use the kernel in Eq. (3.2.10) to discuss the propagation of a collimated Gaussian beam, launched on-axis such that its intensity peaks at the central axis of the GRIN medium. Under such conditions, the input field at $z = 0$ has the form

$$E(x, y, 0) = A_0 \exp\left(-\frac{x^2 + y^2}{2w_0^2}\right), \tag{3.2.22}$$

where A_0 is the peak amplitude and w_0 is the spot size (1/e width) of the Gaussian beam. We may use Eq. (3.2.15) as such a beam exhibits radial symmetry. However, we choose to work in the Cartesian coordinates because the problem is simpler for them.

We find the electric field at any point $\mathbf{r} = (x, y, z)$ inside the GRIN medium by using Eq. (3.2.22) in Eq. (3.2.5) together with the propagation kernel in Eq. (3.2.10). The result can be written as

$$E(\mathbf{r}) = \frac{A_0 k}{2\pi i} \left(\frac{be^{i\psi}}{\sin(bz)}\right) H(x, x') H(y, y'), \tag{3.2.23}$$

where the function $H(x, x')$ is defined as

$$H(x, x') = \int_{-\infty}^{\infty} \exp\left(-\frac{x'^2}{2w_0^2} + \frac{ikb}{2\sin(bz)}[\cos(bz)x'^2 - 2xx']\right) dx'. \tag{3.2.24}$$

The integral can be performed using the known result [16]

$$\int_{-\infty}^{\infty} \exp(-px^2 + qx) \, dx = \sqrt{\frac{\pi}{p}} \exp\left(\frac{q^2}{4p}\right). \tag{3.2.25}$$

After some algebra, the electric field of an initially Gaussian beam at a distance z inside the GRIN medium takes the form [18]

$$E(\mathbf{r}) = A_0 F(\mathbf{r}) \exp[i\phi(\mathbf{r})], \qquad (3.2.26)$$

where the beam's amplitude retains its Gaussian form with a different spot size:

$$F(\mathbf{r}) = \frac{w_0}{w(z)} \exp\left[-\frac{x^2 + y^2}{2w^2(z)}\right], \qquad (3.2.27)$$

and its phase $\phi(\mathbf{r})$ is given by

$$\phi(\mathbf{r}) = \frac{k}{2w}\frac{dw}{dz}(x^2 + y^2) + kz + \tan^{-1}(C_f \tan bz). \qquad (3.2.28)$$

The spot size $w(z)$ in Eq. (3.2.27) evolves with z in a periodic fashion as

$$w(z) = w_0[\cos^2(2\pi z/L_p) + C_f^2 \sin^2(2\pi z/L_p)]^{1/2}, \qquad C_f = \frac{L_p}{2\pi k w_0^2}. \qquad (3.2.29)$$

The parameter C_f is the same as defined in Eq. (2.5.34). As a reminder, L_p is the pitch of the GRIN medium given in Eq. (3.2.20).

We note that the variational solution obtained in Section 2.5.3 and the eikonal-based solution obtained in Section 2.5.4 agree fully with the exact solution given in Eq. (3.2.26). The amplitude, the width, and the phase of a Gaussian beam evolve in a periodic fashion such that the beam recovers all of its input features periodically at distances $z = mL_p/2$ (m is any positive integer). When $C_f < 1$, the beam's width takes its minimum value, $w_0 C_f$ in the middle of each period, i.e., the value of C_f governs the extent of beam compression (or focusing) during each cycle. This reduction in width is similar to the focusing provided by a lens. The focal length L_f of a GRIN lens is defined as the shortest distance at which the beam first comes to a focus. Recall that the L_f is just one quarter of the self-imaging period L_p. If we use $a = 1$ mm and $\Delta = 0.05$ as typical values for a GRIN lens, the numerical value of L_f is close to 5 mm.

As an example, Figure 3.3 shows the evolution of a Gaussian beam over one self-imaging period using $C_f = 0.4$. For this value of C_f, the beam width is reduced by a factor of 2.5 at the point of maximum compression, and its peak intensity is enhanced by a factor of 6.25. Compression by a factor of 10, and intensity enhancement by a factor of 100, can be realized by making $C_f = 0.1$. As seen in Eq. (3.2.29), C_f depends on two lengths, L_p and the diffraction length defined as $L_d = kw_0^2$. If we use $w_0 = 0.1$ mm and $k = n_0(2\pi/\lambda)$ with $n_0 = 1.5$ and $\lambda = 1.55$ μm, we obtain $L_d = 6$ cm. In this case, the value of C_f is close to 0.05, and the Gaussian beam can be focused down to a spot size of about 5 μm. Such a GRIN lens is quite suitable for coupling light into a single-mode fiber. Figure 3.4 shows how the width ratio w/w_0 varies over one period for several values of the parameter C_f.

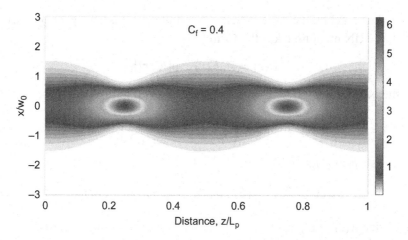

Figure 3.3 Evolution of an on-axis Gaussian beam inside a GRIN fiber over one self-imaging period. Beam's cross section along the $y = 0$ plane is shown for $C_f = 0.4$.

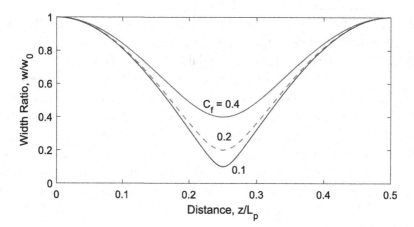

Figure 3.4 Beam-width ratio w/w_0 plotted over one self-imaging period for several values of C_f.

The Gaussian beam described by Eq. (3.2.22) is launched such that its center coincides with the GRIN rod's axis at $x = y = 0$, and its intensity profile is radially symmetric around this axis. In the modal picture, only radially symmetric modes are excited by such an input beam. The propagation kernel in Eq. (3.2.10) allows us to consider the impact of off-axis launch for which the beam's center is shifted from the GRIN rod's axis at $x = y = 0$. If the peak of the Gaussian beam is shifted by a distance s along the x axis, the input field in Eq. (3.2.22) is replaced with

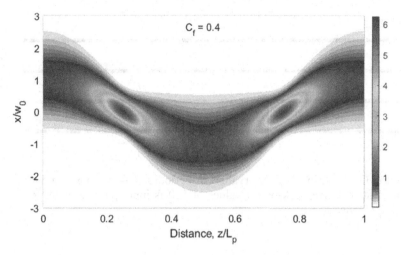

Figure 3.5 Evolution of an off-axis Gaussian beam (centered at $x = w_0$) inside a GRIN fiber over one self-imaging period. Beam's cross section along the $y = 0$ plane is shown for $C_f = 0.4$.

$$E(x, y, 0) = A_0 \exp\left(-[(x - s)^2 + y^2]/2w_0^2\right). \tag{3.2.30}$$

The optical field at any point $\mathbf{r} = (x, y, z)$ is found using Eq. (3.2.30) in Eq. (3.2.5) together with the kernel in Eq. (3.2.10). The integrals can again be done using the result in Eq. (3.2.25). The final result has the form given in Eq. (3.2.26), but the beam's amplitude is now governed by [18]

$$F(\mathbf{r}) = \frac{w_0}{w(z)} \exp\left[-\frac{(x - s\cos bz)^2 + y^2}{2w^2(z)}\right], \tag{3.2.31}$$

where $w(z)$ varies with z in a periodic fashion as indicated in Eq. (3.2.29).

Figure 3.5 shows how an off-axis Gaussian beam evolves over one self-imaging period by plotting its intensity variations after choosing $s = w_0$ and $C_f = 0.4$. Similar to the on-axis launch, the beam's width oscillates with z, compressing and recovering its input value periodically. The new feature is that the beam's center also oscillates around the fiber's axis. This is in agreement with the quantum approach of Section 2.5, where such periodic oscillations were predicted for any beam shape. Even though the Gaussian beam in Figure 3.5 recovers its initial width w_0 at $z = L_p/2$, its center is located on the opposite side at $x = -s$, resulting in a flipped image. The self-imaging period for a shifted Gaussian beam doubles compared to a beam launched on-axis because its intensity distribution is not radially symmetric around the GRIN rod's axis.

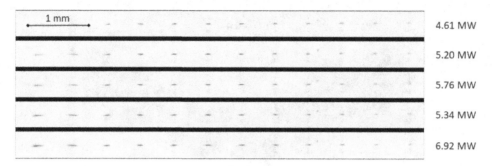

Figure 3.6 Experimental observation of periodic self-imaging in a 5-cm-long GRIN fiber at five peak-power values of 120-fs input pulses. (After Ref. [20] ©2020 Optica.)

3.2.4 Experiments on Self-Imaging

It is not easy to observe periodic self-imaging occurring inside a GRIN fiber because one does not have direct access to its core. As only the intensity of the optical field exiting the fiber can be measured, one can only use GRIN fibers of different lengths and see how the intensity patterns change at the output end of each fiber [19].

In a 2020 experiment, the nonlinear phenomenon of multiphoton absorption was exploited to observe periodic self-imaging all along the length of a GRIN fiber [20]. The core of a silica-based GRIN fiber contains defects, forming because of germanium doping. When intense femtosecond pulses are launched into such a fiber at a wavelength in the near-infrared region, multiphoton absorption occurs mostly in the vicinity of the focal regions, where the intensity is the highest. As a result, defects in this region radiate some of the absorbed energy through photoluminescence (at shorter wavelengths in the blue-violet region). This radiation scatters out of the fiber's cladding all along the fiber's length [21]. Figure 3.6 shows the observed patterns at several input peak-power levels when 120-fs pulses were launched into a 5-cm-long GRIN fiber in the form of a Gaussian beam [20]. It follows from the discussion in this section that the spacing between two neighboring bright spots corresponds to one half of the self-imaging period L_p that depends on the parameters of the GRIN fiber as indicated in Eq. (3.2.20). Using the known values, $a = 25$ μm and $\Delta = 0.01$, the predicted self-imaging period of 1.1 mm agrees with the experimental data.

It is interesting to note that the observed self-imaging period does not change with the peak power of input pulses. This may appear surprising at first because silica glass exhibits the optical Kerr effect that should lead to nonlinear self-focusing. This issue is discussed in Section 5.1.2, where we study the impact of self-focusing on the self-imaging phenomenon.

3.3 Focusing and Imaging by a GRIN Lens

The self-imaging property of a GRIN medium reveals that optical beams of any shape and size reproduce their input spatial distribution after each self-imaging period L_p, which is related to the parameters a and Δ as indicated in Eq. (3.2.20). In particular, any collimated optical beam with a plane wavefront initially undergoes a focusing phase and acquires a minimum spot size at a distance of $L_f = L_p/4$. A GRIN rod whose length is chosen to be L_f (or smaller) acts as a lens and can be used for focusing and imaging, with the added benefit of being both compact and flat. The same device can be used in reverse to collimate a narrow optical beam coming out of a single-mode fiber.

GRIN lenses were commercialized as early as 1970 under the tradename *Selfoc lens* [22]. Their focusing and imaging properties were studied during the 1970s in the geometrical-optics approximation through ray tracing and aberration theory [23]. Around the same time, Maxwell's equations were used to find the modes of a GRIN medium [24]. By 1974, a kernel governing propagation of optical beams inside a GRIN beam has been obtained that included all excited modes of a GRIN fiber automatically [9]. It was used in 1976 to analyze the imaging characteristics of a GRIN lens [15]. A review in 1980 was devoted to the theory of GRIN imaging [11]. In this section we use the propagation kernel obtained in Section 3.2 to discuss the focusing and imaging properties of a GRIN lens.

3.3.1 Focusing by a GRIN Lens

The propagation kernel given in Eq. (3.2.10) can be used to find the spot size at the focal point of a GRIN lens. For this purpose, we consider a collimated circular beam with a constant intensity within a circle of radius ρ_0. As this circular beam propagates down a GRIN rod, its diffractive spreading is opposed by the focusing action of the index gradient. As a result, the beam develops a converging wavefront and narrows down until it attains a minimum spot size at the focal point L_f at a distance of $L_p/4$.

Let us assume that the center of the circular beam at $z = 0$ coincides with the axis of the GRIN rod. In this situation, the input field exhibits radial symmetry that is maintained as the beam propagates down the GRIN rod. For such a beam, we can use Eq. (3.2.15) with the input $E(\rho', \phi', 0) = A_0$ for $\rho' \leq \rho_0$ and zero otherwise. Using $b = 2\pi/L_p$ in Eq. (3.2.15), the electric field at a distance z can be written as [25]

$$E(\rho, z) = \frac{A_0 e^{ikz}}{iC_f \sin(2\pi z/L_p)} \exp\left[\frac{i\rho^2}{2C_f \rho_0^2}\cot(2\pi z/L_p)\right]F(\rho, z), \qquad (3.3.1)$$

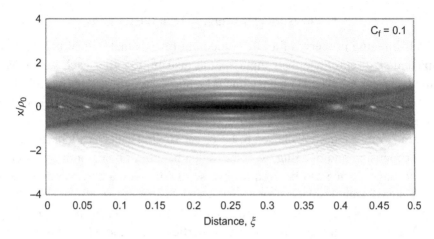

Figure 3.7 Evolution of a circular beam inside a GRIN rod over a distance of $L_p/2$.

where $F(\rho, z)$ is given by

$$F(\rho, z) = \int_0^1 J_0\left(\frac{u\rho}{C_f \rho_0 \sin(2\pi z/L_p)}\right) \exp\left[\frac{iu^2}{2C_f}\cot(2\pi z/L_p)\right] u \, du, \qquad (3.3.2)$$

and C_f is defined similar to Eq. (3.2.29):

$$C_f = \frac{L_p}{2\pi k \rho_0^2} = \frac{a}{k\rho_0^2 \sqrt{2\Delta}}. \qquad (3.3.3)$$

We expect C_f to set the minimum spot size at the focal point of a GRIN rod, just as it did for a Gaussian beam.

The integral in Eq. (3.3.2) requires a numerical approach. As an example, Figure 3.7 shows how an initially circular beam evolves along the GRIN rod's length by plotting $|E(x, 0, z)|^2$ as a function of z/L_p for a fixed value $C_f = 0.1$. Similar to a Gaussian beam, a circular beam also evolves periodically with the period $z_p = L_p/2$. During each period, it undergoes a focusing phase and acquires a minimum spot size at its midpoint. The main difference compared to a Gaussian beam is the appearance of a fringe pattern that is related to the diffractive effects associated with the circular edge of the beam.

Figure 3.8 shows the diffractive effects more clearly by plotting the intensity distribution $|E(x, 0, z)|^2$ in the $y = 0$ plane at three locations within one focal length. The beam expands initially and develops an oscillatory structure because of diffraction. However, it comes to a focus at $z = L_f$ with a minimum spot that is surrounded by several low-intensity rings at a level of 1% (-20 dB) of the central maximum. At this location, the integral in Eq. (3.3.2) simplifies considerably and takes the form

$$F(\rho, L_p/4) = \int_0^1 J_0\left(\frac{u\rho}{C_f \rho_0}\right) u \, du. \qquad (3.3.4)$$

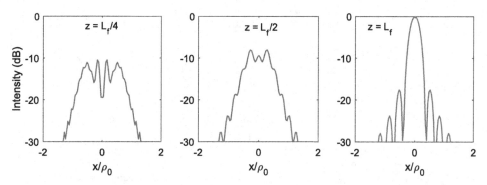

Figure 3.8 Intensity patterns of an initially circular beam inside a GRIN rod at distances $z/L_f = 0.25, 0.5,$ and 1.

This integral can be done analytically using the known identity [26]

$$\int_0^v J_0(u)\, u\, du = v J_1(v). \tag{3.3.5}$$

Using it, the intensity at the focal point, $L_f = L_p/4$, is found to be

$$I(\rho) = |E(\rho, L_f)|^2 = \left[A_0 C_f \frac{\rho_0}{\rho} J_1\left(\frac{\rho}{C_f \rho_0}\right)\right]^2. \tag{3.3.6}$$

The function $[J_1(u)/u]^2$ has a central bright region that is surrounded by multiple rings, resulting in the so-called Airy pattern seen in Figure 3.8 in the focal plane of the GRIN lens. The first zero of this function occurs for $u = 3.8317$. Using this value, the full width at half-maximum (FWHM) of the bright spot can be approximated as $w_m = 3C_f \rho_0$. Thus, a circular beam with a FWHM of 1 mm can be focused down to 3 μm when $C_f = 0.001$. We can use Eq. (3.3.3) to realize such small values of C_f. As an example, $C_f = 0.001$ at wavelengths near 1 μm when $a = 2$ mm, $\Delta = 0.02$, and $\rho_0 = 1$ mm. The focal length of such a GRIN lens is about 1.5 cm.

As expected, diffractive effects limit the focal spot size close to the wavelength of the incident radiation. A second limitation is imposed by the degree to which a parabolic index profile can be realized inside the GRIN rod used as a lens. A simple way to include small departures from an ideal parabolic shape is to modify Eq. (1.2.2) for the refractive index as

$$n(\rho) = n_0 \left[1 - \Delta\left(\frac{\rho}{a}\right)^2 + h_4 \Delta\left(\frac{\rho}{a}\right)^4 + \cdots\right], \tag{3.3.7}$$

where the h_4 term represents the largest perturbation. Even though an analytic solution of the wave equation becomes difficult, considerable progress can be made using the ray equation in the geometrical-optics approximation [11]. The results

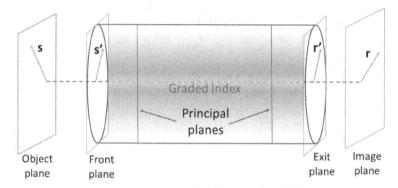

Figure 3.9 Geometry and the notation used for a GRIN-lens-based imaging system.

show that the focal spot size increases when h_4 is finite because of aberrations produced by a non-parabolic index profile.

The length of a GRIN lens can be less than one quarter of the pitch ($L_f = L_p/4$), when the objective is to bring the beam to focus at some distance from the output end of the lens. In that situation, the beam exits GRIN lens with a curved wavefront that brings the beam to focus as it propagates in air. The radius of curvature of the phase front at the output end of GRIN lens of length L is found from Eq. (3.3.1) to be

$$R_c = 2C_f \rho_0^2 / \cot(2\pi L/L_p), \qquad (3.3.8)$$

where L_p is the pitch of the GRIN lens. We can use R_c to find the distance at which the exiting beam will come to a focus in air and acquire its smallest spot size.

3.3.2 Imaging by a GRIN Lens

A GRIN lens can be used to form the image of an object, located at some distance from its front end, on the opposite side of the lens because, similar to a conventional lens, a GRIN lens also curves the wavefront of an incoming beam. The imaging characteristics of GRIN lenses were studied during the 1970s with the ray-tracing techniques [23]. The propagation kernel given in Eq. (3.2.10) was first used in 1976 for the imaging problem [15]. In this section, we use the wave-optics approach to discuss the imaging characteristics of a GRIN lens.

Figure 3.9 shows the geometry and the notation employed for a GRIN lens-based imaging system. The object is placed at a distance u from the front end of a GRIN lens of length L, and its image is formed at a distance v from the other end of the lens. We need to find the optical field $E_i(\mathbf{r})$ in the image plane from its known value $E_o(\mathbf{s})$ in the object plane. The problem requires three propagation steps: (i) propagation in air over a distance u, propagation inside the GRIN lens over a distance L, and propagation in air again over a distance v. Each step involves a convolution with the appropriate kernel. The final result can be written as

$$E_i(\mathbf{r}) = \int K(\mathbf{r}, \mathbf{s}) E_o(\mathbf{s}) \, d\mathbf{s}, \qquad (3.3.9)$$

where $\mathbf{r}$ and $\mathbf{s}$ are two-dimensional vectors in the image and object planes, respectively. The surface integration, denoted by a single integral sign, is carried out over the entire object plane.

The composite kernel $K(\mathbf{r}, \mathbf{s})$ involves three kernels, corresponding to the three propagation stages involved, and has the form (see Figure 3.9)

$$K(\mathbf{r}, \mathbf{s}) = \iint K_h(\mathbf{r}, \mathbf{r'}) K_g(\mathbf{r'}, \mathbf{s'}) K_h(\mathbf{s'}, \mathbf{s}) \, d\mathbf{s'} \, d\mathbf{r'}, \qquad (3.3.10)$$

where K_h and K_g represent the kernels for the homogeneous and GRIN media, respectively. The form of these kernels is given in Section 3.3. In the present notation, the homogeneous kernel has the form

$$K_h(\mathbf{r}, \mathbf{s}) = \frac{k e^{ik_0 z}}{2\pi i z} \exp\left(\frac{ik}{2z} [(r^2 + s^2) - \frac{ik}{z} \mathbf{r} \cdot \mathbf{s}] \right). \qquad (3.3.11)$$

The kernel for the GRIN lens is more complicated and is given by

$$K_g(\mathbf{r}, \mathbf{s}) = \frac{k e^{ikL}}{2\pi i} \left(\frac{b}{\sin(bL)} \right) \exp\left[\frac{ikb}{2} \cot(bL)(r^2 + s^2) - \frac{ikb}{\sin(bL)} \mathbf{r} \cdot \mathbf{s} \right]. \qquad (3.3.12)$$

where b is related to the pitch of the GRIN lens. The two surface integrals in Eq. (3.3.10) can be done analytically. After some algebra, the final result is given by [15]

$$K(\mathbf{r}, \mathbf{s}) = \frac{k e^{ikL}}{2\pi i \xi} \left(\frac{b}{\sin(bL)} \right) \exp[ik(u + v + L)]$$

$$\exp\left[\frac{ikb}{2\xi} [(\cot bL - bu)r^2 + (\cot bL - bv)s^2] - \frac{ikb}{\xi \sin(bL)} \mathbf{r} \cdot \mathbf{s} \right], \qquad (3.3.13)$$

where ξ is defined as

$$\xi = 1 + b(u + v)\cot(bL) - b^2 uv. \qquad (3.3.14)$$

The curvature of the wavefront in the image plane is governed by the term containing r^2 in the phase factor in Eq. (3.3.13). It can be written as

$$R_c = \frac{\xi}{b} [\cot(bL) - bu]^{-1}. \qquad (3.3.15)$$

The wavefront is planar when the radius of curvature is infinity. This happens when the object is at a distance $u = f = \cot(bL)/b$. This distance corresponds to the focal length of a GRIN lens of length L. If $u > f$, the wavefront will converge, and a real image will be formed in the image plane. In contrast, a virtual image is formed when $u < f$.

Let us investigate the condition under which $E_i(\mathbf{r})$ in Eq. (3.3.9) resembles the object field $E_o(\mathbf{s})$. Clearly, this can only happen when the kernel $K(\mathbf{r}, \mathbf{s})$ is of the form

$$K(\mathbf{r}, \mathbf{s}) = C\delta(\mathbf{r} - M\mathbf{s}), \tag{3.3.16}$$

where C is a complex constant and M governs the image's magnification. It follows from Eq. (3.3.9) that $K(\mathbf{r}, \mathbf{s})$ takes of the form in Eq. (3.3.16) when ξ given in Eq. (3.3.14) vanishes. This requirement indicates that an image is formed when the following imaging condition is satisfied

$$\frac{1}{u} + \frac{1}{v} = \frac{1}{f}[1 - (b^2 uv)^{-1}]. \tag{3.3.17}$$

The same condition is obtained in the geometrical-optics approximation [27]. The constants M and C in the imaging situation are found to be

$$M = \sin(bL)[\cot(bL) - bu], \qquad C = M \exp[ik(u + v + L) - ikr^2/2(f - v)].$$

An imaging system is specified in terms of its focal length and the position of the focal planes, principal planes, and nodal planes. The focal length, $f = \cot(bL)/b$, fixes the position of the focal plane in the object space. The focal plane in the image space is also situated at a distance f measured from the output end of the GRIN lens. This follows directly from the symmetry of the kernel with respect to u and v in Eq. (3.3.13). As the focal planes are symmetrically situated in the object and the image space, the principal planes, and the nodal planes coincide. The location f_0 of the principal planes can be found using Newton's formula [2], $(u - f)(v - f) = f_0^2$, and is given by

$$f_0 = \sqrt{f^2 + 1/b^2} = [b \sin(bL)]^{-1}, \qquad f = \cot(bL)/b. \tag{3.3.18}$$

As $f_0 > f$, the principal planes are located inside the GRIN lens. If we fix the position of these planes with respect to the center of the GRIN lens, we find that they are situated on both sides at a distance

$$l_p = \frac{L}{2} + f - f_0 = \frac{L}{2}\left(1 - \frac{\tan(bL/2)}{bL/2}\right). \tag{3.3.19}$$

It is common to measure the distances of the object and the image from the corresponding principal planes. If we define these distances as u_0 and v_0, we find that the imaging condition given in Eq. (3.3.17) is reduced to the standard lens formula,

$$\frac{1}{u_0} + \frac{1}{v_0} = \frac{1}{f_0}. \tag{3.3.20}$$

The magnification factor also takes the much simpler form $M = (1 - v_0/f_0)$ in terms of f_0 and v_0.

A conventional lens produces the Fourier transform of the field when the object and image planes coincide with the corresponding focal planes. We expect a GRIN lens to behave in the same way. If we substitute $u = v = f$ in Eq. (3.3.13), the kernel takes the form

$$K(\mathbf{r}, \mathbf{s}) = \frac{kb \sin(bL)}{2\pi i} \exp[ik(u + v + L) - ikb \sin(bL)\mathbf{r} \cdot \mathbf{s}]. \qquad (3.3.21)$$

Using this result in Eq. (3.3.9), we find that, up to a constant factor, the field in the back focal plane is indeed the Fourier transform of the field at the front focal plane. This property implies that two identical GRIN lenses can be used as a 4f-imaging system for applications related to image processing [17].

3.4 Applications of GRIN Devices

GRIN rods were developed around 1970 to serve as a compact flat lens for applications where the use of a traditional lens is undesirable [22]. They were commercialized and have been used for a variety of applications [28]. GRIN fibers were developed during the 1970s [10] and used for the first generation of optical communication systems [29]. More recent applications of GRIN devices are in the areas of optical sensing [30] and medical imaging [31].

3.4.1 GRIN Rods as a Planar Lens

As we saw in Section 3.3.1, a GRIN rod, with a length that is one quarter of its pitch L_p, acts as a lens that can focus an incoming collimated beam to a spot size comparable to its wavelength. The same device can work in reverse as a collimator. The focal length of such a flat lens depends on the two GRIN parameters a and Δ as indicated in Eq. (3.1.14):

$$L_f = \frac{1}{4}L_p = \frac{\pi a}{2\sqrt{2\Delta}}. \qquad (3.4.1)$$

Choosing $a = 1$ mm and $\Delta = 0.05$ as typical values for a GRIN rod, the numerical value of L_f is about 5 mm. A GRIN rod with a length $L < L_f$ also acts as a lens that focuses the beam after it exits the rod.

The physical size of such GRIN lenses is relatively small (diameter close to 2 mm and length close to 5 mm). In contrast to a conventional lens, such a compact device is also flat at both its ends. These features are useful in many technological areas including medical imaging and optical communications. As early as 1980, it was pointed out that GRIN lenses can be used as connectors, attenuators, directional couplers, and channel multiplexers [32]. A GRIN lens was also used for coupling light from a semiconductor laser into a single-mode fiber [33].

As we saw in Section 3.3.2, length of a GRIN lens used for imaging an object can be shorter than L_f. For a grin rod of length L, two principal planes are located symmetrically inside the GRIN lens at a distance from the center given in Eq. (3.3.19). If the object and its image are located at distances u_0 and v_0, respectively, from the corresponding principal planes, they are related by the standard lens formula given in Eq. (3.3.20), where the focal length (measured from a principal plane) is related to the pitch L_p as

$$f_0 = \frac{L_p}{2\pi} \sin(2\pi L/L_p). \tag{3.4.2}$$

This relation shows that the focal length can be controlled by adjusting the device length L. As an example, when L is chosen to be $L_p/12$, the focal length $L_p/4\pi$ is a small fraction of L_p. Two imaging applications that attracted considerable attention during the 1980s were the use of an array of GRIN lenses in photocopy machines [34] and the development of a GRIN objective lens for compact-disk (CD) players [35]. More recently, GRIN lenses are being used for medical imaging, a topic discussed next.

3.4.2 Medical Imaging

An endoscope is an imaging device in the form of a thin long tube that is used to image inner parts of a human body in a noninvasive manner. Hopkins and Storz considered as early as 1966 the use of GRIN lenses and fibers for endoscopy [36]. In their proposed device, GRIN lenses were sandwiched between fiber bundles to form an endoscope. Since then, considerable advances have improved fiber-optic endoscopes so much that they are used routinely for medical diagnosis.

Imaging requires a microscope through which a laser beam is focused onto the tissue and absorbed by it. Scanning of the focal spot allows one to use fluorescence from the tissue to form an image of the tissue on a CCD camera, whose resolution depends on the lens used to focus the laser beam. By 2000, the advent of nonlinear microscopy led to considerable improvement in the resolution of such images. In nonlinear microscopy, energetic laser pulses are focused inside the tissue at a wavelength such that a single photon is not energetic enough to be absorbed by the tissue. However, two-photon absorption can still occur in a narrow region where pulses have the highest peak intensities. In a 2012 experiment, a single GRIN lens was used for improving the performance of a multiphoton microscope capable of imaging in vivo [37]. The GRIN lens was 1.7-mm-long (diameter 0.5 mm), had

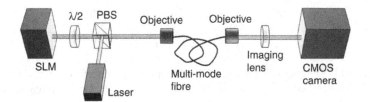

Figure 3.10 Schematic of the setup used for training a neural network employed for reconstructing images transmitted through a multimode fiber. SLM: spatial light modulator, PBS: polarizing beam splitter. (After Ref. [42]; ©CC BY.)

a numerical aperture of 0.6, and was designed to balance the spherical aberration of the microscope's objective lens. Its use allowed to realize a lateral resolution of 618 nm and an axial resolution of 5.5 μm.

A bundle of optical fibers has been used for imaging since the 1970s such that each fiber in the bundle acts as a picture element (pixel). In 2004, a flexible fiber bundle with 30,000 pixels was combined with a GRIN lens and used with a two-photon microscope [38]. In an alternative approach, a single multimode fiber is used for forming the image. The main issue with the use of a multimode fiber is related to the distortion of images resulting from modal interference and random mode coupling along the fiber's length (see Section 2.4). Several techniques have been adopted in recent years for overcoming this distortion [39], including phase conjugation, holography, and adaptive phase compensation.

In a 2016 experiment, a thin rigid endoscope was made using a single GRIN fiber with an external diameter of only 125 μm [40]. It made use of two-photon absorption within the tissue by employing short intense pulses. Image distortion was overcome by measuring the transmission matrix connecting the input and output modes of the fiber. The device was capable of forming images of three-dimensional samples through two-photon fluorescence. However, the use of short pulses limited the length of the endoscope to about 10 cm to avoid excessive broadening of pulses.

Endoscopes can be longer when computational neural network (or deep learning) techniques are employed to remove the distortions induced by a GRIN fiber. In a 2018 experiment, images of objects transmitted through a 1-km-long GRIN fiber could be reconstructed successfully with such a technique [41]. Since then, deep-learning techniques have been used extensively for imaging purposes [42–45]. Figure 3.10 shows a typical experimental setup used for training the neural network. A spatial light modulator (SLM) is employed to create optical beams that are transmitted through a multimode fiber and their images recorded by a camera. In a 2021 experiment, a 10-m-long GRIN fiber was used to train a neural network [44]. The fidelity of the reconstructed images was high enough to be useful for endoscopic applications.

Figure 3.11 Schematic of a sensor based on modal interference inside GRIN fiber. LED: light-emitting diode; SMF: single-mode fiber; OSA: optical spectrum analyzer.

3.4.3 Fiber-Optic Sensing

Optical fibers have been used for making a variety of sensors, especially when a Bragg grating is fabricated inside their core to provide wavelength selectivity [46–48]. GRIN fibers have been found useful for some such sensors [30]. A GRIN fiber was used in 2007 to make a low-loss, low-cost sensor capable of measuring strain and temperature with high sensitivity [49]. Its operation was based on the modal interference occurring inside a GRIN fiber (see Section 2.4.2). Figure 3.11 shows the concept schematically. The sensor is formed by sandwiching a GRIN fiber between two single-mode fibers (SMFs) such that the two splicing areas act as mode couplers. Radiation from a light-emitting diode (LED) is transmitted through the first SMF and is coupled into several different modes of the GRIN fiber. At the other end, the GRIN fiber's output is coupled to the second SMF, which sends it to an optical spectrum analyzer. Experiments showed that only two modes were involved in creating the interference pattern used by such a sensor.

The operation of the sensor can be understood from Eq. (2.4.14) containing a sinusoidal term resulting from the interference between the two excited modes. For a fiber of length L, we can write it as

$$I = I_1 + I_2 + 2\sqrt{I_1 I_2}\cos[2\pi(\delta n)L/\lambda], \qquad (3.4.3)$$

where I_1 and I_2 are the modal intensities and the relation $k_0 = 2\pi/\lambda$ was used to show the wavelength dependence of the intensity explicitly. The relative phase difference between the mode is assumed to remain constant. The spectrum of the output power exhibits a multi-peak structure as the cosine term changes in a periodic fashion over the bandwidth of the LED's spectrum. Any change in the core's refractive index, induced by changes in the local temperature or strain, produces a shift in this pattern, which can be used to deduce the amount of change in the temperature or strain. In a 2007 experiment, the observed sensitivity was close to 58 pm for a 1 °C change in temperature [49].

In a different approach to making sensors, two fibers are combined with a thin air gap between them that acts as a Fabry–Perot resonator. In a 2010 study, periodic self-imaging of GRIN fibers was employed to improve the performance of such a sensor [50]. The sensor was used to measure the refractive index. It was capable

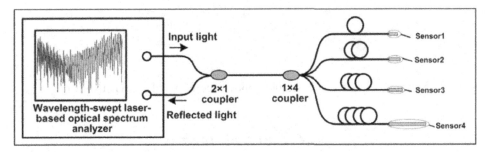

Figure 3.12 Scheme for multiplexing several fiber-based sensors in parallel. Each sensor is made by combining a GRIN fiber and another fiber with a thin gap to form a Fabry-Perot resonator. (After Ref. [51]; ©CC BY.)

of detecting temperature changes with a sensitivity of 11.5 pm for a 1 °C change in temperature. In a later experiment [51], several such sensors were combined in parallel using the setup shown in Figure 3.12. The spectrum of light reflected from all sensors was analyzed using a Fourier-transform technique. The position and the amplitude of the four peaks were used to measure changes in the refractive index and to deduce changes in temperature.

A GRIN lens has also been used in a low-cost instrument designed for pH sensing. Recall that the pH level is a measure of the acidity of a solution. In a 2012 study, the sensor was made by connecting a plastic fiber to a GRIN lens [52]. Light from a laser was reflected by a dichroic mirror and focused through a microscope to illuminate one end of the plastic fiber. At the other end of the fiber, the GRIN lens focused light onto the sample. The resulting fluorescence was sent through the plastic fiber to the same microscope objective, and a CCD camera was used to form an image that was analyzed by a spectrometer.

Sensors have also been made by forming a long-period grating within a GRIN fiber. In a 2021 experiment, such a grating was formed by alternating 1-mm-long sections of a single-mode fiber with 0.7-mm-long sections of a GRIN fiber, resulting in a grating period of 1.7 mm [53]. The length of each GRIN section was close to the self-imaging period to ensure that little distortion of the optical beam occurred inside it. The spectrum of light transmitted through the sensor exhibited a resonance dip, whose position was found to shift with temperature with a sensitivity of 90.8 pm/°C. The wavelength of the resonance peak was relatively insensitive to strain and bending of the fiber.

GRIN fibers can also be used for probing animal tissue *in vivo*. Microprobes made by fusing a short piece of GRIN fiber to the end of a single-mode fiber have been used for some time as a non-invasive probe of animal tissue [54–56]. The same probe can also act as a lens and help in imaging of the tissue.

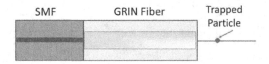

Figure 3.13 Schematic of a fiber-optic tweezer based on a GRIN fiber.

3.4.4 Fiber-Optic Tweezers

An optical tweezer is used for trapping and manipulating tiny particles [57]. Step-index fibers were used for this purpose as early as 1993 [58]. The device was the fiber version of a three-dimensional light-force trap and was used to manipulate small dielectric spheres and living yeast. Since then, many fiber-based devices have been developed [60]. Recently, the use of GRIN fibers has been proposed for tweezing applications [59–63]. Figure 3.13 shows the design of such a device schematically. Such tweezers are relatively small, have low costs, and are easy to use.

The physical mechanism behind the trapping of particles is a gradient force, resulting from a nonuniform intensity distribution of the incident light [57]. This force pulls the particle to a high-intensity region. In the case of a GRIN fiber, the force depends on the intensity distribution of the modes that are excited within the fiber. As was discussed in Section 2.4, when the incident field is radially symmetric, only the LP_{0m} modes with $\ell = 0$ are excited inside a GRIN fiber. The two-mode case is the simplest, for which the intensity distribution at the exit of the fiber can be written from Eq. (2.4.12) as [62]

$$I(x, y, 0) = K|U_{01}(x, y) + se^{i\psi_0}U_{02}(x, y)|^2, \qquad (3.4.4)$$

where K is a constant, s is the amplitude ratio, and ψ_0 is the relative phase. The known mode profiles for the two modes can be used to include diffraction outside the fiber. After propagating a distance z in air, the intensity is found to be

$$I(x, y, z) = \frac{K}{w^2}\exp\left(-\frac{\rho^2}{w^2}\right)\left|1 + se^{i\psi}\left(1 - \frac{\rho^2}{w^2}\right)\right|^2, \qquad (3.4.5)$$

where the spot size $w(z)$ varies with z as

$$w(z) = w_g\sqrt{1 + (z/L_d)^2}, \qquad L_d = kw_g^2, \qquad (3.4.6)$$

and the initial spot size w_g is related to the fundamental mode of the GRIN fiber as $w_g = a/\sqrt{V}$. The phase ψ also varies with z because of a curved wavefront.

Equation (3.4.5) can be used to find the gradient of the on-axis intensity by setting $\rho = 0$, which in turn is used to deduce the trapping distance. Following Section 2.4.1, the analysis can even be extended to include all LP_{0m} modes excited by a

Gaussian beam within the GRIN fiber [62]. This approach can be used to find the distance where the on-axis intensity peaks, and the particle is most likely to be trapped. This distance is known as the working distance.

In practice, fiber-optic tweezers are made by splicing a single-mode fiber to the input end of a short GRIN fiber (see Figure 3.13). In a 2014 experiment, a longer working distance was realized by introducing an air gap between the two fibers [59]. Both fibers were inserted into a silica capillary, and the gap between them was carefully adjusted. The output end of the GRIN fiber was also tapered down to below 10 μm. Even though the optical beam coming out of the single-mode fiber diffracted inside the air gap, it was periodically focused by the GRIN fiber because of self-imaging phenomenon discussed in Section 3.2. Such a device was observed to trap the yeast cells easily. In a 2016 experiment, optofluidic manipulation of a microsphere was realized by tuning the strain applied to a GRIN fiber [61]. It was not even necessary to taper the GRIN fiber.

References

[1] A. Ghatak and K. Thyagarajan, *An Introduction to Fiber Optics* (Cambridge University Press, 1998).

[2] M. Born and E. Wolf, *Principles of Optics*, 7th ed. (Cambridge University Press, 1999).

[3] C. Gomez-Reino, M. V. Perez, and C. Bao, *Gradient Index Optics: Fundamentals and Applications* (Springer, 2003).

[4] F. P. Kapron, *J. Opt. Soc. Am.* **60**, 1433 (1970); https://doi.org/10.1364/JOSA.60.001433.

[5] N. F. Borrelli, *Microoptics Technology: Fabrication and Applications of Lens Arrays and Devices*, 2nd ed. (CRC Press, 2004), 69-128.

[6] D. Stoler, *J. Opt. Soc. Am.* **71**, 334 (1981); https://doi.org/10.1364/JOSA.71.000334.

[7] M. Nazarathy and J. Shamir, *J. Opt. Soc. Am.* **72**, 356 (1982); https://doi.org/10.1364/JOSA.72.000356.

[8] J. N. McMullin, *Appl. Opt.* **25**, 2184 (1986); https://doi.org/10.1364/AO.25.002184.

[9] G. P. Agrawal, A. K. Ghatak, and C. L. Mehta, *Opt. Commun.* **12**, 333 (1974); https://doi.org/10.1016/0030-4018(74)90028-5.

[10] R. Olshansky, *Rev. Mod. Phys.* **51**, 341 (1979); https://doi.org/10.1103/RevModPhys.51.341.

[11] K. Iga, *Appl. Opt.* **19**, 1039 (1980); https://doi.org/10.1364/AO.19.001039.

[12] F. G. Mehler, *J. für die Reine und Angewandte Mathematik* (in German) **66**, 161 (1866); https://en.wikipedia.org/wiki/Mehler_kernel.

[13] W. Pauli, *Wave Mechanics*, Vol. 5 of Pauli Lectures on Physics (Dover, 2000), sec. 44.

[14] H. M. Ozaktas, Z. Zalevsky, and M. A. Kutay, *The Fractional Fourier Transform* (Wiley, 2001).

[15] G. P. Agrawal, *Nouv. Revue d'Optique* **7**, 299 (1976); https://doi.org/10.1088/0335-7368/7/5/303.

[16] I. S. Gradshteyn and I. M. Ryzhik, Eds., *Table of Integrals, Series, and Products*, 6th ed. (Academic Press, 2000).

[17] J. W. Goodman, *Introduction to Fourier Optics*, 4th ed. (W. H. Freeman, 2017).

[18] G. P. Agrawal, *Opt. Fiber Technol.* **50**, 309 (2019); https://doi.org/10.1016/j.yofte .2019.04.012.

[19] S. W. Allison and G. T. Gillies, *Appl. Opt.* **33**, 1802 (1994); https://doi.org/10.1364/ AO.33.001802.

[20] T. Hansson, A. Tonello, T. Mansuryan et al., *Opt. Express* **28**, 24005 (2020); https: //doi.org/10.1364/OE.398531.

[21] M. Ferraro, F. Mangini, M. Zitelli et al., *Photon. Res.* **9**, 2443 (2021); https://doi.org/ 10.1364/PRJ.425878.

[22] T. Uchida, M. Furukawa, I. Kitano, K. Koizumi, and H. Matsumura, *IEEE J. Quantum Electron.* **6**, 606 (1970); https://doi.org/10.1109/JQE.1970.1076326.

[23] E. W. Marchand, *Gradient Index Optics* (Academic Press, 1978).

[24] D. Gloge and E. A. J. Marcatili, *Bell Syst. Tech. J.* **52**, 1563 (1973); https://doi.org/ 10.1002/j.1538-7305.1973.tb02033.x.

[25] A. S. Ahsan and G. P. Agrawal, *J. Opt. Soc. Am. B* **37**, 858 (2020); https://doi.org/ 10.1364/JOSAB.379253.

[26] M. Abramowitz and I. A. Stegun, Eds., *Handbook of Mathematical Functions* (Dover, 1983).

[27] Y. Aoki and M. Suzuki, *IEEE Trans. Microw. Theory Techn.* **15**, 2 (1967); https: //doi.org/10.1109/TMTT.1967.1126362.

[28] I. Kitano, *Appl. Opt.* **29**, 3992 (1990); https://doi.org/10.1364/AO.29.003992.

[29] G. P. Agrawal, *Fiber-Optic Communication Systems*, 5th ed. (Wiley, 2021).

[30] A. G. Mignani, A. Mencaglia, M. Brenci, and A. Scheggi, Radially Gradient-Index Lenses: Applications to Fiber Optic Sensors, in *Diffractive Optics and Optical Microsystems*, Ed. S. Martellucci and A. N. Chester (Springer, 1997), 311-325; https://doi.org/10.1007/978-1-4899-1474-3_26.

[31] J. Son, B. Mandracchia, and S. Jia, *Biomed. Opt. Express* **11**, 7221 (2020); https: //doi.org/10.1364/BOE.410605.

[32] W. J. Tomlinson, *Appl. Opt.* **19**, 1128 (1980); https://doi.org/10.1364/AO.19.001128.

[33] I. Kitano, H. Ueno, and M. Toyama, *Appl. Opt.* **25**, 3336 (1986); https://doi.org/ 10.1364/AO.25.003336.

[34] J. D. Rees and W. Lama, *Appl. Opt.* **23**, 1715 (1984); https://doi.org/10.1364/AO.22 .001715.

[35] H. Nishi, H. Ichikawa, M. Toyama, and I. Kitano, *Appl. Opt.* **25**, 3340 (1986); https: //doi.org/10.1364/AO.25.003340.

[36] P. C. De Groen, "History of the Endoscope," *Proc. IEEE* **105**, 1987 (2017); https: //doi.org/10.1109/JPROC.2017.2742858.

[37] T. A. Murray and M. J. Levene, *J. Biomed. Opt.* **17**, 021106 (2012); https://doi.org/ 10.1117/1.JBO.17.2.021106.

[38] W. Göbel, J. N. D. Kerr, A. Nimmerjahn, and F. Helmchen, *Opt. Lett.* **29**, 2521 (2004); https://doi.org/10.1364/OL.29.002521.

[39] T. Cizmár and K. Dholakia, *Opt. Express* **19**, 18871 (2011); https://doi.org/10.1364/ OE.19.018871.

[40] S. Sivankutty, E. R. Andresen, R. Cossart et al., *Opt. Express* **24**, 825 (2016); https: //doi.org/10.1364/OE.24.000825.

[41] N. Borhani, E. Kakkava, C. Moser, and C. Psaltis, *Optica* **5**, 960 (2018); https: //doi.org/10.1364/OPTICA.5.000960.

[42] P. Caramazza, O. Moran, R. Murray-Smith, and D. Faccio, *Nat. Commun.* **10**, 2029 (2019); https://doi.org/10.1038/s41467-019-10057-8.

[43] T. Zhao, S. Ourselin, T. Vercauteren, and W. Xia, *Opt. Express* **28**, 20978 (2020); https://doi.org/10.1364/OE.396734.

[44] J. Zhao, X. Ji, M. Zhang et al., *J. Phys. Photon.* **3**, 015003 (2021); https://doi.org/10.1088/2515-7647/abcd85.

[45] Z. Liu, L. Wang, Y. Meng et al., *Nat. Commun.* **13**, 1433 (2022); https://doi.org/10.1038/s41467-022-29178-8.

[46] S. Yin, P. B. Ruffin, and F. T. S. Yu, *Fiber Optic Sensors*, 2nd ed. (CRC Press, 2017)

[47] M. Ding and G. Brambilla, Optical Fiber Sensors, in *Biomedical Optical Sensors*, Ed. R. De La Rue, H. P. Herzig, and M. Gerken (Springer, 2020), 155-179.

[48] S. C. Tjin and L. Wei, *Fiber Optic Sensors and Applications* (MDPI, 2020).

[49] Y. Liu and L. Wei, *Appl. Opt.* **46**, 2516 (2007); https://doi.org/10.1364/AO.46.002516.

[50] Y. Gong, Y. Guo, Y.-J. Rao, T. Zhao, and Y Wu, *IEEE Photon. Technol. Lett.* **22**, 1708 (2010); https://doi.org/10.1109/LPT.2010.2082518.

[51] L. Liu, Y. Gong, Y. Wu et al., *Sensors* **12**, 12377 (2012); https://doi.org/10.3390/s120912377.

[52] J. Wang and L. Wang, *Appl. Spectroscopy* **66**, 300 (2012); https://doi.org/10.1366/11-06492.

[53] H. Niu, W. Chen, Y. Liu et al., *Opt. Express* **29**, 22922 (2021); https://doi.org/10.1364/OE.430168.

[54] W. A. Reed, M. F. Yan, and M. J. Schnitzer, *Opt. Lett.* **27**, 1794 (2002); https://doi.org/10.1364/OL.27.001794.

[55] D. Lorenser, X. Yang, and D. D. Sampson, *Opt. Lett.* **37**, 1616 (2012); https://doi.org/10.1364/OL.37.001616.

[56] M. Sato, K. Shouji, D. Saito, and I. Nishidate, *Appl. Opt.* **55**, 3207 (2016); https://doi.org/10.1364//AO.55.003297.

[57] A. Ashkin, *IEEE J. Sel. Topics Quantum Electron.* **6**, 841 (2000); https://doi.org/10.1109/2944.902132.

[58] A. Constable, J. Kim, J. Mervis, F. Zarinetchi, and M. Prentiss, *Opt. Lett.* **18**, 1867 (1993); https://doi.org/10.1364/OL.18.001867.

[59] Y. Gong, W. Huang, Q.-F. Liu et al., *Opt. Express* **22**, 25267 (2014); https://doi.org/10.1364/OE.22.025267.

[60] R. S. R. Ribeiro, O. Soppera, A. G. Oliva, A. Guerreiro, and P. A. S. Jorge, *J. Lightwave Technol.* **33**, 3394 (2015); https://doi.org/10.1109/JLT.2015.2448119.

[61] C.-L. Zhang, Y. Gong, Q.-F. Liu et al., *IEEE Photon. Technol. Lett.* **28**, 256 (2016); https://doi.org/10.1109/LPT.2015.2494583.

[62] E. Mobini and A. Mafi, *J. Lightwave Technol.* **35**, 3854 (2017); https://doi.org/10.1109/JLT.2017.2718501.

[63] Y. Lou, D. Wu, and Y. Pang, *Adv. Fiber Mater.* **1**, 83 (2019); https://doi.org/10.1007/s42765-019-00009-8.

4

Dispersive Effects

Graded-index (GRIN) fibers were developed during the 1970s and used by 1980 for the first generation of optical communication systems. They differ from GRIN rods only in the relative size of the core region; the core's radius a is reduced to below 40 μm in a GRIN fiber. As a result, the number of modes supported by such a fiber is much smaller compared to a GRIN rod. This number can be as small as one or two for GRIN fibers designed with $a \leq 5$ μm. For some applications, optical beams launched into a GRIN fiber contain a train of short pulses. Such pulses are affected considerably by the dispersive effects resulting from the frequency dependence of the refractive index.

This chapter is devoted to the study of various dispersive effects that affect short pulses inside a GRIN fiber. In Section 4.1, we obtain an equation governing the evolution of optical pulses inside a GRIN medium. The dispersion parameters appearing in this equation change, depending on which mode is being considered. Section 4.2 focuses on the distortion of optical pulses resulting from differential group delay (DGD) and group-velocity dispersion (GVD). Section 4.3 deals with the effects of linear coupling among the modes, occurring because of random variations in the core's shape and size along a fiber's length. A non-modal approach is developed in Section 4.4 for the propagation of short optical pulses inside a GRIN medium. The focus of Section 4.5 is on the applications where optical pulses are sent through a GRIN rod or fiber.

4.1 Impact of Chromatic Dispersion

Dispersion-induced pulse broadening was studied during the 1970s for GRIN fibers with different power-law profiles [1–3]. This study made use of the approximate propagation constants found in Section 2.3 in the WKB approximation. In this chapter, we focus on GRIN fibers designed with the parabolic index profile. The propagation constants found in Section 2.2 are employed to derive a time-domain propagation equation satisfied by the amplitudes of different modes. This equation is used to study how optical pulses are distorted inside a GRIN fiber.

4.1.1 Propagation Equation for Mode Amplitudes

Consider a GRIN fiber whose front end is located at $z = 0$ and let the z axis coincide with its central axis. A pulsed beam with the electric field $E(x, y, 0, t)$ excites multiple modes of this fiber. As the modes in Section 2.2 were found in the frequency domain, we work with the Fourier transform of the electric field:

$$\tilde{E}(\mathbf{r}, \omega) = \int_{-\infty}^{\infty} E(\mathbf{r}, t)\, e^{i\omega t}\, dt. \tag{4.1.1}$$

At the input end of the GRIN fiber, we use the modal expansion in the form

$$\tilde{E}(x, y, 0, \omega) = \sum_j \tilde{A}_j(0, \omega) U_j(x, y), \tag{4.1.2}$$

where we simplify the notation by using a single index j for the LP_{lm} modes of the fiber. Recall from Section 2.2 that all modes are normalized such that

$$\iint_{-\infty}^{\infty} U_j^*(x, y) U_k(x, y)\, dx\, dy = \delta_{jk}. \tag{4.1.3}$$

The expansion coefficients appearing in Eq. (4.1.2) are found by multiplying this equation with $U_k^*(x, y)$ and integrating over the entire transverse plane. After using Eq. (4.1.3), the result is given by

$$\tilde{A}_j(0, \omega) = \iint_{-\infty}^{\infty} U_j^*(x, y)\tilde{E}(x, y, 0, \omega)\, dx\, dy. \tag{4.1.4}$$

As all modes propagate without change in their spatial profiles and only acquire a phase shift inside the GRIN medium, $\tilde{E}(\mathbf{r}, \omega)$ at a distance z is obtained from Eq. (4.1.2) by multiplying $U_j(x, y)$ with a mode-dependent phase factor:

$$\tilde{E}(\mathbf{r}, \omega) = \sum_j \tilde{A}_j(0, \omega) U_j(x, y) \exp(i\beta_j z), \tag{4.1.5}$$

where $\beta_j(\omega)$ is the propagation constant of the jth mode. The electric field $E(\mathbf{r}, t)$ can now be calculated by taking the inverse Fourier transform as

$$E(\mathbf{r}, t) = \frac{1}{2\pi} \int_{-\infty}^{\infty} \tilde{E}(\mathbf{r}, \omega)\, e^{-i\omega t}\, d\omega. \tag{4.1.6}$$

If the fiber support only a few modes, it is useful to work with the mode amplitudes $A_j(z, t)$ and ask how they evolve inside a GRIN fiber. For this purpose, we combine Eqs. (4.1.5) and (4.1.6) and write the electric field as

$$E(\mathbf{r}, t) = \frac{1}{2\pi} \sum_j U_j(x, y) \int_{-\infty}^{\infty} \tilde{A}_j(0, \omega) \exp(i\beta_j z - i\omega t)\, d\omega. \tag{4.1.7}$$

We assume that the spectrum of input beam is centered at the frequency ω_0 and is relatively narrow. Writing the electric field in the form,

$$E(\mathbf{r}, t) = \sum_j U_j(x, y) A_j(z, t) \exp(i\beta_{0j}z - i\omega_0 t), \tag{4.1.8}$$

we obtain the slowly varying amplitude of the jth mode:

$$A_j(z, t) = \frac{1}{2\pi} \int_{-\infty}^{\infty} \tilde{A}_j(0, \omega) \exp[i(\beta_j - \beta_{0j})z - i(\omega - \omega_0)t] \, d\omega, \tag{4.1.9}$$

where $\tilde{A}_j(0, \omega)$ depends on the input field as indicated in Eq. (4.1.4). The preceding integral depends on the dispersive properties of a GRIN fiber through the frequency dependence of the propagation constant $\beta_j(\omega)$.

In practice, one is interested in knowing how the shape of optical pulses is affected by the dispersion. For this purpose, we consider the temporal power profile obtained using

$$P(z, t) = \iint_{-\infty}^{\infty} |E(\mathbf{r}, t)|^2 \, dx \, dy, \tag{4.1.10}$$

Substituting $E(\mathbf{r}, t)$ from Eq. (4.1.8) and using orthogonality of the modes in Eq. (4.1.3), we obtain the simple relation

$$P(z, t) = \sum_j P_j(z, t) = \sum_j |A_j(z, t)|^2. \tag{4.1.11}$$

It shows that the temporal power profile at any distance z is found by simply summing $|A_j|^2$ over all modes excited at the input end of a GRIN medium. It is thus sufficient to obtain a propagation equation for the amplitudes $A_j(z, t)$.

To obtain this propagation equation, we use the narrow-bandwidth condition $\Delta\omega \ll \omega_0$. We expand $\beta_j(\omega)$ in Eq. (4.1.9) in a Taylor series around the central frequency ω_0 of the spectrum and retain terms up to the second order. It follows from Eq. (2.4.17) that

$$\beta_j(\omega) = \beta_{0j} + \beta_{1j}\Omega + \tfrac{1}{2}\beta_{2j}\Omega^2, \tag{4.1.12}$$

where $\Omega = \omega - \omega_0$. Recall from Section 2.4.2 that β_{1j} is the group delay for the jth mode and β_{2j} is its GVD parameter. Using Eq. (4.1.12) in Eq. (4.1.9), we obtain

$$A_j(z, t) = \frac{1}{2\pi} \int_{-\infty}^{\infty} \tilde{A}_j(0, \Omega) \exp\left[i(\beta_{1j}\Omega + \tfrac{1}{2}\beta_{2j}\Omega^2)z - i\Omega t\right] d\Omega. \tag{4.1.13}$$

Thus, we can find $A_j(z, t)$ at any distance z inside the GRIN fiber by performing an integration over the spectrum of the input pulse.

It is useful to find the equation whose solution Eq. (4.1.13) represents. Calculating $\partial A_j/\partial z$ and replacing Ω with $i(\partial A_j/\partial t)$ in the time domain, Eq. (4.1.13) can be converted into the following time-domain equation [4]:

$$\frac{\partial A_j}{\partial z} + \beta_{1j}\frac{\partial A_j}{\partial t} + \frac{i}{2}\beta_{2j}\frac{\partial^2 A_j}{\partial t^2} = 0. \qquad (4.1.14)$$

This equation governs evolution of the jth mode inside a GRIN medium and includes the effects of fiber's dispersion up to the second order.

4.1.2 DGD-Induced Broadening of Pulses

If pulses launched into a GRIN fiber are relatively broad or propagate close to its zero-dispersion wavelength, we can neglect the β_2 term in Eq. (4.1.14), resulting in the solution $A_j(z, t) = A_j(0, t - \beta_{1j}z)$. Except for the group delay by an amount $\beta_{1j}z$, the initial amplitude $A_j(0, t)$ remains unchanged. However, the pulse shape can still change if this group delay is different for different modes. Using the preceding solution in Eq. (4.1.11), the pulse shape at the end of a fiber of length L is obtained from

$$P(L, t) = \sum_j |A_j(0, t - \beta_{1j}L)|^2 = \sum_j |A_j(0, t - \tau_{1j}L)|^2, \qquad (4.1.15)$$

where τ_{1j} is the DGD, as defined in Eq. (2.4.20), and we used the first mode as a reference ($\tau_{11} = 0$) for removing a constant time delay.

The simplest situation occurs when the input beam excites just a few modes. As discussed in Section 2.4.1, this can occur when a Gaussian beam is launched with its center aligned with the fiber's axis. In this case, only a few radially symmetric LP_{0m} modes are excited with coupling efficiencies η_{0m}. We can use Eq. (4.1.15) to write the output power in the form

$$P(L, t) = P_0 \sum_m \eta_{0m} S[t + (m - 1)\tau L], \qquad (4.1.16)$$

where $P(0, t) = P_0 S(t)$, P_0 is the input peak power, and $S(t)$ describes the shape of input pulses, normalized such that $S(0) = 1$. Also, we used Eq. (2.4.20) to introduce a single parameter $\tau = \tau_{12}$ representing the DGD between the first two mode groups.

Figure 4.1 shows the distortion induced by DGD for values τ in the range 0–0.3 ps/m, assuming 30-ps-wide Gaussian pulses are launched into a 100-m-long GRIN fiber. If we use $S(t) = \exp(-t^2/T_0^2)$, the full width at half maximum (FWHM) of the pulse is related to T_0 as

$$T_{\text{FWHM}} = 2(\ln 2)^{1/2} T_0 \approx 1.665 T_0. \qquad (4.1.17)$$

For a 30-ps-wide Gaussian pulse, $T_0 = 18$ ps. The coupling efficiencies were obtained from Eq. (2.4.9) with $qw_0 = 2$ and the first five modes were included in the sum. As seen in Figure 4.1, an initially Gaussian pulse becomes not only broader but also asymmetric in shape after propagating just 100 m inside a GRIN fiber. These distortions occur because different modes arrive at slightly different times at the end of the fiber. Distortions occur toward the front end of the pulse because higher-order

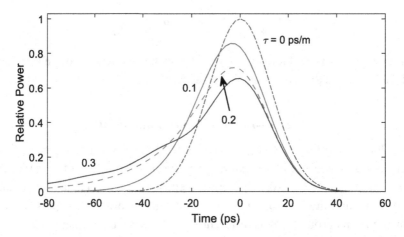

Figure 4.1 DGD-induced distortion of 30-ps-wide Gaussian pulses inside a 100-m-
long GRIN fiber for τ values in the range of 0-0.3 ps/m. The pulse is not distorted
for $\tau = 0$.

modes travel faster and arrive earlier compared to the LP_{01} mode of the fiber that
carries more of the pulse energy.

4.1.3 GVD-Induced Pulse Broadening

For shorter input pulses, one must include the effects of GVD governed by the β_{2j}
term in Eq. (4.1.14). This equation can be solved analytically by using the Fourier
transform of $A_j(z, t)$, and the solution is given in Eq. (4.1.13).

As an example, consider the case of a chirped Gaussian pulse with the initial
amplitude,

$$A_j(0, t) = A_{0j} \exp\left[-\frac{1 + iC}{2}\left(\frac{t}{T_0}\right)^2\right], \tag{4.1.18}$$

where A_{0j} is the peak amplitude and T_0 is related to the FWHM of the pulse as
indicated in Eq. (4.1.17). The parameter C governs the frequency chirp imposed
on the pulse. A pulse is said to be chirped if its frequency changes with time. The
frequency change is related to the phase derivative and is given by

$$\delta\omega(t) = -\frac{\partial\phi}{\partial t} = \frac{C}{T_0^2}t, \tag{4.1.19}$$

where ϕ is the phase of $A_j(0, t)$. A time-dependent frequency shift is called the *chirp*.
The spectrum of a chirped pulse is broader than that of an unchirped pulse. This
can be seen by taking the Fourier transform of Eq. (4.1.18) to obtain

$$\tilde{A}_j(0, \omega) = A_0 \left(\frac{2\pi T_0^2}{1 + iC}\right)^{1/2} \exp\left[-\frac{\omega^2 T_0^2}{2(1 + iC)}\right]. \tag{4.1.20}$$

The spectral half-width (at $1/e$ intensity point) is given by

$$\Delta\omega_0 = (1 + C^2)^{1/2} T_0^{-1}. \tag{4.1.21}$$

In the absence of frequency chirp ($C = 0$), the spectral width satisfies the relation $\Delta\omega_0 T_0 = 1$. Such a pulse has the narrowest spectrum and is called *transform-limited*. The spectral width is enhanced by a factor of $(1 + C^2)^{1/2}$ in the presence of linear chirp.

Using Eq. (4.1.20) in Eq. (4.1.13), integration can be performed analytically using the known formula [5]

$$\int_{-\infty}^{\infty} e^{-ax^2} e^{ibx} dx = \sqrt{\pi/a} \, \exp(-b^2/4a). \tag{4.1.22}$$

The final result can be written in the form [6]

$$A_j(z, t) = \frac{A_{0j}}{\sqrt{Q(z)}} \exp\left[-\frac{(1 + iC)(t - \beta_{1j}z)^2}{2T_0^2 Q(z)}\right], \tag{4.1.23}$$

where $Q(z) = 1 + (C - i)\beta_{2j}z/T_0^2$. This equation shows that a pulse maintains its Gaussian shape inside the GRIN fiber, but its width, chirp, and amplitude change because of the dependence of $Q(z)$ on β_{2j}. The peak of the pulse is shifted in time by $\beta_{1j}z$ as the pulse takes some time to arrive at its current location.

We can find from Eq. (4.1.23) the width and chirp of the pulse in the jth mode at a distance z. Denoting them with T_1 and C_1 respectively, we obtain

$$T_1(z) = |Q(z)|T_0, \qquad C_1(z) = C + s(1 + C^2)\xi, \tag{4.1.24}$$

where $s = \text{sgn}(\beta_{2j})$, $\xi = z/L_D$, and the dispersion length is defined as $L_D = T_0^2/|\beta_{2j}|$. Changes in the pulse width are quantified through the broadening factor

$$T_1/T_0 = [(1 + sC\xi)^2 + \xi^2]^{1/2}. \tag{4.1.25}$$

Figure 4.2 shows (a) the broadening factor and (b) the chirp as a function of ξ in the case of anomalous GVD ($s = -1$). An unchirped pulse ($C = 0$) broadens monotonically by a factor of $(1 + \xi^2)^{1/2}$ and develops a negative chirp such that $C_1 = -\xi$ (the dotted curves). Chirped pulses, on the other hand, may broaden or compress depending on whether β_{2j} and C have the same or opposite signs. For $\beta_{2j}C < 0$, pulse's width decreases initially and becomes $T_1^{\min} = T_0/(1 + C^2)^{1/2}$ at a distance $\xi = |C|L_D/(1 + C^2)$. This is the temporal analog of lens-induced focusing. The dispersion-induced chirp counteracts the initial chirp, and the net chirp vanishes at the distance where the pulse is the shortest.

Let us consider the power distribution $P(z, t)$ of the output pulse given in Eq. (4.1.11) as a sum over all excited modes. Energy of the input pulse is distributed over multiple modes with the amplitudes $A_j(0, t)$. Each of these amplitudes changes

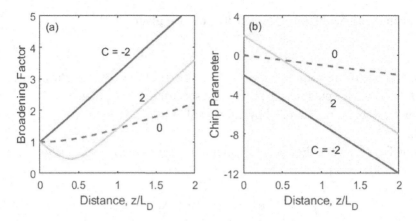

Figure 4.2 (a) Broadening factor and (b) frequency chirp as a function of ξ when a Gaussian pulse propagates in a specific mode of a GRIN fiber in the region of anomalous dispersion ($s = -1$).

as indicated in Eq. (4.1.23). In the absence of GVD, each mode is delayed by a different amount, but its temporal profile $|A_j(0,t)|^2$ does not change. In the presence of GVD, the temporal profiles of modes change differently because their dispersion lengths are not the same when the values of β_{2j} are different for them. Compared to the pulse shapes seen in Figure 4.1, output pulse becomes broader and more distorted when the effects of GVD are included.

4.2 Average Pulse Broadening

For some applications, it is useful to have a single measure of pulse's broadening, averaged over all modes excited by the pulse. Such an approach was used as early as 1976 to estimate the extent of temporal broadening of a pulse [3]. It is based on the concept of the variance of a random variable and makes use of its moments, similar to those introduced in Section 2.5.2 in the context of the spatial width of an optical beam. In the case of optical pulses, the moments are defined as

$$M_n(z) = \int_{-\infty}^{\infty} t^n P(z, t)dt, \tag{4.2.1}$$

where n is an integer and $P(z, t)$ is the temporal power profile at a distance z. Its modal distribution at a given time t is indicated in Eq. (4.1.11). It is easy to conclude that M_0 corresponds to pulse's energy, M_1 is related to the average delay of the pulse, and M_2 can be used to calculate the variance providing a reasonable measure of pulse broadening.

4.2.1 Impulse-Based Model

In a simple model, the pulses are assumed to be so short that each can be treated as an impulse, whose power is distributed over a wide range of frequencies such that

$$P(z, t) = \int_0^\infty \tilde{P}(z, t, \omega) d\omega. \tag{4.2.2}$$

The power of each spectral component is distributed over many modes of the fiber. Noting that impulse response of each mode depends on the group delay of that mode, this power can be written as [3]

$$\tilde{P}(z, t, \omega) = \sum_j P_j(z, \omega) \delta[t - z\tau_j(\omega)], \tag{4.2.3}$$

where $z\tau_j(\omega)$ is the group delay of the jth mode. The frequency dependence of $\tau_j(\omega)$ accounts for the fiber's dispersion. Using Eqs. (4.2.1) through (4.2.3), we obtain

$$M_n(z) = z^n \int_0^\infty \left[\sum_j \tau_j^n(\omega) P_j(z, \omega) \right] d\omega. \tag{4.2.4}$$

The frequency dependence of $P_j(z, \omega)$ stems from the source used to launch the impulse. We can write it as $P_j(z, \omega) \approx S(\omega) p_j(z)$, where $S(\omega)$ is the spectrum of the source normalized such that $\int S(\omega) d\omega = 1$ and $p_j(z)$ is the modal distribution of the power at a distance z inside the GRIN fiber. The central frequency and the bandwidth of the spectrum are related to $S(\omega)$ as

$$\omega_0 = \int_0^\infty \omega S(\omega) d\omega, \qquad \sigma_\omega^2 = \int_0^\infty (\omega - \omega_0)^2 S(\omega) d\omega. \tag{4.2.5}$$

The moments in Eq. (4.2.4) can be calculated approximately by expanding the delay $\tau_j(\omega)$ in a Taylor series around ω_0 and retaining only the first two terms:

$$\tau_j \approx \beta_{1j} + (\omega - \omega_0) \beta_{2j}. \tag{4.2.6}$$

Using this form of τ_j in Eq. (4.2.4), we define the modal average of the variable B as

$$\langle B \rangle = \frac{1}{P_0} \sum_j p_j B_j. \tag{4.2.7}$$

where the total power $P_0 = \sum_j p_j(z)$ is related to M_0. The first moment M_1 is related to the average delay as $T_{av}(z) = M_1/M_0$. This delay depends linearly on the distance as $T_{av}(z) = z\langle \beta_1 \rangle$.

The variance of power distribution can now be calculated using

$$\sigma_p^2(z) = M_2(z)/M_0 - T_{av}^2(z), \tag{4.2.8}$$

where σ_p is the root-mean-square (RMS) width of the pulse. The variance can be divided into two parts, $\sigma_p^2 = \sigma_1^2 + \sigma_2^2$, where σ_1 is the intermodal part and σ_2 is the intramodal part. Both vary linearly with distance and are given by [3]

$$\sigma_1(z) = z[\langle \beta_1^2 \rangle - \langle \beta_1 \rangle^2]^{1/2}, \qquad \sigma_2(z) = z\sigma_\omega \langle \beta_2^2 \rangle^{1/2}. \tag{4.2.9}$$

It follows that the RMS width of the pulse increases linearly with distance.

The intermodal part in Eq. (4.2.9) depends on the standard deviation of β_1, while the intramodal part depends on $\langle \beta_2^2 \rangle$. Effects of third-order dispersion β_3 can be included by keeping an additional term in the Taylor expansion in Eq. (4.2.6). In practice, the intermodal part σ_1 dominates. This is consistent with the results in Section 4.1, where DGD resulting from different values of β_{1j} produced most pulse broadening. Whereas, the results of Section 4.1 must be used when one is interested in actual pulse shapes, the results of this section provide a quick way to judge the impact of fiber's dispersion on short pulses. They also allow one to go beyond the parabolic index profile and have been applied to a power-law profile [3].

4.2.2 Extension to Gaussian Pulses

It was noticed in 1977 that the impulse-based model is oversimplified and can lead to inconsistent results [7]. A delta function in time should have an infinitely wide spectrum, but the spectrum in Eq. (4.2.5) clearly has a finite bandwidth. Also, any input pulse will have a finite duration, and we saw in Section 4.1 that the width of output pulse depends on this duration. This is not the case in Eq. (4.2.9).

In the approach used in Ref. [7], the initial electric field is written in the form

$$E(x, y, 0, t, \omega') = F(x, y, 0)g(t)S(\omega'), \tag{4.2.10}$$

where $F(x, y, 0)$ is the spatial distribution, $g(t)$ specifies the pulse shape and $S(\omega')$ is the spectrum of the source. When we take its Fourier transform with respect to time, we obtain

$$\tilde{E}(x, y, 0, \omega, \omega') = F(x, y, 0)\tilde{g}(\omega)S(\omega'), \tag{4.2.11}$$

This equation makes it clear that the pulse's spectrum has two parts: $\tilde{g}(\omega)$ is the Fourier spectrum of the pulse, whereas $S(\omega')$ is the spectrum of the source.

Following Eq. (4.1.5), we expand $F(x, y, 0)$ in terms of fiber's modes and find the spatial distribution at a distance z using

$$F(x, y, z) = \sum_j C_j U_j(x, y) \exp(i\beta_j z), \tag{4.2.12}$$

where the expansion coefficients C_j are known in terms of the $F(x, y, 0)$. The electric field in the time domain is found to be

$$E(\mathbf{r}, t, \omega') = \sum_j C_j U_j(x, y) S(\omega') \int_{-\infty}^{\infty} \tilde{g}(\omega) \exp[i\beta_j(\omega)z) - i\omega t] \, d\omega. \quad (4.2.13)$$

The integral in the preceding equation can be done for Gaussian input pulses for which $\tilde{g}(\omega)$ is a Gaussian function. The procedure involves expanding $\beta_j(\omega)$ in a Taylor series up to the second order, as was done in Section 4.1. Integrating the result over the transverse area, the power profile is obtained as

$$P(z, t, \omega') = \iint_{-\infty}^{\infty} |E(\mathbf{r}, t, \omega')|^2 \, dx \, dy. \quad (4.2.14)$$

This result is used to calculate the moments defined in Eq. (4.2.4) and to find the variance in Eq. (4.2.8). The result can be written in the form [7]

$$\sigma_p^2(z) = \sigma_0^2 + \sigma_1^2 + \sigma_2^2, \quad (4.2.15)$$

where σ_0 is the RMS width of the input Gaussian pulse, σ_1 is identical to that in Eq. (4.2.9), and σ_2 is modified as

$$\sigma_2(z) = z \langle \beta_2^2 \rangle^{1/2} [\sigma_\omega^2 + (2\sigma_0)^{-2}]^{1/2}. \quad (4.2.16)$$

Equation (4.2.15) does not reduce to Eq. (4.2.9) in the limit of an impulse ($\sigma_0 = 0$). However, it does reduce to Eq. (4.2.9) when the source bandwidth is so large that $\sigma_\omega \gg 1/\sigma_0$. In the opposite limit, $\sigma_\omega \ll 1/\sigma_0$, Eq. (4.2.15) corresponds to the situation discussed in Section 4.1.3.

4.3 Random Mode Coupling

It has been assumed so far that each mode of a GRIN fiber propagates independently without coupling to other modes. This assumption holds for an ideal fiber and follows from the orthogonality condition in Eq. (4.1.3). In practice, optical fibers are far from being ideal. Any fiber suffers from random variations in the shape and size of its core along its length that occur invariably during its fabrication. These unintended variations lead to random coupling among a fiber's modes that must be considered for a realistic description of light propagation in long fibers [8–10].

Mode coupling also occurs if the fiber is bent or suffers from micro-bending. In this case, guided modes of a fiber couple to the cladding modes (or radiation modes) of the same fiber. Such coupling manifests as a bending-induced power loss for the guided mode. This type of mode coupling for a GRIN fiber was considered in the 1970s and was used to calculate the fiber's bending loss [11, 12].

4.3.1 Coupled-Mode Equations

A simple way to include random coupling among a fiber's modes is to add a small perturbation $\delta\epsilon(x, y, z)$ to the index profile of a GRIN fiber in Eq. (1.2.1). The perturbation varies with z in a random fashion because of fluctuations in the core's shape and size (or stress) along the fiber. Following the procedure of Section 4.1, we find that perturbation adds a coupling term to the mode's propagation equation given in Eq. (4.1.3) such that [4]

$$\frac{\partial A_j}{\partial z} + \beta_{1j}\frac{\partial A_j}{\partial t} + \frac{i}{2}\beta_{2j}\frac{\partial^2 A_j}{\partial t^2} = i\sum_{k \neq j} \kappa_{jk}A_k \exp[i(\beta_{0k} - \beta_{0j})z]. \tag{4.3.1}$$

where the coupling coefficient is defined as

$$\kappa_{jk}(z) = \frac{\omega_0}{c}\iint_{-\infty}^{\infty} \delta\epsilon(x, y, z)U_k^*(x, y,)U_j(x, y,)\,dx\,dy. \tag{4.3.2}$$

The magnitude of $\kappa_{jk}(z)$ depends on the local perturbation $\delta\epsilon(x, y, z)$, in addition to the spatial overlap of the two modes involved.

It is natural to ask under what conditions coupling between the modes of a fiber can be ignored. If we consider only two modes ($j = 1, 2$) excited by a CW beam (no time variations) and use $A_j = B_j \exp(-i\delta_a z)$ in Eq. (4.3.1) with $\delta_a = (\beta_{01} - \beta_{02})/2$, we obtain two coupled linear equations:

$$\frac{dB_1}{dz} = i\delta_a B_1 + i\kappa_{12}B_2, \qquad \frac{dB_2}{dz} = -i\delta_a B_2 + i\kappa_{21}B_1. \tag{4.3.3}$$

It is relatively easy to solve these equations when κ_{12} does not depend on z. Even though this is not the case for long fibers, κ_{12} can be treated as a constant over a short section of the fiber. The two equations in Eq. (4.3.3) can then be combined to obtain the following single equation:

$$\frac{d^2 B_2}{dz} + \kappa_e^2 B_2 = 0, \qquad \kappa_e = \sqrt{\kappa_{12}\kappa_{21} + \delta_a^2}. \tag{4.3.4}$$

Assuming that only the first mode is excited at $z = 0$ so that $B_2(0) = 0$ initially, the solution is found to be

$$B_2(z) = (i\kappa_{21}/\kappa_e)B_1(0)\sin(\kappa_e z). \tag{4.3.5}$$

This result shows that power is transferred to the second mode in a periodic fashion. Maximum power transfer occurs for the first time at a distance, $L_c = \pi/(2\kappa_e)$, known as the coupling length.

The fraction of power transferred after one coupling length is found to be

$$\eta = \left|\frac{B_2(L_c)}{B_1(0)}\right|^2 = \frac{|\kappa_{21}|^2}{\kappa_{12}\kappa_{21} + \delta_a^2} = \left(1 + \frac{\delta_a^2}{|\kappa_{21}|^2}\right)^{-1}, \tag{4.3.6}$$

where we used the relation $\kappa_{12} = \kappa_{21}^*$ that follows from Eq. (4.3.2). This fraction represents the coupling efficiency, and it depends only on the ratio $\delta_a/|\kappa_{21}|$. Clearly, η is small if two modes are weakly coupled because of a small value of κ_{21}. However, it also becomes small when δ_a is much larger than $|\kappa_{21}|$. The coupling efficiency is 100% only when the two modes have the same propagation constants ($\delta_a = 0$). This is the reason why mode coupling is relatively strong for two degenerate or nearly degenerate modes.

The preceding analysis shows that degenerate mode pairs ($\delta_a = 0$) of a mode group are most strongly coupled in any GRIN fiber. Random fluctuations in the fiber's birefringence lead to strong coupling between the orthogonally polarized versions of the same mode because of their degenerate nature. The coupling length is also short for them (< 1 m) because of their total spatial overlap. Similarly, the even and odd versions of any LP mode with $l > 0$ become strongly coupled, but coupling length is larger for them (> 10 m). Two neighboring mode groups may exhibit coupling over long distances (> 100 m), depending on the magnitude of δ_a for them.

4.3.2 Polarization-Mode Dispersion

A potential source of pulse broadening is related to polarization-mode dispersion (PMD). As discussed earlier, random changes in the shape and size of a fiber's core occur for any fiber, including the GRIN fibers. Such variations break the degeneracy of the orthogonally polarized versions of each mode, that is, their propagation constants become slightly different, and this difference varies in a random fashion along the fiber's length. As the two modes travel at different speeds, they arrive at different times at the fiber's output end. This situation is similar to the DGD between two spatial modes but with one major difference. Whereas the DGD is constant for two spatial modes, PMD varies along a fiber's length in an unpredictable manner.

PMD was measured as early as 1978 for single-mode fibers [13]. It has attracted most attention for such fibers because of their use in optical communication systems [14–16]. In the case of GRIN fibers, PMD occurs for all modes, but its impact does not vary much from one mode to the next. We can quantify PMD for any mode using the difference in the group delays between its orthogonally polarized versions: $\delta\beta_1 = \beta_{1x} - \beta_{1y}$, where $\delta\beta_1$ varies randomly along the fiber's length. The relative delay ΔT between the two polarization components is obtained by integrating $\delta\beta_1$ over the fiber's length L:

$$\Delta T = \int_0^L \delta\beta_1(z)\, dz. \tag{4.3.7}$$

When $\delta\beta_1$ is constant (e.g., for a polarization-maintaining fiber), the time delay, $\Delta T = \delta\beta_1 L$, scales linearly with the fiber's length. In contrast, when $\delta\beta_1$ fluctuates along z, ΔT scales with length as $\sqrt{L}$. This is so because the PMD problem is similar in nature to a random walk in one dimension. The x-polarized component of a mode may move faster in some sections of the fiber, but it may also move slower in other sections. It is intuitively clear that the state of polarization (SOP) of any mode of the fiber will also evolve randomly along the length of a GRIN fiber because of PMD. In the case of optical pulses, the SOP will also be different for different spectral components of the pulse.

The analytical treatment of PMD is complicated in general because of its statistical nature [14–16]. A simple model divides the fiber into a large number of segments. Both the degree of birefringence and the orientation of the principal axes remain constant in each section, but they change randomly from section to section. In effect, each fiber section is treated as a phase plate using a Jones matrix. Propagation of each frequency component associated with an optical pulse is then governed by the composite Jones matrix obtained by multiplying individual Jones matrices for each fiber section. The composite matrix shows that two principal states of polarization exist for any fiber. When a pulse is polarized along them, the SOP at the fiber's output becomes frequency independent to first order. These states are analogous to the slow and fast axes associated with a polarization-maintaining fiber.

The principal states of polarization provide a convenient basis for calculating the moments of ΔT. PMD-induced pulse broadening is characterized by the root-mean-square (RMS) value of ΔT, obtained after averaging over random birefringence changes. The variance σ_T^2 for a fiber of length L is found to be [17]

$$\sigma_T^2 = 2(\Delta\beta_1)^2 l_c^2 [\exp(-L/l_c) + L/l_c - 1], \tag{4.3.8}$$

where l_c is the correlation length, defined as the length over which two polarization components remain correlated; its values are ~ 10 m for most fibers. For a long fiber such that $L \gg l_c$, σ_T can be approximated as

$$\sigma_T \approx \Delta\beta_1 \sqrt{2l_c L} \equiv D_p\sqrt{L}, \tag{4.3.9}$$

where D_p is known as the PMD parameter. Measured values of D_p vary from fiber to fiber but are ~ 1 ps/km$^{1/2}$. Because of the $\sqrt{L}$ dependence, PMD-induced pulse broadening is relatively small compared with the DGD effects. However, PMD can become a limiting factor for telecommunication systems designed to operate over long distances at high bit rates [6].

4.3.3 Intermodal Power Transfer

Equation (4.3.1), governing the evolution of mode amplitudes inside an optical fiber, includes random mode coupling through its last term. With some simplifications, we can use it to deduce how the average powers of modes vary inside the fiber [10]. We assume that the perturbation $\delta\epsilon$ in Eq. (4.3.1) can be factored as $\delta\epsilon(x,y)h(z)$, where $h(z)$ is a random function of z such that it vanishes on average. If we focus on the case of a CW beam so that the two dispersion terms can be ignored, Eq. (4.3.1) is reduced to

$$\frac{dA_j}{dz} = i \sum_{k \neq j} \kappa_{jk} h(z) A_k \exp[i(\beta_{0k} - \beta_{0j})z], \tag{4.3.10}$$

where κ_{jk} is a constant because its randomness is included through $h(z)$. We use this equation to find average modal powers, $P_j = \langle A_j^* A_j \rangle$, where $\langle \cdots \rangle$ denotes averaging over the random process $h(z)$. The result is

$$\frac{dP_j}{dz} = \sum_{k \neq j} \kappa_{jk} \langle A_j^*(z) A_k(z) h(z) \rangle \exp[i(\beta_{0k} - \beta_{0j})z] + \text{c.c.}, \tag{4.3.11}$$

where c.c. stands for the complex conjugate term.

To calculate the average $\langle A_j^*(z) A_k(z) h(z) \rangle$, we integrate Eq. (4.3.10) from z' to z to obtain

$$A_j(z) = A_j(z') + \sum_{n \neq j} i \kappa_{jn} h(z) A_n(z') \int_{z'}^{z} h(u) \exp[i(\beta_{0n} - \beta_{0j})u] \, du. \tag{4.3.12}$$

Substituting this solution in Eq. (4.3.11), we obtain

$$\frac{dP_j}{dz} = \sum_{k \neq j} \sum_{n \neq j} \kappa_{jk} \kappa_{jn}^* \langle A_n^*(z') A_k(z') \rangle \exp[i(\beta_{0k} - \beta_{0n})z]$$

$$\times \int_{z'}^{z} \langle h(u)h(z) \rangle \exp[i(\beta_{0n} - \beta_{0j})u] \, du + \text{c.c.}, \tag{4.3.13}$$

where we used $\langle A_n^*(z') A_k(z) h(z) \rangle = \langle A_n^*(z') A_k(z) \rangle \langle h(z) \rangle = 0$.

This equation is simplified considerably using $\langle A_n^*(z) A_k(z) \rangle = \langle |A_n|^2 \rangle \delta_{kn}$ because only the diagonal terms in the double sum contribute significantly. The correlation function, $R(z-u) = \langle h(u)h(z) \rangle$, for fluctuations is finite only when u is close z. With these simplifications, Eq. (4.3.13) can be written in the simple form [10]

$$\frac{dP_j}{dz} = \sum_{k=1}^{M} h_{jk}(P_k - P_j), \tag{4.3.14}$$

where M is the total number of modes and h_{jk} is defined as

$$h_{jk} = |\kappa_{jk}|^2 \int_{-\infty}^{\infty} R(u)\exp[i(\beta_{0k} - \beta_{0j})u]\,du. \tag{4.3.15}$$

This deterministic set of coupled equations governs how the average power is distributed among different modes of the fiber because of random mode coupling among them. For long fibers, we should add a loss term to the right side of Eq. (4.3.14) to obtain

$$\frac{dP_j}{dz} = \sum_{k=1}^{M} h_{jk}(P_k - P_j) - \alpha_j P_j, \tag{4.3.16}$$

where α_j is the power-loss coefficient for the jth mode. Notice that mode coupling increases the power of the jth mode when $P_j < P_k$, and the opposite happens when $P_j > P_k$. This is also expected intuitively.

Equation (4.3.14) applies to the CW situation in which the mode powers do not change with time. It can be extended to the pulsed case by including different group delays for different modes governed by the dispersion parameters β_{1j}. Pulse broadening induced by β_{2j} cannot be included in this discussion based on modal powers. Let $P_j(z,t)$ denote the time-dependent power for the jth mode at a distance z. Different speeds of different modes are included by replacing Eq. (4.3.16) with

$$\frac{\partial P_j}{\partial z} + \beta_{1j}\frac{\partial P_j}{\partial t} = \sum_{k=1}^{M} h_{jk}(P_k - P_j) - \alpha_j P_j. \tag{4.3.17}$$

The preceding set of coupled equations can be solved because of its linear nature. By taking the Fourier transform with respect to time, we obtain

$$\frac{d\tilde{P}_j}{dz} - i\omega\beta_{1j}\tilde{P}_j = \sum_{k=1}^{M} h_{jk}(\tilde{P}_k - \tilde{P}_j) - \alpha_j\tilde{P}_j, \tag{4.3.18}$$

where a tilde is used for the frequency-domain variables. We solve this set of equations by using $\tilde{P}_j(z) = B_j e^{-sz}$ where B_j and s are to be determined. The resulting algebraic equations can be written in a matrix form, $GB = 0$, where B is a column vector and the elements of the matrix G are given by

$$G_{jk} = h_{jk} + (s + i\omega\beta_{1j} - \alpha_j - H_j)\delta_{jk}, \tag{4.3.19}$$

with $H_j = \sum_{k=1}^{M} h_{jk}$. The eigenvalues $s^{(n)}$ ($n = 1$ to M) are found by setting determinant of G to zero. Corresponding eigenvectors $B^{(n)}$ are used to write the solution as a linear combination of M eigenvectors:

$$\tilde{P}_j(z,\omega) = \sum_{n=1}^{M} c_n B_j^{(n)}(\omega)\exp[-s^{(n)}(\omega)z]. \tag{4.3.20}$$

The coefficient c_n is found from the initial power at $z = 0$ using the orthogonality relation $\sum_j B_j^{*(m)} B_j^{(n)} = \delta_{mn}$:

$$c_n = \sum_{j=1}^{M} B_j^{*(n)} \tilde{P}_j(0, \omega).$$

(4.3.21)

By taking the inverse Fourier transform in Eq. (4.3.20), we obtain

$$P_j(z, t) = \frac{1}{2\pi} \sum_{n=1}^{M} \int_{-\infty}^{\infty} c_n B_j^{(n)}(\omega) \exp[-s^{(n)}(\omega)z - i\omega t] \, d\omega.$$

(4.3.22)

Equation (4.3.22) provides the average power in the jth mode at a distance z, when its initial value $P_j(0, t)$ is known. Modal powers can be calculated for any input pulse using the technique discussed in Section 2.4.1. For a Gaussian input pulse, we obtain

$$P_j(0, t) = \eta_j P_0 \exp[-(t/T_0)^2],$$

(4.3.23)

where η_j is the excitation efficiency of the jth mode and P_0 is the peak power for a pulse of duration $2T_0$. The Fourier transform of $P_j(0, t)$ yields

$$\tilde{P}_j(0, \omega) = \sqrt{\pi} \eta_j P_0 T_0 \exp[-(\omega T_0/2)^2].$$

(4.3.24)

If the eigenfunctions $B_j^{(n)}$ are known, Eqs. (4.3.21) and (4.3.22) can be used to calculate c_n and $P_j(z, t)$, respectively. In general, a numerical approach is needed for the frequency integration in Eq. (4.3.22).

In the case of weak coupling, a perturbative approach can be employed [10]. To the lowest order, the eigenvectors can be assumed to be frequency independent, but the eigenvalues depend on frequency as $s(\omega) = s_0 - i\omega s_1$. In this case, Eq. (4.3.22) takes the form

$$P_j(z, t) = \frac{1}{2\sqrt{\pi}} \eta_j P_0 T_0 \sum_{n=1}^{M} B_j^{(n)} \exp(-s_0^{(n)} z) \int_{-\infty}^{\infty} \exp[-(\omega T_0/2)^2 - i\omega(t - s_1^{(n)} z)] \, d\omega.$$

(4.3.25)

The frequency integral can now be done to obtain

$$P_j(z, t) = \eta_j P_0 \sum_{n=1}^{M} B_j^{(n)} \exp(-s_0^{(n)} z) \exp[-(t - s_1^{(n)} z))^2 / T_0)^2].$$

(4.3.26)

This equation shows that the Gaussian pulse in each mode is delayed by an amount that is different from $\beta_{1j} z$ expected in the absence of mode coupling. Whereas the DGD-induced pulse broadening scales linearly with the fiber's length L in the absence of mode coupling (see Section 4.1.2), it scales as $\sqrt{L}$ when mode coupling is taken into account.

4.3.4 Spatiotemporal Fluctuations

As discussed in Section 2.4.2, multimode interference leads to spatial intensity fluctuations, called speckles, at the output of a GRIN fiber when a CW beam is incident on it. Such spatial fluctuations occur even when a pulsed beam is incident on the fiber. In addition, temporal intensity fluctuations occur at any spatial point because of random mode coupling along the fiber [18–20]. In effect, the spatial pattern of speckles fluctuates with time in the pulsed case.

When a pulsed input beam excites multiple modes of a GRIN fiber, the intensity, $I = |E|^2$, at a distance z can be written from Eq. (4.1.8) as

$$I(\mathbf{r}, t) = \left| \sum_j U_j(x, y) A_j(z, t) \exp(i\beta_{0j}z - i\omega_0 t) \right|^2, \qquad (4.3.27)$$

where the sum is limited to the modes excited by the input beam. As the values of β_{0j} differ for different modes, the intensity pattern is affected by the interference among different modes. We can see this more clearly by writing Eq. (4.3.27) in the form

$$I(\mathbf{r}, t) = \sum_j |U_j(x, y) A_j(z, t)|^2 + \sum_j \sum_{k \neq j} U_j(x, y) U_k^*(x, y) A_j(z, t) A_k^*(z, t)$$

$$\exp[i(\beta_{0j} - \beta_{0k})z], \qquad (4.3.28)$$

where $A_j(z, t)$ depends on the dispersion parameters, as indicated in Eq. (4.1.23) in the case of Gaussian-shape pulses. In the presence of mode coupling, the intensity fluctuates because $A_j(z, t)$ varies in a random fashion. We can calculate the average intensity $\langle I(\mathbf{r}, t) \rangle$ by averaging over such fluctuations.

The interference term in Eq. (4.3.28) depends on the relative phase difference, $\phi_j - \phi_k$ between two different modes. This phase difference can fluctuate because of random environmental changes occurring inside the fiber. The net result is that the intensity at any spatial point fluctuates in time, in addition to the DGD-induced pulse broadening. This becomes an issue when multimode fibers are used for delivering short optical pulses to a specific location. In a 2003 study, spatial amplitude and phase variations were measured at the output of a 6-m-long GRIN fiber using a heterodyne interferometer [18]. A short pulse with a variable delay line was used as a reference for the cross-correlation measurements. This technique allows one to measure the spatial phase distribution after integration over a large number of pulses.

Techniques have been developed in recent years to compensate for the spatio-temporal speckles and to deliver undistorted short pulses at the output end of a fiber. In a 2015 study, phase conjugation of selective modes was used to form a focused spot with minimal temporal broadening [19]. Spectral phase shaping was employed in a 2018 experiment to produce spatiotemporal focusing of short pulses distorted by modal dispersion [20]. It was found that temporal speckles at different spatial locations were uncorrelated. This feature was used to focus a short pulse at one spatial location. In another approach, optical pulses were reshaped at the input end such that the pulse appears at a predefined focus at the output of a multimode fiber [21].

4.4 Non-modal Approach

When a pulsed beam excites a large number of modes inside a GRIN medium, the modal technique becomes impractical. We can use a non-modal approach to solve Eq. (2.1.13), as was done in Section 2.5 for CW beams. The starting point is Eq. (2.5.1). In the case of an optical pulse, this equation is satisfied for each frequency component of the pulse such that [22, 23]

$$2i\beta(\omega)\frac{\partial \tilde{A}}{\partial z} + \frac{\partial^2 \tilde{A}}{\partial x^2} + \frac{\partial^2 \tilde{A}}{\partial y^2} - \beta^2 b^2(x^2 + y^2)\tilde{A}(\mathbf{r}, \omega) = 0. \tag{4.4.1}$$

Two notational changes were made. We replaced $A(\mathbf{r})$ with $\tilde{A}(\mathbf{r}, \omega)$ and k with $\beta(\omega) = n_0(\omega)\omega/c$ to indicate their frequency dependence explicitly. Equation (4.4.1) should be solved to find $\tilde{A}(\mathbf{r}, \omega)$ at a distance z in terms of its initial value at $z = 0$. The inverse Fourier transform of the result,

$$A(\mathbf{r}, t) = \frac{1}{2\pi} \int_{-\infty}^{\infty} \tilde{A}(\mathbf{r}, \omega)e^{-i(\omega - \omega_0)t} d\omega, \tag{4.4.2}$$

then provides us with the spatial as well as temporal dependence of the slowly varying amplitude of a pulsed beam.

4.4.1 Spatiotemporal Propagation Equation

An alternative approach converts Eq. (4.4.1) to the time domain using Eq. (4.1.6). Recalling that $\tilde{E} = \tilde{A}\exp[i\beta(\omega)]z$, we can write the electric field as

$$E(\mathbf{r}, t) = \frac{1}{2\pi} \int_{-\infty}^{\infty} \tilde{A}(\mathbf{r}, \omega)\exp[i\beta(\omega)z - i\omega t] \, d\omega. \tag{4.4.3}$$

The frequency integral is complicated because β is frequency dependent. We can use its value β_0 at the central frequency ω_0 of the pulsed beam to obtain $E = A(\mathbf{r}, t)\exp(i\beta_0 z - i\omega_0 t)$, where the slowly varying amplitude $A(\mathbf{r}, t)$ satisfies

$$A(\mathbf{r}, t) = \frac{1}{2\pi} \int_{-\infty}^{\infty} \tilde{A}(\mathbf{r}, \omega)\exp[i(\beta - \beta_0)z - i(\omega - \omega_0)t]d\omega. \tag{4.4.4}$$

We use this equation to calculate the spatial derivative,

$$\frac{\partial A}{\partial z} = \frac{1}{2\pi} \int_{-\infty}^{\infty} \left[\frac{\partial \tilde{A}}{\partial z} + i(\beta - \beta_0)\tilde{A} \right] \exp[i(\beta - \beta_0)z - i(\omega - \omega_0)t]d\omega. \quad (4.4.5)$$

We substitute $\partial \tilde{A}/\partial z$ from Eq. (4.4.1) and approximate $\beta(\omega)$ with $\beta(\omega) = \beta_0 + \beta_1\Omega + \frac{1}{2}\beta_2\Omega^2$ from Eq. (4.1.12), where $\Omega = \omega - \omega_0$. With this simplification, the frequency integration can be done, and we obtain the following four-dimensional equation:

$$\frac{\partial A}{\partial z} + \frac{1}{2i\beta_0}\left(\frac{\partial^2 A}{\partial x^2} + \frac{\partial^2 A}{\partial y^2}\right) + \beta_1\frac{\partial A}{\partial t} + \frac{i\beta_2}{2}\frac{\partial^2 A}{\partial t^2} + \frac{i\beta_0}{2}b^2(x^2 + y^2)A = 0. \quad (4.4.6)$$

Equation (4.4.6) governs the spatiotemporal evolution of a pulsed optical beam inside a GRIN medium. We made both the paraxial and the narrow-bandwidth approximations in its derivation. We replaced β with β_0 in the diffraction and GRIN terms in Eq. (4.4.1). We also neglected the frequency dependence of Δ, which can be included by expanding Δ in a Taylor series, as was done for $\beta(\omega)$. The main advantage of Eq. (4.4.6) is that it makes no reference to the modes. One can say that it includes all modes of a GRIN medium automatically. Equation (4.4.6) can be simplified somewhat in a moving frame with $t' = t - \beta_1 z$ because the β_1 term then disappears. The single GVD parameter in the resulting equation can be interpreted as the value of β_2 averaged over the modes. One can also use the β_2 value of the fundamental mode. Even though Eq. (4.4.6) is linear, its four-dimensional nature and the potential for spatiotemporal coupling make it difficult to solve this equation analytically. Numerical solutions of Eq. (4.4.6) are also time-consuming for the same reason.

4.4.2 Simplified Time-Domain Equation

To simplify the four-dimensional problem governed by Eq. (4.4.6), we consider the length scales associated with the spatial and temporal evolutions of a pulsed optical beam. We have seen in Section 3.2 that the length scale of spatial changes is governed by the self-imaging period L_p given in Eq. (3.2.20), with values of ~ 1 mm for GRIN fibers. We have also seen in Section 4.1.3 that the length scale of temporal changes is governed by the dispersion length L_D defined as $L_D = T_0^2/|\beta_2|$, where T_0 is a measure of pulse duration. When $T_0 \sim 1$ ps, this length is ~ 100 m for typical values of β_2 at wavelengths near 1 μm.

The vast disparity in the two length scales implies that spatial evolution of a pulsed beam is not much affected by the GVD effects when its bandwidth satisfies the condition $\Delta\omega \ll \omega_0$. In contrast, the spatial self-imaging effects are likely to influence the temporal effects. A simple way to implement this physical reasoning is to decouple the spatial effects in Eq. (4.4.2) with the approximation

$$\tilde{A}(\mathbf{r}, \omega) \approx F(\mathbf{r}, \omega_0)\tilde{A}(z, \omega), \qquad (4.4.7)$$

where $F(\mathbf{r}, \omega_0)$ governs spatial variations of the beam at the central frequency ω_0. This approximation is justified for narrow-bandwidth pulses because the spatial self-imaging features do not vary much over the bandwidth [24].

Using Eq. (4.4.7) in Eq. (4.4.4), the electric field becomes

$$E(\mathbf{r}, t) = F(\mathbf{r}, \omega_0)\frac{1}{2\pi}\int_{-\infty}^{\infty}\tilde{A}(z, \omega)\exp[i\beta(\omega)z - i\omega t]\, d\omega. \qquad (4.4.8)$$

Writing it in the form, $E = F(\mathbf{r}, \omega_0)A(z, t)\exp(i\beta_0 z - i\omega_0 t)$, the slowly varying amplitude $A(z, t)$ satisfies

$$A(z, t) = \frac{1}{2\pi}\int_{-\infty}^{\infty}\tilde{A}(z, \omega)\exp[i(\beta - \beta_0)z - i(\omega - \omega_0)t]d\omega. \qquad (4.4.9)$$

Notice that this equation does not depend on x and y because the spatial part of the optical field has been factored out as indicated in Eq. (4.4.8). In effect, the four-dimensional problem has been reduced to a simpler two-dimensional problem.

By expanding $\beta(\omega)$ in a Taylor series, $\beta(\omega) = \beta_0 + \beta_1\Omega + \frac{1}{2}\beta_2\Omega^2$, and following the procedure used in Section 4.1.1, we can convert Eq. (4.4.9) int the following pulse-propagation equation

$$\frac{\partial A}{\partial z} + \beta_1\frac{\partial A}{\partial t} + \frac{i}{2}\beta_2\frac{\partial^2 A}{\partial t^2} = 0. \qquad (4.4.10)$$

This equation is identical to Eq. (4.1.14) and can be solved with the Fourier-transform method discussed in Section 4.1.3. When the input beam contains chirped Gaussian pulses such that

$$A(0, t) = A_0\exp\left[-\frac{1 + iC}{2}\left(\frac{t}{T_0}\right)^2\right], \qquad (4.4.11)$$

the solution of Eq. (4.4.10) is given by

$$A(z, t) = \frac{A_0}{\sqrt{Q(z)}}\exp\left[-\frac{(1 + iC)(t - \beta_1 z)^2}{2T_0^2 Q(z)}\right], \qquad (4.4.12)$$

where $Q(z) = 1 + (C - i)\beta_2 z/T_0^2$. The peak of the pulse is delayed in time by $\beta_1 z$ because it takes some time for it to arrive at its current location. In contrast with the discussion in Section 4.1.3, the solution for the electric field,

$$E(\mathbf{r}, t) = F(\mathbf{r}, \omega_0)A(z, t)\exp(i\beta_0 z - i\omega_0 t), \qquad (4.4.13)$$

is obtained without using the concept of modes. For a Gaussian beam, the spatial part $F(\mathbf{r}, \omega_0)$ is given in Eq. (3.2.27). If the pulse shape is also Gaussian, $A(z, t)$ has the form given in Eq. (4.4.12).

One should consider the conditions under which the solution in Eq. (4.4.13) is valid. As a modal approach was not used, it contains only two dispersion parameters, β_1 and β_2, representing values averaged over the modes at the frequency ω_0. Also, the spatial and temporal features evolve independently, without any coupling between them. In spite of these simplifications, it represents a valid solution (i) when the input beam excites a large number of modes, (ii) bandwidth of pulses is small relative to the central frequency of the optical spectrum, and (iii) mode coupling is negligible.

4.4.3 Frequency-Domain Solution

An exact analytic approach to propagation of pulsed optical beams in GRIN media makes use of the Fourier transform of the electric field in the frequency domain, $\tilde{E}(\mathbf{r}, \omega)$, as indicated in Eq. (4.1.6). At the input end at $z = 0$, $\tilde{E}(\mathbf{r}, \omega)$ takes the form

$$\tilde{E}(x, y, 0, \omega) = F(x, y, 0)S(\omega), \tag{4.4.14}$$

where $F(x, y, 0)$ is the spatial distribution of the input beam and $S(\omega)$ is the spectral content of each pulse. Using it, the electric field in the time domain becomes

$$E(\mathbf{r}, t) = \frac{1}{2\pi} \int_{-\infty}^{\infty} S(\omega)F(\mathbf{r}, \omega) \exp[i\phi(\mathbf{r}, \omega) - i\omega t)] \, d\omega, \tag{4.4.15}$$

where $F(\mathbf{r}, \omega)$ is the spatial distribution and $\phi(\mathbf{r}, \omega)$ is the phase at a distance z. Their functional forms were obtained in Section 3.2.3 for a CW Gaussian beam and are given in Eq. (3.2.27) and Eq. (3.2.28). We can use these equations after replacing the propagation constant k with $\beta(\omega)$:

$$F(\mathbf{r}, \omega) = \frac{w_0}{w(z)} \exp\left[-\frac{x^2 + y^2}{2w^2(z)}\right], \tag{4.4.16}$$

$$\phi(\mathbf{r}, \omega) = \beta \frac{dw^2}{dz}(x^2 + y^2) + \beta(\omega)z + \tan^{-1}(C_f \tan bz). \tag{4.4.17}$$

The spot size $w(z)$ evolves with z in a periodic fashion as

$$w(z) = w_0[\cos^2(bz) + C_f^2 \sin^2(bz)]^{1/2}, \tag{4.4.18}$$

where $C_f = [\beta(\omega)bw_0^2]^{-1}$. If we neglect the dispersion of the parameter Δ, b can be treated as a constant, but C_f depends on frequency through β. As before, we can include this frequency dependence by expanding $\beta(\omega)$ in a Taylor series around the central frequency ω_0 of the pulse's spectrum $S(\omega)$ as

$$\beta(\omega) = \beta_0 + \beta_1\Omega + \tfrac{1}{2}\beta_2\Omega^2, \qquad \Omega = \omega - \omega_0. \tag{4.4.19}$$

Equation (4.4.15) provides an exact solution to the pulse-propagation problem. The frequency integral can be done numerically if the spectrum $S(\omega)$ of the pulsed Gaussian beam is known. One can draw several general conclusions from it. First, for short optical pulses containing only a few optical cycles, the spectrum will be so broad that the spatial intensity distribution of an initially Gaussian beam will not remain Gaussian inside a GRIN fiber. This behavior is similar to that occurring for short pulses propagating inside a homogeneous medium [25]. Second, the periodic self-imaging of the beam is not affected much by dispersion. Third, the minimum spot size occurring during each period is affected by the GVD because of the frequency dependence of C_f, but the change is expected to be relatively small. Fourth, the phase-front curvature is also affected by GVD because of the presence of $\beta(\omega)$ in Eq. (4.4.17).

The situation changes for wider pulses with a relatively narrow spectrum. In this case, the beam's amplitude $F(\mathbf{r}, \omega)$ is relatively unaffected by the spectrum, and we can take it outside the integral in Eq. (4.4.15) using its value at the center frequency ω_0. However, the frequency dependence of $\beta(\omega)$ must be retained in the phase term. Equation (4.4.15) in this case takes the form

$$E(\mathbf{r}, t) = F(\mathbf{r}, \omega_0)A(\mathbf{r}, t) \exp[i(\beta_0 z - i\omega(t - \beta_1 z) + i\beta_0(dw^2/dz)(x^2 + y^2)], \quad (4.4.20)$$

where the pulse shape is governed by

$$A(\mathbf{r}, t) = \frac{1}{2\pi} \int_{-\infty}^{\infty} S(\Omega) \exp[i(\beta_1\Omega + \tfrac{1}{2}\beta_2\Omega^2)(dw^2/dz)(x^2 + y^2) + i\beta_2\Omega^2 z - i\omega t) \, d\Omega.$$

$$(4.4.21)$$

This equation shows the spatiotemporal coupling induced by the GVD of a GRIN fiber through the phase-front curvature. Only when we neglect the spatial phase term in Eq. (4.4.21), we can separate the spatial and temporal parts in Eq. (4.4.20).

4.5 Applications

Transmission of short optical pulses through a GRIN rod or a GRIN fiber occurs in a variety of applications. In this section, we focus on two applications related to optical communications and biomedical imaging.

4.5.1 Optical Communication Systems

The use of GRIN fibers for optical communication systems attracted considerable during the 1970s because DGD-induced distortion of pulses is less for them compared to step-index fibers [26]. Indeed, GRIN fibers were used successfully beginning in the year 1980 for the first generation of telecommunication systems [6]. They were also used for making useful devices such as couplers and connectors [27].

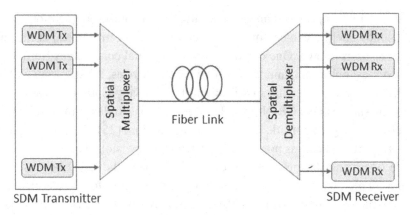

Figure 4.3 Schematic of an SDM system showing how multiple WDM signals can be combined using a spatial multiplexer, transmitted over the same fiber link, and separated with a spatial demultiplexer at the receiver end.

Even though the use of single-mode fibers became prevalent for long-haul systems after 1985, plastic-based GRIN fibers are still used for transferring data between computers. After the year 1995, capacity of telecom systems could be enhanced through wavelength-division multiplexing (WDM), a technique in which multiple optical channels at different wavelengths are launched into the same fiber.

Around the year 2010, it was realized that the telecom systems based on single-mode fibers were approaching their capacity limit set by the nonlinear effects inside such fibers. One way to solve the capacity issue is to make use of multimode fibers such that different data streams are transmitted using different spatial modes of the same fiber. Such an approach is referred to as mode-division multiplexing and is an example of space-division multiplexing (SDM), a technique that exploits parallel spatial paths to enhance the capacity of an optical communication system [6]. GRIN fibers have become relevant for SDM systems because of their lower DGDs compared to that of step-index fibers.

Figure 4.3 shows the schematic of an SDM system. Its transmitter is more complex because it combines multiple WDM data streams using a spatial multiplexer and launches them into a multimode GRIN fiber providing parallel paths. At the receiving end, a spatial demultiplexer is used to direct the output from different modes to different receivers. It should be evident that the advent of SDM required development of many new components. In addition to spatial multiplexers, multimode amplifiers have been developed that can amplify the entire telecom signal, multiplexed using both the WDM and SDM techniques, at the same time using a single device.

As seen in Figure 4.3, a spatial multiplexer takes multiple signals from several single-mode fibers at its input end and processes them such that each signal is coupled into a different mode of a GRIN fiber. The same device can also be used for demultiplexing spatial channels because most optical devices function the same way when light propagates backward. Early spatial multiplexers were bulky because they made use of discrete components such as mirrors, prisms, lenses, and beam splitters. More compact multiplexers were developed later using optical fibers and are known as photonic lanterns [28].

The use of GRIN fibers for SDM suffers from intermodal crosstalk, resulting from random linear coupling along the fiber's length. Such a system can still work if a digital coherent receiver reduces the degradation caused by linear crosstalk through digital signal processing (DSP). This technique is known as the MIMO, short for multiple-input multiple-output [29]. It makes use of application-specific microelectronic chips, designed to recover the signals launched into each mode from the digitized symbol streams received in all spatial channels. A MIMO chip must deal with all spatial channels simultaneously. Even when only the first two mode groups of a GRIN fiber are used, the total number of modes is six when modal degeneracies are taken into account. A second hurdle for GRIN fibers is related to the DGD between different spatial modes. The total delay among the SDM channels arriving at the receiver should not exceed a few nanoseconds for most MIMO chips used in practice.

With continuing progress in the MIMO technology [29], it has become possible to increase the number of spatial modes to beyond six. Fifteen spatial modes of a conventional GRIN fiber (core diameter 50 μm) used in a 2015 experiment required 30×30 MIMO processing. The transmission distance was only 22.8 km because of the DGD and other practical limitations [30]. A conventional GRIN fiber was also used in 2016 to realize transmission in 10 spatial modes over a distance of 40 km [31]. By 2018, total capacity of 257 Tb/s was realized over 48 km using just 10 spatial modes such that each mode carried 336 WDM channels at 12 Gbaud [32]. These results indicate that the use of GRIN fibers for long-haul communication is likely to remain attractive.

4.5.2 *Biomedical Imaging*

Some medical applications require short optical pulses that must be transported to a specific place where the imaging object is located. The use of GRIN fibers for this purpose has attracted attention in recent years [33–36]. In a 2016 study, a 95-m-long GRIN fiber was used to produce a multi-wavelength source of nanosecond pulses used for the purpose of photoacoustic microscopy [34]. The Q-switched pump laser, operating at 1047 nm, injected pulses with energies of up to 140 mJ into a GRIN fiber. The GRIN fiber not only transported these pulses but also produced additional

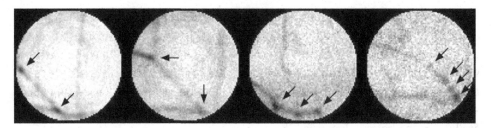

Figure 4.4 Images of neurons taken by inserting a fiber endoscope 2 to 3 mm deep into a mouse's brain. The arrows mark branching points and synaptic boutons. (After Ref. [35]; ©2018 CC BY.)

pulses at several longer wavelengths through the process of cascaded stimulated Raman scattering (see Section 5.4). This feature allowed multi-spectral imaging of a lipid-rich tissue.

In another application, a GRIN fiber with a GRIN lens at its distal end is used for making an endoscope that can be employed for imaging biological tissue inside a human body. A pulsed beam is injected into the endoscope and focused onto the tissue, where it is absorbed and creates a signal at a different wavelength through fluorescence that propagates backward through the GRIN fiber and is used to form an image. However, as we have seen in Section 3.4.2, multimode interference and intermodal coupling within the GRIN fiber can degrade the quality of a pulsed beam to such an extent that it cannot be focused on a narrow spot in the tissue. Several techniques have been developed in recent years to solve this problem. A holographic phase-conjugation technique was employed in a 2015 experiment to image small-size beads through two-photon fluorescence by focusing 120-fs pulses that were transported through the endoscope [33]. In the case of two-photon imaging, the central region of the focused spot, where the intensity is large enough for simultaneous absorption of two photons, creates fluorescence, resulting in improved spatial resolution of the image ($\sim 1~\mu$m).

In the holographic phase-conjugation technique [33], a modal interference pattern at the output of a GRIN fiber is recorded during the calibration stage as a hologram using a spatial light modulator (digital phase conjugation). During the imaging step, this hologram is used to eliminate the degradation caused by modal interference and coupling within the GRIN fiber and to focus the pulsed beam sharply on the sample to be imaged. Two-photon imaging is carried out by scanning the focused spot over the sample. By 2019, this technique was used for the ablation of biological tissue [36]. The GRIN fibers used in this experiment had core diameters of up to 400 μm. Despite such a large size of the core, nonlinear effects still played a considerable role because of high pulse energies involved. The endoscope was still able to image cochlear hair cells with a resolution of about 1 μm. As an example, Figure 4.4

shows the in-vivo images of a mouse's brain recorded with a fiber-based endoscope [35].

Computational neural networks have also been employed recently to remove the distortions induced by a GRIN fiber [37–40]. This technique works by training the neural network to construct a transmission matrix relating the modal expansion coefficients at the input and output end of the GRIN fiber. The inverse of this matrix is used to remove all degradations. In a 2018 experiment, a 1-km-long GRIN fiber was used to reconstruct images of different objects transmitted through it [37]. Since then, deep-learning techniques have been employed for imaging various objects. In a 2021 experiment, a 10-m-long GRIN fiber was used to train a neural network [40]. The fidelity of the reconstructed images was high enough to be useful for endoscopic applications.

References

[1] D. Gloge and E. A. J. Marcatili, *Bell Syst. Tech. J.* **52**, 1563 (1973); https://doi.org/10.1002/j.1538-7305.1973.tb02033.x.

[2] M. Ikeda, *IEEE J. Quantum Electron.* **10**, 362 (1974); https://doi.org/10.1109/JQE.1974.1068137.

[3] R. Olshansky and D. B. Keck, *Appl. Opt.* **15**, 483 (1976); https://doi.org/10.1364/AO.15.000483.

[4] G. P. Agrawal, *Nonlinear Fiber Optics*, 6th ed. (Academic Press, 2019), 621-684.

[5] I. S. Gradshteyn and I. M. Ryzhik, Eds., *Table of Integrals, Series, and Products*, 6th ed. (Academic Press, 2000).

[6] G. P. Agrawal, *Fiber-Optic Communication Systems*, 5th ed. (Wiley, 2021).

[7] K. Thyagarajan and A. K. Ghatak, *Appl. Opt.* **16**, 2583 (1977); https://doi.org/10.1364/AO.16.002583.

[8] D. Marcuse and R. M. Derosier, *Bell Syst. Tech. J.* **48**, 3217 (1969); https://doi.org/10.1002/j.1538-7305.1969.tb01743.x.

[9] A.W. Snyder, *IEEE Trans. Microw. Theory Techn.* **18**, 608 (1970); https://doi.org/10.1109/TMTT.1970.1127296.

[10] D. Marcuse, *Theory of Dielectric Optical Waveguides*, 2nd ed. (Academic Press, 1991).

[11] D. Gloge, *Appl. Opt.* **11**, 2506 (1972); https://doi.org/10.1364/AO.11.002506.

[12] R. Olshansky and D. B. Keck, *Appl. Opt.* **14**, 935 (1975); https://doi.org/10.1364/AO.14.000935.

[13] S. C. Rashleigh and R. Ulrich, *Opt. Lett.* **3**, 60 (1978); https://doi.org/10.1364/OL.3.000060.

[14] C. D. Poole and J. Nagel, Polarization Effects in Lightwave Systems in *Optical Fiber Telecommunications*, Vol. 3A, Ed. I. P. Kaminow and T. L. Koch (Academic Press, 1997), 114-155.

[15] G. J. Foschini, R. M. Jopson, L. E. Nelson, and H. Kogelnik, *J. Lightwave Technol.* **17**, 1560 (1999); https://doi.org/10.1109/50.788561.

[16] B. Huttner, C. Geiser, and N. Gisin, *IEEE J. Sel. Topics Quantum Electron.* **6**, 317 (2000); https://doi.org/10.1109/2944.847767.

[17] C. D. Poole, *Opt. Lett.* **13**, 687 (1988); https://doi.org/10.1364/OL.13.000687.

[18] R. Rokitski and S. Fainman, *Opt. Express* **23**, 1497 (2003); https://doi.org/10.1364/OE.11.001497.

[19] E. E. Morales-Delgado, S. Farahi, I. N. Papadopoulos, D. Psaltis, and C. Moser, *Opt. Express* **23**, 9109 (2015); https://doi.org/10.1364/OE.23.009109.

[20] B. Liu and A. M. Weiner, *Opt. Lett.* **43**, 4675 (2018); https://doi.org/10.1364/OL.43.004675.

[21] M. C. Velsink, L. V. Amitonova, and P. W. H. Pinkse, *Opt. Express* **29**, 272 (2021); https://doi.org/10.1364/OE.412714.

[22] M. Karlsson, D. Anderson, and M. Desaix, *Opt. Lett.* **17**, 22 (1992); https://doi.org/10.1364/OL.17.000022.

[23] T. Hansson, A. Tonello, and T. Mansuryan, *Opt. Express* **28**, 24005 (2020); https://doi.org/10.1364/OE.398531.

[24] M. Conforti, C. M. Arabi, A. Mussot, and A. Kudlinski, *Opt. Lett.* **42**, 4004 (2017); https://doi.org/10.1364/OL.42.004004.

[25] G. P. Agrawal, *Opt. Commun.* **157**, 52 (1998); https://doi.org/10.1016/S0030-4018(98)00481-7.

[26] R. Olshansky, *Rev. Mod. Phys.* **51**, 341 (1979); https://doi.org/10.1103/RevModPhys.51.341.

[27] W. J. Tomlinson, *Appl. Opt.* **19**, 1128 (1980); https://doi.org/10.1364/AO.19.001128.

[28] T. A. Birks, I. Gris-Sánchez, S. Yerolatsitis, S. G. Leon-Saval, and R. R. Thomson, *Adv. Opt. Photon.* **7**, 107 (2015); https://doi.org/10.1364/AOP.7.000107.

[29] K. Shibahara, T. Mizuno, D. Lee, and Y. Miyamoto, *J. Lightwave Technol.* **36**, 336 (2018); https://doi.org/10.1109/JLT.2017.2764928.

[30] N. K. Fontaine, R. Ryf, H. Chen et al., *Opt. Fiber Commun. Conf.*, paper Th5C.1 (Optica Publishing Group, 2015); https://doi.org/10.1364/OFC.2015.Th5C.1.

[31] J. J. A. van Weerdenburg, A. M. V. Benitez, R. G. H. van Uden et al., *Opt. Fiber Commun. Conf.*, Paper Th4C.3 (Optica Publishing Group, 2016); https://doi.org/10.1364/OFC.2016.Th4C.3.

[32] D. Soma, S. Beppu, Y. Wakayama et al., *J. Lightwave Technol.* **36**, 1375 (2018); https://doi.org/10.1109//JLT.2018.2792484.

[33] E. E. Morales-Delgado, D. Psaltis, and C. Moser, *Opt. Express* **23**, 32158 (2015); https://doi.org/10.1364/OE.23.032158.

[34] T. Buma, J. L. Farland, and M. R. Ferrari, *Photoacoustics* **4**, 83 (2016); https://doi.org/10.1016/j.pacs.2016.08.002.

[35] S. Turtaev, I. T. Leite, T. Altwegg-Boussac et al., *Light Sci. Appl.* **7**, 92 (2018); https://doi.org/10.1038/s41377-018-0094-.

[36] E. Kakkava, M. Romito, D. B. Conkey et al., *Biomed. Opt. Express* **10**, 423 (2019); https://doi.org/10.1364/BOE.10.000423.

[37] N. Borhani, E. Kakkava, C. Moser, and C. Psaltis, *Optica* **5**, 960 (2018); https://doi.org/10.1364/OPTICA.5.000960.

[38] P. Caramazza, O. Moran, R. Murray-Smith, and D. Faccio, *Nat. Commun.* **10**, 2029 (2019); https://doi.org/10.1038/s41467-019-10057-8.

[39] T. Zhao, S. Ourselin, T. Vercauteren, and W. Xia, œ28, 20978 (2020); https://doi.org/10.1364/OE.396734.

[40] J. Zhao, X. Ji, M. Zhang et al., *J. Phys. Photon.* **3**, 015003 (2021); https://doi.org/10.1088/2515-7647.

5

Nonlinear Optical Phenomena

As we saw in Section 1.3.3, the refractive index of any material increases for intense beams because of the optical Kerr effect. In this chapter, we focus on such intensity-dependent changes in the refractive index of a GRIN medium and discuss how they lead to a variety of interesting nonlinear effects. In Section 5.1, we consider self-focusing of a continuous-wave (CW) beam inside a GRIN medium. Pulsed beams are the focus of Section 5.2, where we derive a nonlinear propagation equation and discuss the phenomena of self-phase modulation (SPM) and cross-phase modulation (XPM). Section 5.3 is devoted to modulation instability and the formation of multimode solitons. Intermodal nonlinear effects are considered in Section 5.4 with emphasis on four-wave mixing and stimulated Raman scattering. Nonlinear applications discussed in Section 5.5 include supercontinuum generation, spatial beam cleanup, and second harmonic generation.

5.1 Kerr Nonlinearity

The Kerr nonlinearity is known to induce self-focusing of optical beams inside a homogeneous medium such that the beam contracts in size, rather than diffracting, as it propagates down the medium [1–3]. If the input power exceeds a critical value, the contraction becomes so extreme that the beam's size becomes comparable to its wavelength. In the case of a GRIN medium, the index gradient acts as a lens and produces periodic focusing of the beam, even at low input powers. Self-focusing should modify this behavior at high power levels. Self-focusing inside a GRIN fiber was observed [4] and studied theoretically during the 1980s using the paraxial approximation [5–7].

5.1.1 Kerr-Induced Self-Focusing

It was found in Section 1.3.3 that the Kerr effect inside a GRIN medium imposes a nonlinear phase shift on a pulsed optical beam, in addition to the linear and

GRIN-induced phase shifts. Using the refractive index given in Eq. (1.3.10), the three phase shifts acquired over a short distance d inside a GRIN medium take the following forms:

$$\phi_L = n_0 k_0 d, \qquad \phi_G = -n_0 k_0 d(b^2 \rho^2), \qquad \phi_{NL} = n_2 k_0 d I(\mathbf{r}, t), \qquad (5.1.1)$$

where $b = \sqrt{2\Delta}/a$ is the index gradient and n_2 is the Kerr coefficient. The linear part ϕ_L is much larger than the other two but can be ignored because it has a constant value. The GRIN part ϕ_G varies spatially and produces a curved wavefront that leads to focusing of a beam, similar to that produced by a convex lens.

The nonlinear part ϕ_{NL} can vary both spatially and temporally because of its dependence on the local intensity $I(\mathbf{r}, t)$ of the optical field. In the case of a CW Gaussian beam, ϕ_{NL} only varies spatially and takes the form

$$\phi_{NL} = n_2 k_0 d I_0 \exp[-(\rho/w)^2] \approx n_2 k_0 d I_0 [1 - (\rho/w)^2], \qquad (5.1.2)$$

where $2w$ can be interpreted as the spatial width of the Gaussian beam. The approximation in Eq. (5.1.2) holds near the beam's center where $\rho \ll w$. In a homogeneous medium with $n_2 > 0$, the nonlinear phase shift produces a curved wavefront that opposes beam's diffraction and leads to self-focusing. The curvature is of opposite kind when $n_2 < 0$. It leads to self-defocusing such that the beam expands more than what would be expected from diffraction alone.

The situation is more complex for a GRIN medium because both ϕ_G and ϕ_{NL} curve the wavefront of the Gaussian beam simultaneously. Writing the combined phase in the form $\exp(-i n_0 k_0 \rho^2 / R_c)$ and ignoring a constant phase, the radius of curvature of the wavefront is given as

$$\frac{1}{R_c} = b^2 d + \frac{n_2 I_0 d}{n_0 w^2}, \qquad (5.1.3)$$

Both terms are positive in the case of a GRIN medium with $n_2 > 0$, indicating that the Kerr nonlinearity adds to the focusing produced by the GRIN part ϕ_G. However, the magnitude of the last term is relatively small at moderate intensity levels, and the first term containing b^2 dominates. In practice, self-focusing remains a minor effect because ϕ_G dominates over ϕ_{NL}, until the intensity becomes so large that the two become comparable.

5.1.2 *Impact of Self-Focusing on Self-Imaging*

In the case of a CW beam, we begin with the paraxial equation for the slowly varying amplitude given in Eq. (2.1.15):

$$2ik\frac{dA}{dz} + \nabla_T^2 A + [n^2(\mathbf{r}, I)k_0^2 - k^2]A = 0, \qquad (5.1.4)$$

where $k = n_0 k_0$ is the propagation constant and the dependence of the refractive index on the local intensity is shown explicitly. The Kerr nonlinearity is included by using the form of $n(\mathbf{r}, I)$ given in Eq. (1.3.10). Using it with $f(\mathbf{r}) = (\rho/a)^2$ for a parabolic index profile, we have

$$n^2(\mathbf{r}, I) \approx n_0^2(1 - b^2\rho^2) + 2n_0 n_2 I(\mathbf{r}).$$ (5.1.5)

Using this form in Eq. (5.1.4) with $I = |A|^2$, we obtain

$$2ik\frac{dA}{dz} + \nabla_T^2 A - k^2 b^2(x^2 + y^2) + 2k^2\frac{n_2}{n_0}|A|^2 A = 0.$$ (5.1.6)

This equation governs the propagation of CW beams inside a nonlinear GRIN medium. The second term accounts for diffraction. The last two terms are responsible for self-imaging and self-focusing, respectively.

One way to solve Eq. (5.1.6) is to use the variational method discussed in Section 2.5.2 in the linear case ($n_2 = 0$). This method was used in 1992 to solve Eq. (5.1.6) and to discuss the effects of self-focusing [7]. The Lagrangian density given in Eq. (2.5.23) has an extra term to account for the nonlinear term in Eq. (5.1.6):

$$\mathscr{L}_d = ik\left(A\frac{\partial A^*}{\partial z} - A^*\frac{\partial A}{\partial z}\right) + \left|\frac{\partial A}{\partial x}\right|^2 + \left|\frac{\partial A}{\partial y}\right|^2 + k^2 b^2(x^2 + y^2)|A|^2 - 2k^2\frac{n_2}{n_0}|A|^2.$$ (5.1.7)

As was done in the linear case, the beam is assumed to retain its Gaussian shape such that

$$A(x, y, z) = A_p \exp\left(-\frac{x^2 + y^2}{2w^2} + \frac{ikh}{2}(x^2 + y^2) + i\psi\right).$$ (5.1.8)

where the four parameters (A_p w, h, and ψ) vary with z.

As we discussed in Section 2.5.2, A_p is related to the width w through $A_p^2 = P_0/(\pi w^2)$, where P_0 is the beam's power. Following the procedure used there, the remaining parameters are found to satisfy the following equations:

$$\frac{dw}{dz} = hw, \qquad \frac{d\psi}{dz} = -\frac{1}{kw^2}\left(1 - \frac{3p}{2}\right),$$ (5.1.9)

$$\frac{dh}{dz} + \left(h^2 + b^2 - \frac{(1-p)}{k^2 w^4}\right) = 0,$$ (5.1.10)

where the dimensionless parameter p is defined as

$$p = \frac{n_2}{2n_0}(kA_0 w_0)^2 = \frac{P_0}{P_c}, \qquad P_c = \frac{2\pi n_0}{n_2 k^2}.$$ (5.1.11)

Here A_0 and w_0 are the initial values of A_p and w at $z = 0$.

The w and h equations can be combined to obtain the following equation for w:

$$\frac{d^2w}{dz^2} + b^2w - \frac{(1-p)}{k^2w^3} = 0. \tag{5.1.12}$$

The Kerr coefficient n_2 enters this equation though the ratio $p = P_0/P_c$, where P_0 is the input power of the beam and P_c is the critical power needed for beam collapse) to occur inside a homogeneous medium. For silica-based GRIN devices, P_c is about 4 MW at wavelengths near 1 μm when we use $n_2 = 2.7 \times 10^{-20}$ m^2/W in Eq. (5.1.11).

The equation for the beam's width, Eq. (5.1.12), is identical to Eq. (2.5.29) obtained in the linear case in Section 2.5.2, except that the last term is multiplied by the factor $1 - p$. The presence of this factor does not change the form of the solution given in Eq. (2.5.33). As a result, the beam's width in the nonlinear case evolves with z as

$$w(z) = w_0[\cos^2(2\pi z/L_p) + C_f^2 \sin^2(2\pi z/L_p)]^{1/2}, \tag{5.1.13}$$

where L_p is the self-imaging period and the parameter C_f is modified as

$$C_f = \frac{w_g^2}{w_0^2}\sqrt{1-p} = \frac{L_p\sqrt{1-p}}{2\pi k w_0^2}. \tag{5.1.14}$$

For $p \ll 1$, we recover the linear result showing that the beam's width oscillates along z in a periodic fashion (see Figure 2.9). This periodic evolution is maintained when p becomes finite but remains below 1, because the period L_p does not depend on the nonlinear parameter n_2. However, the minimum width during each cycle becomes smaller because C_f is reduced by a factor of $\sqrt{1-p}$. When $P_0 = P_c$ or $p = 1$, the parameter C_f vanishes, and the width evolves with z as $w(z) = w_0 \cos(2\pi z/L_p)$, that is, the width is reduced to zero at the first focal point located at $z = L_p/4$. This is indicative of the beam's collapse induced by self-focusing. In practice, width never becomes zero because the paraxial approximation breaks down when a beam's width becomes comparable to its wavelength.

The collapse of the beam occurs for any $P_0 > P_c$ (or $p > 1$). For such values of p, C_f^2 in Eq. (5.1.13) becomes negative, and $w(z)$ vanishes at a distance such that $z < L_p/4$. Figure 5.1 shows how the beam's width is reduced to zero for three values of the power ratio P_0/P_c when a Gaussian beam with $w_0 = 10$ μm is launched into a GRIN fiber at a wavelength near 1 μm. In the case of a nonlinear GRIN medium, the distance z_c at which the beam collapse occurs is always shorter compared to a homogeneous medium. It can be found by setting $w(z_c) = 0$ in Eq. (5.1.13) and is given by

$$z_c = \frac{L_p}{2\pi}\cot^{-1}|C_f|. \tag{5.1.15}$$

When $P_0 = P_c$ (or $p = 1$), C_f becomes zero and $z_c = L_p/4$, as expected. When P_0 exceeds P_c, the collapse distance shrinks, as seen in Figure 5.1.

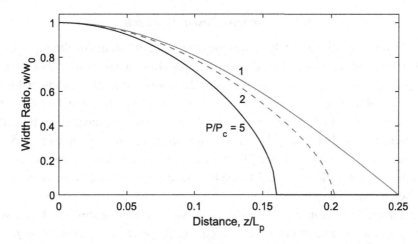

Figure 5.1 Variations of the width of a Gaussian beam inside a GRIN medium when its input power exceeds the self-focusing critical power P_c. The width ratio $w(z)/w_0$ is plotted as a function z/L_p for three values of P_0/P_c.

The remaining parameters of the Gaussian beam are also affected by the Kerr nonlinearity. The parameter h is related inversely to the radius of curvature of the wavefront. From Eq. (5.1.9), it is given by

$$h(z) = \frac{1}{2w^2}\frac{dw^2}{dz} = \left(\frac{\pi}{L_p}\right)\frac{(C_f^2 - 1)\sin(4\pi z/L_p)}{1 + (C_f^2 - 1)\sin^2(2\pi z/L_p)}, \tag{5.1.16}$$

where we used w^2 from Eq. (5.1.13). This equation shows that $h = 0$ when $C_f = 1$, as expected for a planar wavefront. Moreover, h changes sign when C_f exceeds 1. The phase ψ can be calculated by integrating Eq. (5.1.9) and is given by

$$\psi(z) = \frac{3p-2}{2kw_0^2}\int_0^z [\cos^2(2\pi z/L_p) + C_f^2\sin^2(2\pi z/L_p)]^{-1}dz. \tag{5.1.17}$$

This integral can be done analytically [8], but the result depends on whether $p < 1$, equals 1, or exceeds 1. In the range $< 0 < p < 1$, the solution is

$$\psi(z) = \frac{(3p-2)L_p}{4\pi kw_0^2|C_f|}\tan^{-1}[C_f\tan(2\pi z/L_p)]. \tag{5.1.18}$$

Its form for $p > 1$ changes $\tan^{-1}(x)$ function to $\tanh^{-1}(x)$ because C_f becomes imaginary for $p > 1$. When $p = 1$, we have $C_f = 0$, and the phase takes the following form:

$$\psi(z) = [L_p/(4\pi kw_0^2)]\tan(2\pi z/L_p). \tag{5.1.19}$$

5.1.3 Moment-Based Approach

The preceding variational analysis has assumed that the shape of the input beam was Gaussian and remained Gaussian with propagation inside a GRIN fiber. Based on the propagation kernel found in Section 3.2, we know that the periodic nature of the beam's evolution applies to any spatial shape of the input beam in the linear case. One may thus expect this periodicity to hold even in the nonlinear case. Indeed, it was found in 2004 that a self-similar solution similar to that given in Eq. (5.1.8) applies to a variety of input beams [9]. In this section, we extend the moment-based quantum approach of Section 2.5.1 to the nonlinear case to deduce results that apply to any beam shape.

To calculate the evolution of a beam's width as it propagates inside a GRIN medium, we calculate the second moment $\langle \rho^2 \rangle$ using $\rho^2 = (x^2 + y^2) = (q_1^2 + q_2^2)/kb$, where q_1 and q_2 are defined in Section 2.5. As $\langle \rho \rangle = 0$ for the on-axis launch of a radially symmetric beam, the variance is calculated using

$$\sigma^2 = \langle q_1^2 + q_2^2 \rangle = \langle q_1^2 \rangle + \langle q_2^2 \rangle. \tag{5.1.20}$$

As was done in Section 2.5.2, we use Heisenberg's equation of motion,

$$i\frac{d\langle O \rangle}{d\xi} = \langle [\hat{O}, \hat{H}] \rangle, \tag{5.1.21}$$

to calculate the average of an operator inside a nonlinear GRIN medium. Here, $\xi = bz$ and the Hamiltonian in Eq. (2.5.5) has an extra nonlinear term:

$$\hat{H} = \frac{1}{2}\left(\hat{p}_1^2 + \hat{p}_2^2\right) + \frac{1}{2}\left(\hat{q}_1^2 + \hat{q}_2^2\right) - \gamma_0 \hat{I}. \tag{5.1.22}$$

This term is intensity dependent and depends on the nonlinear parameter γ_0 defined as $\gamma_0 = n_2 k/(n_0 b)$. It turns out that the Hamiltonian that remains invariant in the nonlinear case is $H_0 = 2H + \gamma_0 \hat{I}$ [10].

We follow the procedure of Section 2.5.2 to calculate σ^2 and obtain

$$\frac{d^2\sigma^2}{d\xi^2} = 2\langle H_0 \rangle - 4\sigma^2. \tag{5.1.23}$$

This equation is easily solved to obtain

$$\sigma^2(\xi) = \frac{1}{2}\langle H_0 \rangle + a_r \cos(2\xi + \phi_r), \tag{5.1.24}$$

where a_r is the amplitude and ϕ_r is the phase of sinusoidal oscillations exhibited by the beam's width. These constants are set by the initial conditions at $z = 0$. Using $\sigma = \sigma_0$ and $d\sigma/d\xi = 0$ at $\xi = 0$, we find $a_r = \sigma_0^2 - \langle H_0 \rangle/2$ and $\phi_r = 0$. With these values,

$$\sigma^2(\xi) = \langle H_0 \rangle/2 + [\sigma_0^2 - \langle H_0 \rangle/2] \cos(2\xi), \tag{5.1.25}$$

where $\langle H_0 \rangle$ depends on spatial distribution of the input beam.

The preceding equation shows that the width of any input beam oscillates inside a GRIN medium with the same period found in Section 2.5.2 for the linear case. It is possible to rewrite it in the form [10]

$$\sigma^2(\xi) = \frac{1}{2}\sigma_0^2[(1 + D) + (1 - D)\cos(2\xi)], \tag{5.1.26}$$

where the parameter $D = \langle H_0 \rangle / \sigma_0^2 - 1$ and σ_0 is the initial value of σ at $\xi = 0$. As the beam width does not change when $D = 1$, this situation corresponds to the excitation of a single mode inside the GRIN medium. For D values in the range $0 < D < 1$, the width follows an evolution pattern that corresponds to the periodic self-imaging. For negative values of D, σ vanishes at a finite distance, indicating the collapse of the beam induced by self-focusing. These conclusions hold for any beam shape, indicating that both the beam collapse and self-imaging are general phenomena for a nonlinear GRIN medium. In the case of a Gaussian beam, Eq. (5.1.26) is reduced to the form in Eq. (5.1.13) when converted to the original variables. Also, D equal C_f^2 with C_f given in Eq. (5.1.14). As one may expect, the moment method provides the same results as the variational method in the case of a Gaussian beam.

5.2 Pulsed Optical Beams

Nonlinear propagation of optical pulses inside multimode fibers was considered as early as 1980 [11], motivated by the formation of solitons that were observed around this time in single-mode fibers [12]. A modal approach was developed in 1982 for the same purpose [13]. It was extended in 2008 to include both the Kerr and Raman contributions to the Kerr parameter n_2 [14]. Since then, the nonlinear effects in GRIN fibers have been studied extensively [15–17]. In this section, we discuss the phenomenon of SPM before solving the wave equation describing the nonlinear evolution of pulses inside GRIN fibers.

5.2.1 Frequency Chirp and Self-Phase Modulation

Consider the nonlinear phase shift ϕ_{NL} given in Eq. (5.1.1). In the case of a pulsed optical beam, this phase depends on time. Assuming a Gaussian form for the intensity of each pulse, the time dependence of ϕ_{NL} has the form

$$\phi_{NL}(t) = n_2 k_0 d I_0 \exp[-(t/T_0)^2] \approx n_2 k_0 d I_0 [1 - (t/T_0)^2], \tag{5.2.1}$$

where I_0 is the peak intensity and $2T_0$ is related to the temporal width of the pulse. The approximation in Eq. (5.2.1) holds in the pulse's central region where $|t| \ll T_0$.

The time dependence of ϕ_{NL} in Eq. (5.1.2) is referred to as SPM because it shows that an optical pulse modulates its own phase. In the vicinity of the pulse's center, the nonlinear phase varies with time in a quadratic fashion. If we add this phase shift to the harmonic part $\exp(-i\omega_0 t)$, the time-dependent part of the phase can be written as

$$\phi_t(t) = \phi_{NL}(t) - \omega_0 t, \tag{5.2.2}$$

where ω_0 is the central frequency of the pulse's spectrum. We can define the instantaneous frequency of the pulse as $\omega(t) = -d\phi_t/dt$. Clearly, $\omega(t) = \omega_0$ when the nonlinear term is absent. However, when it is present, frequency changes with time as

$$\omega(t) = \omega_0 + \delta\omega(t), \qquad \delta\omega(t) = -d\phi_{NL}/dt = -n_2 k_0 d(dI/dt). \tag{5.2.3}$$

The frequency shift $\delta\omega(t)$ is known as the nonlinear frequency chirp because of its variations with time. It is common to say that the pulse becomes chirped as it passes through a nonlinear medium.

It is evident from Eq. (5.2.3) that both ϕ_{NL} and frequency chirp $\delta\omega(t)$ depend on the shape of pulses. As an example, consider the super-Gaussian shape for which the intensity varies with time as

$$I(t) = I_0 \exp[-(t/T_0)^p], \tag{5.2.4}$$

where p does not have to be an integer. The Gaussian shape is a special case for which $p = 2$. As p increases, the pulse becomes flatter and turns into a rectangular-shape pulse in the limit $p \to \infty$. The SPM-induced chirp for super-Gaussian pulses can be written in the following analytic form

$$\delta\omega(t) = \phi_{max} \frac{p}{T_0} \left(\frac{t}{T_0}\right)^{p-1} \exp\left[-\left(\frac{t}{T_0}\right)^p\right], \tag{5.2.5}$$

where $\phi_{max} = n_2 k_0 d I_0$ is the maximum phase shift at the pulse's center located at $t = 0$. Figure 5.2 shows how (a) the nonlinear phase shift ϕ_{NL} and (b) the frequency chirp $\delta\omega$ vary with time for the Gaussian ($p = 2$) and super-Gaussian ($p = 6$) pulses using $\phi_{max} = 1$. As ϕ_{NL} is proportional to $I(t)$ in Eq. (5.1.1), the nonlinear phase shift mimics the shape of the pulse.

The SPM-induced frequency chirp shown in Figure 5.2(b) has several interesting features. Notice that $\delta\omega$ is negative near the leading edge (a red shift) and becomes positive near the trailing edge (a blue shift) of the pulse. In the case of a Gaussian pulse, the chirp is linear over most of the central region. This feature is useful for compressing a pulse to reduce its width. Super-Gaussian pulses behave quite differently compared to Gaussian pulses. The maximum value of the chirp is considerably larger for them because of their steeper leading and trailing edges. However, the chirp occurs only near the pulse edges and vanishes in the central flat part of the

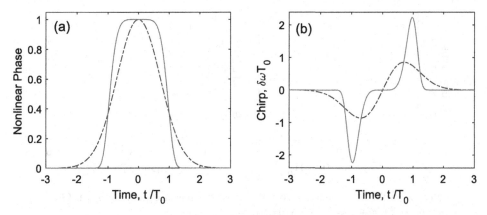

Figure 5.2 Time dependence of (a) SPM-induced phase shift and (b) frequency chirp for Gaussian (dashed curve) and super-Gaussian (solid curve) pulses.

pulse. The main point to note is that chirp variations depend considerably on the exact shape of optical pulses launched into a GRIN fiber.

Since chirping of a pulse affects only the phase of a pulse, its temporal intensity profile does not change inside a nonlinear medium in the absence of the dispersive effects. This does not mean that the SPM effects are not measurable. The SPM-induced chirp produces spectral broadening that can be observed easily. The magnitude of SPM-induced spectral broadening can be estimated from the peak value of $\delta\omega$ in Figure 5.2. We can calculate the peak value by maximizing $\delta\omega(t)$ in Eq. (5.2.5). By setting its time derivative to zero, the maximum value of $\delta\omega$ is found to be

$$\delta\omega_{max} = [pf(p)/T_0]\phi_{max}, \tag{5.2.6}$$

where $f(p)$ is defined as

$$f(p) = (1 - 1/p)^{1-1/p} \exp[-(1 - 1/p)]. \tag{5.2.7}$$

The numerical value of f depends on p only slightly; $f = 0.43$ for $p = 2$ and tends toward 0.37 for large values of p. To obtain the spectral broadening factor, we relate T_0 to the initial spectral width $\Delta\omega_0$ of a Gaussian pulse (at $1/e$ level) as $\Delta\omega_0 = 2/T_0$. The spectral width after a distance d becomes

$$2\delta\omega_{max} = 0.86\,\Delta\omega_0\phi_{max}. \tag{5.2.8}$$

The spectrum broadens by a factor approximately given by the numerical value of the maximum phase shift ϕ_{max}.

The actual shape of the optical spectrum $S(\omega)$ is found by taking the Fourier transform of the electric field. Using $E(t) = \sqrt{I(t)}\exp[i\phi_t(t)]$, where $I(t)$ is the input intensity, we obtain

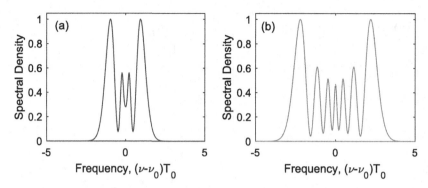

Figure 5.3 SPM-broadened spectra of Gaussian pulses inside a nonlinear GRIN medium for (a) $\phi_{max} = 10$ and (b) $\phi_{max} = 20$.

$$S(\omega) = \left| \int_{-\infty}^{\infty} \sqrt{I(t)} \exp[i\phi_{NL}(t) + i(\omega - \omega_0)T] \, dt \right|^2. \qquad (5.2.9)$$

Figure 5.3 shows the predicted spectra of a Gaussian pulse for two values of ϕ_{max}. The most notable feature is that the SPM-induced chirp produces an oscillatory structure over the entire spectral range. The spectrum consists of several peaks such that the outermost peaks are the most intense. The number of peaks depends on the value of ϕ_{max} and increases linearly with it.

The origin of the oscillatory structure can be understood by referring to Figure 5.2, where the time dependence of the SPM-induced frequency chirp is shown. The same chirp occurs twice, indicating that the pulse has the same instantaneous frequency at two different times. Qualitatively speaking, these two points represent two waves of the same frequency that can interfere constructively or destructively depending on their relative phase difference. The multipeak structure is a result of this interference. Mathematically, the Fourier integral in Eq. (5.2.9) gets dominant contributions from two values of t at which the chirp is the same. These contributions, being complex quantities, may add up in phase or out of phase. One can use the method of stationary phase to obtain an analytic expression of $S(\omega)$ that is valid for large values of ϕ_{max}.

5.2.2 Nonlinear Propagation Equation for Pulses

Equation (4.4.6) of Section 4.4.1 governs the spatiotemporal evolution of a pulsed optical beam inside a linear GRIN medium. We extend this equation to the nonlinear case by comparing it to Eq. (5.1.6) in the CW limit for which we can ignore both time derivatives and replace β_0 with k. It follows that Eq. (4.4.6) can be extended to the nonlinear case by simply adding a nonlinear term to it as

$$\frac{\partial A}{\partial z} + \frac{\nabla_T^2 A}{2i\beta_0} + \beta_1 \frac{\partial A}{\partial t} + \frac{i\beta_2}{2}\frac{\partial^2 A}{\partial t^2} + \frac{i\beta_0}{2}b^2(x^2 + y^2)A = ik_0 n_2 |A|^2 A. \qquad (5.2.10)$$

This equation governs the spatiotemporal evolution of optical pulses inside a nonlinear GRIN medium and includes simultaneously the effects of diffraction, dispersion, and the Kerr nonlinearity. It is referred to as being (3+1)-dimensional, because it has three second-derivative terms (with respect to x y, and t), in addition to the first-derivative term in the z direction. The β_1 term does not play an important role because it can be removed by a simple scaling of time as $T = t - \beta_1 z$ to account for the group delay of the pulse.

It is not easy to solve Eq. (5.2.10) analytically, and its numerical solutions are also time-consuming because of its multidimensional nature. To get some physical insight, consider first the case of a homogeneous medium by setting $b = 0$ in Eq. (5.2.10). If we also neglect the diffraction term, assuming an optical beam with a large spatial size, we obtain the (1+1)-dimensional nonlinear Schrödinger (NLS) equation:

$$\frac{\partial A}{\partial z} + \frac{i\beta_2}{2}\frac{\partial^2 A}{\partial T^2} = ik_0 n_2 |A|^2 A. \qquad (5.2.11)$$

This well-known equation has been studied in many different contexts and can be solved exactly with the inverse scattering method to find specific solutions known as solitons [18–20]. In the optical context, solitons represent pulses whose shape either does not change with propagation or follows a periodic evolution pattern [12]. Their formation requires negative values of the group-velocity dispersion (GVD) parameter β_2.

When diffraction terms are added to Eq. (5.2.11), we obtain Eq. (5.2.10) without the GRIN term ($b = 0$). The soliton-like solutions of this equation were investigated in the 1990s and were called spatiotemporal solitons, or optical bullets, because of their confinement in both space and time [20, 21]. Similar to the case of Eq. (5.2.11), the formation of such solitons requires anomalous GVD. However, unlike Eq. (5.2.11), they are often not stable.

5.2.3 Spatiotemporal Solitons in GRIN Media

One can ask whether Eq. (5.2.10), describing nonlinear evolution inside a GRIN medium, also has solutions in the form of a spatiotemporal soliton. Such solutions were investigated during the 1990s by solving Eq. (5.2.10) with the variational method [22, 23]. For this purpose, it is useful to write Eq. (5.2.10) using dimensionless variables defined as

$$X = qx, \quad Y = qy, \quad Z = bz, \quad \tau = (t - \beta_1 z)/\sqrt{|\beta_2|/b}, \qquad (5.2.12)$$

where $q = \sqrt{\beta_0 b}$. If we also scale the amplitude A as $U = \sqrt{k_0|n_2|/b}\,A$, Eq. (5.2.10) takes the form

$$2i\frac{\partial U}{\partial Z} + \left(\frac{\partial^2 U}{\partial X^2} + \frac{\partial^2 U}{\partial Y^2}\right) - \delta\frac{\partial^2 U}{\partial \tau^2} - (X^2 + Y^2)U + 2v|U|^2 U = 0. \qquad (5.2.13)$$

where $\delta = \text{sign}(\beta_2)$ and $v = \text{sign}(n_2)$. These parameters take values ± 1 and allow us to consider self-focusing or self-defocusing ($v = \pm 1$) with normal or anomalous GVD ($\delta = \pm 1$).

To apply the variational method, we follow the approach used in Section 2.5.2. The Lagrangian density for Eq. (5.2.13) is found to be

$$\mathscr{L}_d = i\left(U\frac{\partial U^*}{\partial Z} - U^*\frac{\partial U}{\partial Z}\right) + \left|\frac{\partial U}{\partial X}\right|^2 + \left|\frac{\partial U}{\partial Y}\right|^2 - \delta\left|\frac{\partial U}{\partial \tau}\right|^2 + (X^2 + Y^2)|U|^2 - v|U|^4,$$

$$(5.2.14)$$

with U^* acting as the generalized coordinate.

The choice of the trial function is crucial to the success of a variational method. For the spatial part, as before, we use the Gaussian shape. However, as we know that the soliton solution of Eq. (5.2.11) has a temporal shape governed by $\text{sech}(t/T_0)$, we adopt this shape and use the ansatz [23]:

$$U(X, Y, Z, \tau) = A_p\text{sech}(\eta\tau)\exp\left(-\frac{X^2 + Y^2}{2w^2}\right)\exp\left[ig\tau^2 + \frac{ih}{2}(X^2 + Y^2) + i\psi\right].$$

$$(5.2.15)$$

We relate A_p to the pulse energy using $E_p = \iiint |U|^2 dX\,dY\,d\tau$ and obtain $A_p^2 = E_p\eta/(2\pi w^2)$. The remaining parameters (η, w, g, h, and ψ) vary with Z as the pulse propagates inside the GRIN medium. The next step is to substitute Eq. (5.2.15) in Eq. (5.2.14) and find the reduced Lagrangian using $\mathscr{L} = \iiint \mathscr{L}_d\,dX\,dY\,d\tau$. All three integrals can be done to obtain

$$\mathscr{L} = E_p\left[\frac{d\psi}{dZ} + \frac{\pi^2}{12\eta^2}\frac{dg}{dZ} - \frac{\delta\eta^2}{6} - \frac{\pi^2\delta g^2}{6\eta^2}\right.$$

$$\left. + \frac{1}{2w^2}\left(1 - \frac{v\eta E_p}{6\pi}\right) + \frac{w^2}{2}\left(\frac{dh}{dZ} + h^2 + 1\right)\right]. \qquad (5.2.16)$$

We next use Eq. (2.5.26) to find how the five parameters in Eq. (5.2.15) evolve with Z. The resulting equations are

$$\frac{dw}{dZ} = hw, \qquad \frac{d\eta}{dZ} = 2\delta\eta g, \qquad (5.2.17)$$

$$\frac{d\psi}{dZ} = \frac{\eta^2}{3} + h^2 w^2 - \frac{v\eta E_p}{24\pi w^2}, \qquad (5.2.18)$$

$$\frac{dg}{dZ} = 2\delta(g^2 - \eta^4/\pi^2) - \frac{v\eta^3 E_p}{2\pi^3 w^2}, \qquad (5.2.19)$$

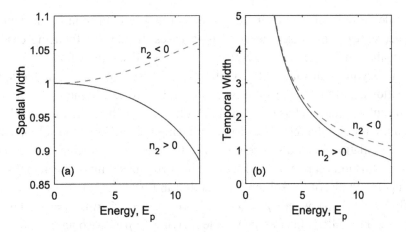

Figure 5.4 Variations of the (a) spatial and (b) temporal widths with pulse's energy E_p for spatiotemporal solitons formed inside a GRIN medium. Solid and dashed curves correspond to the self-focusing and self-defocusing cases, respectively.

$$\frac{dh}{dZ} = \frac{1}{w^4} - (h^2 + 1) - \frac{v\eta E_p}{6\pi w^4}. \tag{5.2.20}$$

These equations require a numerical solution, but they can be solved much faster than the original (3+1)-dimensional partial differential equation. As we are interested in a spatiotemporal soliton whose parameters do not change with Z, we set all Z derivatives to zero, except in the phase equation for ψ. It follows from Eq. (5.2.17) that a chirp-free spatiotemporal soliton exists when $g = h = 0$. From Eq. (5.2.19), the spatial and temporal widths of this soliton are related as

$$w^2\eta = -v\delta E_p/(4\pi). \tag{5.2.21}$$

The spatial width is found using this relation in Eq. (5.2.20) and solving the following cubic equation for w^2:

$$w^6 - w^2 - \delta E_p^2/(24\pi^2) = 0. \tag{5.2.22}$$

The phase ψ changes with Z and can be calculated once η and w are known.

The relation in Eq. (5.2.21) shows that a positive value of η is possible only when $\delta = -1$ with $v = 1$ or $\delta = 1$ with $v = -1$. The first case requires anomalous GVD with self-focusing, and it corresponds to the normal situation for soliton formation. The second case is more interesting because the soliton forms in a self-defocusing nonlinear medium ($n_2 < 0$) exhibiting normal GVD ($\beta_2 > 0$).

Figure 5.4 shows how the spatial width w and the temporal width η^{-1} of the spatiotemporal soliton vary with pulse's energy E_p in the preceding two cases. As one may expect, spatial width decreases with increasing energy in the self-focusing case, while the opposite occurs in the self-defocusing case. In contrast, the temporal width decreases with pulse energy in both cases. In the self-focusing case, the cubic equation in Eq. (5.2.22) have two physically valid solutions for $0 < E_p < 13.5$, and none beyond that range [22]. Only one of these solutions is stable against perturbations, and that is the solution shown in Figure 5.4. In the self-defocusing case, the cubic equation has only one real solution, which is neutrally stable [23]. This feature implies that both the spatial and temporal widths of the soliton exhibit periodic oscillations with distance if the steady state is perturbed in any manner.

Both widths in Figure 5.4 are displayed in normalized units. The actual spatial and temporal widths of the soliton are found from Eq. (5.2.12) to be $w_s = w/q$ and $T_s = (|\beta_2|\beta_0)^{1/2}/(q\eta)$, where $q = \sqrt{V}/a$. If we use $a = 25$ μm, $\beta_2 = -20$ ps^2/km, and $V = 25$ for a GRIN fiber at the wavelength of 1.55 μm, we obtain $w_s = 5w$ μm and $T_s \approx 3/\eta$ fs. At low pulse energies ($E_p \sim 1$), the spatial width of the soliton is close to 5 μm, and its temporal width exceeds 20 fs.

It is interesting to compare the case of a GRIN medium with that of a homogeneous medium [20]. As q vanishes when $b = 0$, no intrinsic spatial scale exists for a homogeneous medium, and we cannot use the preceding analysis. Even though temporal solitons are known to form when a waveguide is used to confine the input beam, no stable spatiotemporal solitons can form in its absence because of the self-focusing-induced beam collapse. Recent work has shown that the use of a GRIN medium allows a variety of stable spatiotemporal solitons to form because the GRIN region acts as a waveguide [24–26]. For example, stable solitons can form, whose donut-like shape resembles a higher-order mode of the GRIN fiber [24]. Even self-imaging spatiotemporal solitons that recover their initial shape periodically have been found to form under suitable conditions.

5.2.4 Effective Time-Domain NLS Equation

It is difficult to observe Spatiotemporal solitons inside a GRIN medium because they must be launched with precise input conditions. Given the self-imaging property of such a medium, we have seen in Section 3.2 that a CW beam evolves in a periodic fashion with a relatively small period (~ 1 mm). This feature suggests that we may able to find solitons that maintain a constant temporal width, but whose spatial features evolve periodically along the GRIN medium's length.

To examine this possibility, we consider a pulsed Gaussian beam launched into a GRIN fiber with a planar wavefront. In the CW case, the solution is given in Eq. (5.1.8) and has the form $A(\mathbf{r}) = A_p F_s(\mathbf{r})$, where the spatial part is given by

$$F_s(\mathbf{r}) = \exp\left[-\frac{x^2 + y^2}{2w^2(z)} + \frac{i}{2}h(z)(x^2 + y^2) + i\psi(z)\right]. \qquad (5.2.23)$$

To apply this CW solution to a pulsed Gaussian beam, we assume that the bandwidth of the pulse is narrow enough that the spatial profile $F_s(\mathbf{r})$ of the beam is not affected much by the nonlinear effects. This assumption is reasonable because the variational analysis of Section 5.1.1 has shown that periodic spatial pattern of the beam in Eq. (5.1.13) maintains the same periodicity when $n_2 > 0$, and its amplitude is only affected through the C_f parameter given in Eq. (5.1.14). As long as the peak power P_0 of the pulse remains below P_c to maintain $p \ll 1$, the pulsed beam should follow the same periodic self-imaging pattern that a CW beam follows inside a linear GRIN medium. With this approximation, the solution of Eq. (5.2.10) is sought in the form [27]

$$A(x, y, z, t) \approx A_t(z, t)F_s(\mathbf{r}), \tag{5.2.24}$$

where $A_t(z, t)$, describing temporal evolution of pulses, does not depend on the spatial coordinates x and y because this dependence is contained in the spatial part $F_s(\mathbf{r})$.

To obtain an equation for $A_t(z, t)$, we follow a procedure well known for single-mode fibers [12]. It is important to realize that $F_s(\mathbf{r})$ does not correspond to any specific mode. When we substitute Eq. (5.2.24) in Eq. (5.2.10), we obtain

$$F_s\left[\frac{\partial A_t}{\partial z} + \beta_1\frac{\partial A_t}{\partial t} + \frac{i\beta_2}{2}\frac{\partial^2 A_t}{\partial t^2} - ik_0 n_2|F_s|^2|A_t|^2 A_t\right],$$
$$+ A_t\left[\frac{\partial F_s}{\partial z} + \frac{\nabla_T^2 F_s}{2i\beta_0} + \frac{i\beta_0}{2}b^2(x^2 + y^2)F_s\right] = 0. \tag{5.2.25}$$

The expression in the second line vanishes because F_s is its solution. We multiply Eq. (5.2.25) with F_s^* and integrate over the transverse coordinates x and y. The final result can be written as [27]

$$\frac{\partial A_t}{\partial z} + \beta_1\frac{\partial A_t}{\partial t} + \frac{i\beta_2}{2}\frac{\partial^2 A_t}{\partial t^2} + = i\bar{\gamma}(z)|A_t|^2 A_t, \tag{5.2.26}$$

where $|A_t|^2$ has units of power and the effective nonlinear parameter is defined as

$$\bar{\gamma}(z) = \frac{k_0 n_2}{A_{\text{eff}}(z)}, \qquad A_{\text{eff}}(z) = \frac{\left(\iint_{-\infty}^{\infty} |F_s(\mathbf{r})|^2 dx\, dy\right)^2}{\iint_{-\infty}^{\infty} |F_s(\mathbf{r})|^4 dx\, dy}. \tag{5.2.27}$$

This is a remarkable result. It shows that the evolution of pulses inside a GRIN fiber can be studied by solving an effective NLS equation in the time domain. Periodic oscillations of the spatial width of the pulsed Gaussian beam, resulting from self-imaging, give rise to an effective nonlinear parameter $\bar{\gamma}(z)$, which is also periodic in z. This is not surprising because the intensity at a given distance z depends on the beam's width, becoming larger when the beam compresses and smaller as it expands. One can interpret this effect as the formation of a Kerr-induced nonlinear index grating within the GRIN medium. The local refractive index of the medium increases in a periodic fashion only near the focal planes where the beam's intensity is enhanced.

The integrals in Eq. (5.2.27) can be carried out using $F_s(\mathbf{r})$ from Eq. (5.2.23). The resulting effective mode area has the form

$$A_{\text{eff}}(z) = A_{\text{eff}}(0)f(z), \qquad f(z) = \cos^2(2\pi z/L_p) + C_f^2 \sin^2(2\pi z/L_p). \qquad (5.2.28)$$

The functional form of $f(z)$ becomes obvious when we note that $A_{\text{eff}}(z)$ at any location scales with the spatial width as $w^2(z)$. We can use this relation to define $\overline{\gamma}(z) = \gamma/f(z)$, where $\gamma = k_0 n_2/A_{\text{eff}}(0)$ is the initial value at $z = 0$. As a final step, we drop the subscript t from A_t and use the reduced time $T = t - \beta_1 z$ to write Eq. (5.2.26) in the form

$$\frac{\partial A}{\partial z} + \frac{i\beta_2}{2}\frac{\partial^2 A}{\partial T^2} = \frac{i\gamma}{f(z)}|A|^2 A. \qquad (5.2.29)$$

This equation reduces to the standard NLS equation when $C_f = 1$.

Similar to the case of single-mode fibers, Eq. (5.2.29) needs to be modified for pulses shorter than a few picoseconds [12]. The reason is related to the nature of the nonlinear response of a material to an electromagnetic field. In general, the molecular vibrations also produce a nonlinear response, in addition to that of electrons. This response, known as the Raman response, is delayed in time, compared to the nearly instantaneous response of electrons, and this delay becomes relevant for short optical pulses. The Raman response is included by replacing the nonlinear index change $(\delta n)_{NL} = n_2|A|^2$ in Eq. (5.2.10) with

$$(\delta n)_{NL} = (1 - f_R)n_2|A|^2 + f_R n_2 \int_0^\infty h_R(t')|A(\mathbf{r}, t - t')|^2 dt', \qquad (5.2.30)$$

where f_R represents the fractional contribution of the delayed response and $h_R(t)$ is the Raman response function of silica molecules, normalized as $\int h_R(t)\, dt = 1$. Its use in Eq. (5.2.10) leads to the following modified form of the effective NLS equation:

$$\frac{\partial A}{\partial z} - i\sum_{n=2}^{M} \frac{i^n \beta_n}{n!}\frac{\partial^n A}{\partial T^n} = \frac{i\gamma}{f(z)}\left[(1 - f_R)|A|^2 A + f_R \int_0^\infty h_R(t')|A(z, T - T')|^2 dT'\right],$$

$$(5.2.31)$$

where the sum includes higher-order dispersion terms in the Taylor expansion of $\beta(\omega)$. The spectral width of ultrashort pulses becomes large enough that their contribution may not remain negligible. Often, it is sufficient to consider the first two terms by choosing $M = 3$, but the β_4 and higher-order terms may also be needed, depending on the situation.

The Raman response $h_R(t)$ in Eq. (5.2.31) affects short pulses through the Raman gain. For pulses with a wide spectrum (> 1 THz), the Raman gain amplifies

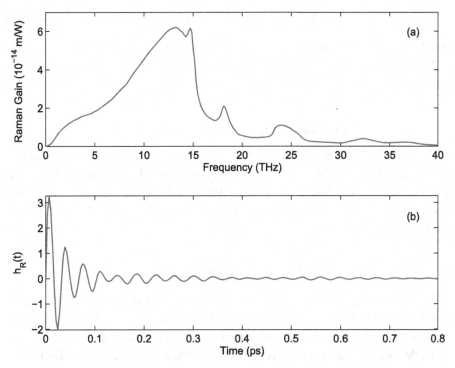

Figure 5.5 (a) Measured Raman gain spectrum of silica fibers; (b) temporal form of the Raman response function deduced from the gain data. (Based on the Raman-gain data provided by R. H. Stolen.)

the low-frequency components of a pulse by transferring energy from the high-frequency components of the same pulse. This phenomenon is called intrapulse Raman scattering. Because of it, the pulse's spectrum shifts toward the low-frequency (red) side as the pulse propagates inside the fiber, a feature referred to as the Raman-induced frequency shift [12].

The functional form of the Raman response $h_R(t)$ depends on the Raman gain spectrum that has been measured for silica fibers. Figure 5.5 shows (a) the experimental gain data and (b) the temporal form of $h_R(t)$ deduced from it [28]. It is useful in practice to have an approximate analytic form for $h_R(t)$. Assuming a single vibrational frequency of silica molecules is involved in the Raman process, $h_R(t)$ can be approximated as [29]

$$h_R(t) = (\tau_1^{-2} + \tau_2^{-2})\tau_1 \exp(-t/\tau_2) \sin(t/\tau_1), \qquad (5.2.32)$$

where τ_1 and τ_2 are related to the frequency and the damping time of vibrations. The commonly used values for silica fibers are $\tau_1 = 12.2$ fs, $\tau_2 = 32$ fs, and $f_R = 0.18$. A modified form of $h_R(t)$ was proposed in 2006 [30] to include the low-frequency hump seen in Figure 5.5:

$$h_R^{\text{new}}(t) = (1 - f_b)h_R(t) + f_b[(2\tau_b - t)/\tau_b^2]\exp(-t/\tau_b), \qquad (5.2.33)$$

where $f_b = 0.21$ and $\tau_b = 96$ fs. This form requires $f_R = 0.245$ and explains better the observed Raman-induced frequency shifts of short optical pulses inside silica fibers.

5.3 Modulation Instability and Solitons

In this section, we use the effective NLS equation given in Eq. (5.2.29) to discuss two important nonlinear phenomena in GRIN fibers, namely the onset of modulation instability and the formation of solitons. Both of these are well known for step-index, single-mode fibers. We refer the reader to Chapters 5 and 14 of Ref. [12] for a detailed discussion.

5.3.1 Modulation Instability

In the case of single-mode fibers, the onset of modulation instability requires anomalous GVD ($\beta_2 < 0$) to initiate temporal oscillations on a CW beam, which are reshaped by the Kerr nonlinearity into a train of optical solitons. It is natural to consider under what conditions modulation instability occurs in a GRIN fiber and produces temporal solitons in spite of different group velocities associated with different modes of the fiber [31, 32].

The stability of a CW Gaussian beam inside a GRIN fiber was studied in 2003 by Longhi [33], starting from Eq. (5.2.10). We follow a simpler approach based on Eq. (5.2.29). A similar equation is encountered in the context of optical communication systems, employing optical amplifiers periodically for compensating fiber losses [34]. In this situation, peak power of optical pulses varies in a periodic fashion, resulting in Eq. (5.2.29) with a different functional form of $f(z)$.

The CW solution of Eq. (5.2.29) is obtained by ignoring the β_2 term and is given by

$$A = \sqrt{P_0}\exp(i\phi_{\text{NL}}), \qquad \phi_{\text{NL}} = \gamma P_0 \int_0^z f^{-1}(z)\,dz, \qquad (5.3.1)$$

where P_0 is the input power launched int the GRIN fiber. The stability of this CW solution is analyzed by perturbing the solution as

$$A = (\sqrt{P_0} + a)\exp(i\phi_{\text{NL}}), \qquad (5.3.2)$$

where $a(z, t)$ is a small perturbation. Linearizing Eq. (5.2.29) in a, we obtain

$$\frac{\partial a}{\partial z} = -\frac{i\beta_2}{2}\frac{\partial^2 a}{\partial T^2} + \frac{i\gamma P_0}{f(z)}(a + a^*). \qquad (5.3.3)$$

This equation can be solved because of its linear nature.

We seek a solution in the form $a = a_1 e^{-i\Omega T} + a_2 e^{i\Omega T}$, where Ω is the modulation frequency. Substituting it in Eq. (5.3.3), the amplitudes a_1 and a_2 are found to satisfy:

$$\frac{\partial a_1}{\partial z} = \frac{i}{2}\beta_2\Omega^2 a_1 + \frac{i\gamma P_0}{f(z)}(a_1 + a_2^*), \tag{5.3.4}$$

$$\frac{\partial a_2}{\partial z} = \frac{i}{2}\beta_2\Omega^2 a_2 + \frac{i\gamma P_0}{f(z)}(a_2 + a_1^*). \tag{5.3.5}$$

Because of the periodic nature of $f(z)$, the two coupled equations can be solved by expanding $f^{-1}(z)$ in a Fourier series as

$$f^{-1}(z) = \sum_{m=-\infty}^{\infty} c_m e^{imKz}, \qquad c_m = \frac{1}{L_p}\int_0^{L_p} f^{-1}(z)e^{-imKz}\, dz, \tag{5.3.6}$$

where $K = 2\pi/L_p$ and L_p is the self-imaging period. The solution is also expanded in a Fourier series, and only the phase-matching terms are retained. It turns out that the nonlinear index grating, produced by periodic self-imaging of the beam, results in an infinite number of spectral sideband pairs located on opposite sides of the CW beam's frequency. The frequencies at which the gain of each sideband pair becomes maximum are found to be [34]

$$\Omega_m = \pm\left(\frac{4\pi m}{\beta_2 L_p} - \frac{2c_0}{\beta_2 L_{\mathrm{NL}}}\right)^{1/2}, \tag{5.3.7}$$

where m is an integer (positive, 0, or negative) and the nonlinear length L_{NL} is defined as $L_{\mathrm{NL}} = (\gamma P_0)^{-1}$. The peak gain of the mth sideband depends on the Fourier coefficient c_m and is given by $g_m = 2\gamma P_0 |c_m|$. The coefficients c_m can be computed from Eq. (5.3.6) by writing it as

$$c_m = \frac{1}{2\pi}\int_0^{2\pi} \frac{\cos(mu) - i\sin(mu)}{(1+C_f^2)+(1-C_f^2)\cos(u)}\, du = \frac{1}{C_f}\left(\frac{C_f - 1}{C_f + 1}\right)^m, \tag{5.3.8}$$

where the imaginary part of c_m vanishes because $\sin(mu)$ is an odd function of u.

The quantity Ω_m^2 must be positive for sidebands to exist. It follows that the $m = 0$ sidebands can exist only if $\beta_2 < 0$, that is, GVD of the GRIN fiber must be anomalous at the input wavelength. These sidebands are identical to those found in the single-mode case [12]. However, periodic self-imaging of the beam produces other sideband pairs for $m \neq 0$ that exist even in the normal-GVD region of the fiber. The gain spectrum of modulation instability exhibits a rich structure containing sideband pairs at frequencies that are not equally spaced. Since periodic spatial variations play a crucial role, this instability is also known as a geometric parametric instability or as a spatiotemporal instability [35–37].

The expression in Eq. (5.3.7) for sideband frequencies can be simplified further. We saw earlier that the self-imaging period L_p is about 1 mm for GRIN fibers. Since

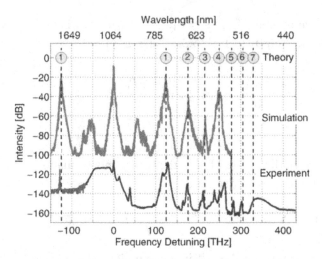

Figure 5.6 Experimental (bottom) and numerically simulated (top) spectra when a 6-m-long GRIN fiber is pumped with 900-ps pulses with 50 kW peak power. Dashed vertical lines and circled numbers show the peak locations predicted by Eq. (5.3.9). (After Ref. [35]; ©2016 APS.)

γ is relatively small (< 0.1 W^{-1}/km), the nonlinear length exceeds 1 meter even at input power levels $\sim$10 kW. The value of c_0 is < 5 for $w_0 < 20$ μm. It follows that the first term in Eq. (5.3.7) dominates, and we can approximate the sideband frequencies as

$$f_m = \Omega_m/2\pi \approx \pm\sqrt{m/(\pi|\beta_2|L_p)}, \quad m = 1, 2, \ldots. \quad (5.3.9)$$

As a specific example, using $\beta_2 = 25$ ps^2/km with $L_p = 1$ mm, the mth sideband is found to be shifted by $127\sqrt{m}$ THz from the pump frequency. For a CW beam with its central frequency at 300 THz (wavelength 1 μm), the $m = 1$ sidebands will be located at the frequencies 173 THz and 427 THz. The former is in the infrared region near 1700 nm, while the latter lies in the visible region near 700 nm. Moreover, higher-order sidebands fall in the ultraviolet and mid-infrared regions. In the case of anomalous GVD ($\beta_2 < 0$) at the beam's wavelength, the $m = 0$ sidebands also form, as seen from Eq. (5.3.7), but their frequencies differ from the incident beam's frequency by < 1 THz. Higher-order sidebands require $m < 0$ in Eq. (5.3.7), and their frequencies are shifted by more than 100 THz from the input frequency.

The sidebands at the frequencies predicted by Eq. (5.3.7) have been observed in several experiments by launching relatively long pulses ($\sim$ 1 ns) into a GRIN fiber to realize a quasi-CW situation [35–37]. In one experiment, 900-ps pulses at 1064 nm in the form of a Gaussian beam (FWHM 35 μm) were launched into a 6-m-long GRIN fiber [35]. The bottom trace in Figure 5.6 shows the observed spectrum when the peak power of input pulses was 50 kW. The top trace shows the numerically simulated spectrum based on Eq. (5.2.10). Dashed vertical lines mark the locations predicted by Eq. (5.3.9). The agreement seen in this figure for the sideband frequencies indicates the usefulness of the NLS equation in Eq. (5.2.29). Its only drawback is that it cannot capture spatial changes that may occur in response to temporal changes.

5.3.2 Graded-Index Solitons

The question whether solitons can form inside multimode fibers attracted attention as early as 1980 [31, 32]. It was realized that different group delays associated with different modes were likely to hinder the formation of such solitons. As intermodal group delays are relatively small for a GRIN fiber, it was natural to consider soliton formation in such a medium. Indeed, theoretical work indicated that formation of multimode solitons was feasible [22, 23], and such solitons were observed in 2013 in an experiment [40].

The variational analysis of Section 5.2.2 has shown that it is not essential to employ the modal description for GRIN fibers. Indeed, one can solve Eq. (5.2.10) either numerically or with a variational technique to find its solutions in the form of a spatiotemporal soliton. In this section, we use Eq. (5.2.29) to find temporal solitons whose spatial shape evolves in a periodic fashion, as dictated by the phenomenon of self-imaging. The presence of $f(z)$ in this NLS equation indicates that it is not integrable by the inverse scattering method. As mentioned earlier, an equation similar to Eq. (5.2.29) is found in the context of single-mode fiber links employing optical amplifiers periodically for compensating fiber losses. It was found in 1990 that a new kind of soliton, called the guiding-center soliton, can form in such fiber links [38]. It is also known as the loss-managed soliton [39].

Based on this analogy, it was found in 2018 that optical pulses can evolve in GRIN fibers such that they preserve their shapes like a soliton in the time domain, even though their spatial width oscillates along the fiber's length [42]. Such pulses are referred to as the GRIN solitons to emphasize that a parabolic index profile is essential for their existence. Before solving Eq. (5.2.29) approximately, it is useful to normalize it using the variables

$$\tau = T/T_0, \quad \xi = z/L_D, \quad U = A/\sqrt{P_0}, \tag{5.3.10}$$

where T_0 and P_0 are the width and the peak power of input pulses and $L_D = T_0^2/|\beta_2|$ is the dispersion length. The resulting equation takes the form

$$i\frac{\partial U}{\partial \xi} + \frac{1}{2}\frac{\partial^2 U}{\partial \tau^2} + \frac{N^2}{f(\xi)}|U|^2 U = 0, \tag{5.3.11}$$

where we assumed $\beta_2 < 0$ and introduced the soliton order as $N = (\gamma P_0 L_D)^{1/2}$. The periodically varying function $f(z)$ given in Eq. (5.2.28) can be written as

$$f(\xi) = \cos^2(2\pi q\xi) + C_f^2 \sin^2(2\pi q\xi), \qquad q = L_D/L_p, \tag{5.3.12}$$

where L_p is the self-imaging period. The numerical value of L_p is about 1 mm for typical GRIN fibers. In contrast, the dispersion length L_D exceeds 1 m for $T_0 > 0.1$ ps if we use $\beta_2 = -20$ ps^2/km, a typical value for silica fibers at wavelengths near 1550 nm. As a result, q is a large number ($q > 100$) under typical experimental conditions. Physically, it represents the number of times self-imaging of the input beam occurs inside a GRIN fiber over a distance of one dispersion length.

The dispersion length provides a scale over which solitons evolve [38]. For this reason, solitons cannot respond to beam-size changes taking place on a scale of 1 mm or less when L_D exceeds 1 meter. If we write the solution of Eq. (5.3.11) as $U = \overline{U} + u$, where $\overline{U}$ represents average over one self-imaging period L_p, perturbations $|u(\xi, \tau)|$ induced by spatial variations remain small enough that they can be neglected (as long as $q \gg 1$). In other words, the dynamics of the soliton can be captured by solving the averaged NLS equation

$$i\frac{\partial \overline{U}}{\partial \xi} + \frac{1}{2}\frac{\partial^2 \overline{U}}{\partial \tau^2} + \overline{N}^2 |\overline{U}|^2 \overline{U} = 0, \tag{5.3.13}$$

where $\overline{N}$ is the effective soliton order defined as

$$\overline{N}^2 = N^2 \langle f^{-1}(\xi)\rangle = N^2 c_0 = N^2/C_f. \tag{5.3.14}$$

As Eq. (5.3.13) is the standard NLS equation, it follows that it has a solution in the form of a fundamental soliton when $\overline{N} = 1$ or $N = \sqrt{C_f}$. This solution exists for a wide range of pulse widths as long as the peak power of pulses is adjusted to satisfy this condition. Recall that C_f represents the fraction by which the beam width shrinks at $z = L_p/4$ before recovering its original width at $z = L_p/2$. Because of periodic self-imaging, the input peak power P_0 must be adjusted to ensure that the condition $\overline{N} = 1$ is maintained on average.

Figure 5.7 compares the input temporal and spectral profiles of a fundamental GRIN soliton before and after it has propagated a distance of $100 L_D$ inside a GRIN fiber. These results were obtained by solving Eq. (5.3.11) numerically with $C_f = 0.2$, $q = 100$, and the initial field $U(0, \tau) = U_0 \text{sech}(\tau)$, where U_0 was chosen to ensure $\overline{N} = 1$. On the logarithmic scale, perturbations induced by spatial oscillations can be seen, but their magnitude remains below the 45 dB level, even after 100 dispersion lengths. Higher-order solitons for which both the spatial and temporal widths evolve periodically can also form inside a GRIN fiber. The period of temporal oscillations is set by the dispersion length and is much longer than the

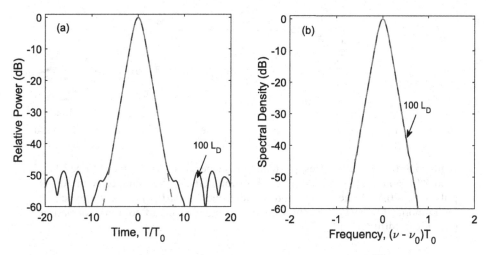

Figure 5.7 Temporal and spectral profiles of a GRIN soliton ($\overline{N} = 1$) before (dashed line) and after it has propagated a distance of $100L_D$ inside a GRIN fiber with $C_g^2 = 0.2$ and $q = 100$.

period of spatial-width oscillations. Higher-order solitons are less stable compared to the fundamental GRIN soliton [42]. They also breakup into multiple fundamental solitons if third-order dispersion is included in Eq. (5.3.11).

A GRIN soliton was observed in a 2013 experiment by launching 300-fs pulses (wavelength near 1550 nm) into a 100-m-long GRIN fiber [40]. At a specific pulse energy of 0.5 nJ, the input and output pulses had nearly the same temporal shapes, but their spectra were different because of different spectral shifts of different modes, which helped in minimizing the DGD effects and allowed a three-mode soliton to form and propagate at a common speed. Figure 5.8 shows (a) spatial profiles, (b) temporal shapes, and (c) optical spectra for the LP_{01}, LP_{02}, and LP_{03} modes that dominated in the formation of the multimode soliton. The (3+1)-dimensional equation in Eq. (5.2.10) was also solved numerically, and its results agreed reasonably well with the experiment.

It is important to emphasize that the observed GRIN soliton is not the spatiotemporal soliton found with the variational method in Section 5.2 because its spatial beam width is not required to remain constant. In the 2013 experiment [40], temporal width of the soliton did reach a constant value of about 200 fs after some initial oscillations. Also, only three modes dominated, and a large faction of soliton's energy was in the fundamental mode of the GRIN fiber. Later numerical and experimental studies included more modes and presented evidence of multimode solitons composed of five or more modes [41]. By 2021, a systematic study of GRIN solitons showed that predictions of Eq. (5.3.13) do not always agree with the

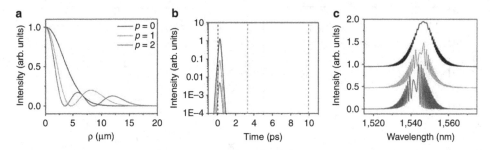

Figure 5.8 (a) Spatial profiles, (b) temporal shapes, and (c) spectra of three LP$_{0m}$ modes ($p = m - 1$) at the end of a 100-m-long GRIN fiber pumped with 300-fs pulses with 0.5 nJ energy. Dashed vertical lines in part (b) mark the expected DGD-induced delays of the three modes in the absence of soliton's formation. (After Ref. [40]; ©2013 CC BY.)

experiments [43–45]. The reason is understood by noting that Eq. (5.3.13) ignores not only the spatiotemporal coupling but also intrapulse Raman scattering that plays a significant role in the case of short optical pulses used in the experiments.

5.3.3 Impact of Intrapulse Raman Scattering

As we discussed in Section 5.2.3, Eq. (5.2.29) can be extended to include intrapulse Raman scattering, resulting in Eq. (5.2.31). If we use dimensionless variables as indicated in Eq. (5.3.10) and average over one self-imaging period, we can write it in the form

$$i\frac{\partial U}{\partial \xi} + \frac{1}{2}\frac{\partial^2 U}{\partial \tau^2} - i\delta_3\frac{\partial^3 U}{\partial \tau^3} + \frac{N^2}{C_f}\left[(1-f_R)|U|^2 A + f_R U\right.$$
$$\left.\int_0^\infty h_R(t')|U(z, T-T')|^2 dT'\right] = 0, \tag{5.3.15}$$

where we used $M = 3$ and included the third-order dispersion through $\delta_3 = \beta_3/(6\beta_2 T_0)$. This equation allows us to consider the Raman-induced spectral shift of short pulses inside GRIN fibers. It can be solved numerically using the form of the Raman response given in Eq. (5.2.33). The Raman-induced spectral shift is found to be enhanced in GRIN fibers, compared to step-index fibers, because of the periodic self-imaging of a pulsed optical beam inside GRIN fibers [46].

Figure 5.9 shows in the (a) time and (b) spectral domains the impact of intrapulse Raman scattering on a 100-fs soliton over a distance of $10L_D$ using $\delta_3 = 0.02$, $N = 1$, and $C_f = 0.5$. Soliton's spectrum in part (b) shifts toward lower frequencies in a continuous fashion because of the Raman-induced transfer of energy from high to low frequencies. The soliton's trajectory in part (a) bends toward the right side

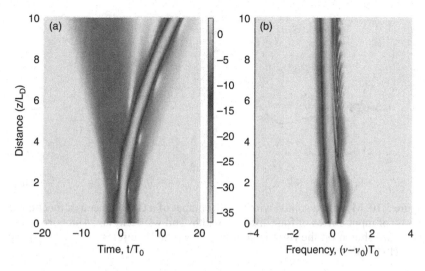

Figure 5.9 Evolution of a soliton over a distance of $10L_D$ in the (a) time and (b) spectral domains using $\delta_3 = 0.02$, $N = 1$, and $C_f = 0.5$. Soliton's trajectory in part (a) bends because of the Raman-induced spectral shift seen in part (b).

because of its continuous slowing down, resulting from this spectral shift. The soliton also sheds some energy as it is perturbed by higher-order dispersive effects. All of these features depend on the spatial compression factor C_f and are enhanced compared to the $C_f = 1$ case for which the spatial width remains constant.

The formation of GRIN solitons was investigated in a 2021 study by launching a pulsed Gaussian beam of 15-μm spot size into a 120-m-long GRIN fiber (core radius 25-μm) such that the first three LP$_{0m}$ modes dominated at the input end [45]. The widths of input pulses were in the range of 60–240 fs and their wavelengths varied from 1300–1700 nm. As seen in Figure 5.10, the width of GRIN solitons at the output end was nearly constant when energy of input pulses was in range 2–6 nJ. Intrapulse Raman scattering played a substantial role in the formation of such solitons, and their wavelengths at the output end were longer by as much as 400 nm than the wavelength of input pulses. Moreover, a large fraction of soliton's energy was in the fundamental mode at the output of the GRIN fiber.

It is well known that higher-order solitons of the NLS equation, forming in single-mode fibers for integer values of $N > 1$, follow a periodic evolution pattern [12] with a period $(\pi/2)L_D$. In the presence of higher-order effects, such as TOD and intrapulse Raman scattering, these solitons undergo a fission process that breaks them into multiple fundamental solitons of different widths. We expect the fission of higher-order solitons to occur even in the case of GRIN fibers because the underlying dynamics remains the same, except for the enhancement of the nonlinear effects when the spatial width of the beam shrinks in each self-imaging period [43].

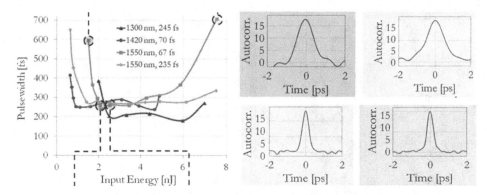

Figure 5.10 Measured soliton widths at the output of a GRIN fiber as a function of energy of input pulses of four different widths and wavelengths. Autocorrelation traces are shown on the right in four cases marked by circles. (After Ref. [45]; ©2021 CC BY.)

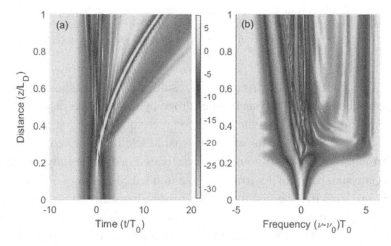

Figure 5.11 (a) Temporal and (b) spectral evolution of a third-order soliton ($N = 3$) over one dispersion length in the (a) time and (b) spectral domains for $\delta_3 = 0.02$ and $C_f = 0.5$. Fission of this soliton broadens the spectrum and results in three fundamental solitons of different widths.

Figure 5.11 shows the evolution of a third-order soliton (width 100 fs) over one dispersion length in the (a) time and (b) spectral domains using $N = 3$, $\delta_3 = 0.02$, and $C_f = 0.5$. Fission of this soliton near $\xi = 0.25$ is initiated by the TOD. The original soliton breaks into three fundamental solitons of different widths. Intrapulse Raman scattering shifts their spectra toward lower frequencies by different amounts and reduces each soliton's speed, resulting in a bent trajectory; the shortest soliton with the largest spectral shift slows down the most. The spectrum in part (b) becomes considerably broader at the output end of the GRIN fiber and exhibits multiple peaks.

A new feature is the appearance of a blue-shifted peak just after the fission occurs. This peak belongs to a dispersive wave, generated because solitons shed radiation when perturbed by the TOD, a well-known feature in single-mode fibers [12]. In the case of GRIN fibers, many dispersive waves are produced at higher input-power levels. This feature is discussed in Section 5.5.1 in the context of supercontinuum generation.

5.4 Modal Approach to Nonlinear Phenomena

So far we have used a multi-dimensional propagation equation to describe the nonlinear effects inside a GRIN medium. This approach is justified for GRIN rods and can be used with success for GRIN fibers supporting a large number of modes. However, its use becomes questionable for devices designed to support a few modes, or when only a few modes of a GRIN fiber are excited intentionally. In this situation, it becomes necessary to adopt a modal approach. A set of coupled multimode nonlinear equations was used in 2008 for studying propagation of ultrashort pulses in multimode fibers [14]. A similar approach was used later in the context of space-division multiplexing [47–51].

5.4.1 Coupled Multimode Nonlinear Equations

The starting point is Eq. (2.1.13) satisfied by the electric field at any point inside a linear GRIN medium. The nonlinear effects are included through the nonlinear contribution $\mathbf{P}_{\mathrm{NL}}(\mathbf{r}, t)$ to the induced polarization. In the frequency domain, the resulting equation takes the form

$$\nabla^2 \tilde{\mathbf{E}} + n^2(\mathbf{r}) \frac{\omega^2}{c^2} \tilde{\mathbf{E}} = -\frac{\omega^2}{\epsilon_0 c^2} \tilde{\mathbf{P}}_{\mathrm{NL}}(\mathbf{r}, \omega), \tag{5.4.1}$$

where $n(\mathbf{r})$ is the linear part of the refractive index. As the nonlinear term on the right side of Eq. (5.4.1) is relatively small, the modes of a GRIN medium found in Section 2.2 without this term can be used for the nonlinear problem as well. We solve Eq. (5.4.1) by expanding $\tilde{\mathbf{E}}$ in terms of these modes as

$$\tilde{\mathbf{E}}(\mathbf{r}, \omega) = \sum_q U_q(x, y) \tilde{\mathbf{A}}_q(z, \omega) \exp[i\beta_q(\omega)z], \tag{5.4.2}$$

where $U_q(x, y)$ governs the shape of the qth spatial mode and β_q is the propagation constant of this mode. As was done in Chapter 4, we use a single subscript q for the LP$_{lm}$ modes.

The polarization degeneracy of the mode is included through the Jones vector, defined as $\mathbf{A}_q = [\hat{x}A_{qx} + \hat{y}A_{qy}]$. The mode amplitudes $\mathbf{A}_q$ vary with z when modes are coupled linearly or nonlinearly. Nonlinear coupling results from the nonlinear term in Eq. (5.4.1). Linear coupling among a fiber's modes occurs because of random variations in the shape and size of a fiber's core along its length. This random coupling can be included by replacing $n^2(\mathbf{r})$ in Eq. (5.4.1) with $n^2(x, y) + \delta\epsilon(x, y, z)$, where $\delta\epsilon$ is a small perturbation that varies with z in a random fashion (see Section 4.3).

Our objective is to derive an equation governing the evolution of mode amplitudes $\tilde{A}_q$ in Eq. (5.4.2). For this purpose, we substitute Eq. (5.4.2) into Eq. (5.4.1) and note that each mode satisfies

$$\nabla_T^2 U_q + n^2(\mathbf{r})k_0^2 U_q = \beta_q^2 U_q, \tag{5.4.3}$$

where the subscript T denotes the transverse part of the ∇^2 operator. Assuming that $\tilde{A}_q$ varies slowly with z and neglecting its second derivative, we obtain

$$\sum_{n=1}^{M} \left(2i\beta_n \frac{\partial \tilde{A}_n}{\partial z} + \delta\epsilon \frac{\omega^2}{c^2} \tilde{A}_n \right) U_n(x, y) e^{i\beta_n z} = -\frac{\omega^2}{\epsilon_0 c^2} \tilde{\mathbf{P}}_{\mathrm{NL}}(\mathbf{r}, \omega). \tag{5.4.4}$$

Multiplying this equation with $U_m^*(x, y)$, integrating over the transverse plane, and using the orthogonal nature of modes, the mode amplitudes are found to satisfy

$$\frac{\partial \tilde{A}_m}{\partial z} = i \sum_n \kappa_{mn} \tilde{A}_n e^{i(\beta_n - \beta_m)z} + \frac{i\omega e^{-i\beta_m z}}{2\epsilon_0 c \bar{n}_m} \iint U_m^*(x, y) \tilde{\mathbf{P}}_{\mathrm{NL}}(\mathbf{r}, \omega) \, dx \, dy, \tag{5.4.5}$$

where $\beta_m = \bar{n}_m \omega / c$ and the coupling coefficients are defined as

$$\kappa_{mn}(z) = k_0 \iint \delta\epsilon(x, y, z) U_m^*(x, y) U_n(x, y) \, dx \, dy. \tag{5.4.6}$$

All modes are normalized such that $\iint U_m^* U_n \, dx \, dy = \delta_{mn}$.

We need to convert Eq. (5.4.5) to the time domain. This is a difficult task without some simplifications. We assume that the dominant contribution to nonlinear polarization comes from the Kerr nonlinearity, for which $\mathbf{P}_{\mathrm{NL}} = \chi^{(3)} \mathbf{EEE}$. The Kerr coefficient n_2 used earlier is related to the third-order susceptibility $\chi^{(3)}$. We write the electric field in the form

$$\mathbf{E}(\mathbf{r}, t) = \sum_m A_m(z, t) U_m(x, y) \exp[i(\beta_{0m} z - \omega_0 t)], \tag{5.4.7}$$

and include the dispersive effects by expanding $\beta_m(\omega)$ in a Taylor series as

$$\beta_m(\omega) = \beta_{0m} + \beta_{1m}(\omega - \omega_0) + \tfrac{1}{2}\beta_{2m}(\omega - \omega_0)^2 + \cdots, \tag{5.4.8}$$

where β_{jm} is the jth derivative of $\beta_m(\omega)$ at the frequency ω_0.

Following the procedure discussed in Ref. [12], we convert Eq. (5.4.5) to the time domain. The final result can be written as [49]

$$\frac{\partial \mathbf{A}_m}{\partial z} + \beta_{1m}\frac{\partial \mathbf{A}_m}{\partial t} + \frac{i}{2}\beta_{2m}\frac{\partial^2 \mathbf{A}_m}{\partial t^2} = i\sum_n \kappa_{mn}\mathbf{A}_n e^{i(\beta_{0n}-\beta_{0m})z}$$

$$+ \frac{i\gamma}{3}\sum_n\sum_p\sum_q f_{mnpq}[2(\mathbf{A}_p^H\mathbf{A}_n)\mathbf{A}_q + (\mathbf{A}_n^T\mathbf{A}_q)\mathbf{A}_p^*]e^{i\Delta\beta_{mnpq}z}, \quad (5.4.9)$$

where the superscripts H and T denote the Hermitian and transpose operations on a Jones vector. The nonlinear parameter γ depends on the effective mode area A_{eff} of the LP_{01} mode as $\gamma = k_0 n_2/A_{\text{eff}}$, where the Kerr ecoefficient n_2 depends on $\chi_{xxxx}^{(3)}$ [12].

The last term involving a triple sum in Eq. (5.4.9) includes three important nonlinear effects related to the Kerr nonlinearity, known as SPM, XPM, and four-wave mixing (FWM). The relative strength of various intermodal terms is governed by the dimensionless modal overlap factor defined as [12]

$$f_{mnpq} = A_{\text{eff}}\iint U_m^*(x,y)U_n(x,y)U_p^*(x,y)U_q(x,y)\,dx\,dy. \quad (5.4.10)$$

As $f_{1111} = 1$ by definition, the value of f_{mnpq} involving four modes represents the strength of each nonlinear term relative to the SPM term for the fundamental mode. The triple sum on the right side of Eq. (5.4.9) takes into account nonlinear coupling among all modes of a GRIN fiber. The magnitude of this coupling depends on f_{mnpq}, or on the overlap of the four modes participating in the corresponding FWM process. It also depends on the phase mismatch governed by

$$\Delta\beta_{mnpq} = \beta_{0n} - \beta_{0m} + \beta_{0q} - \beta_{0p}. \quad (5.4.11)$$

The form of Eq. (5.4.9) indicates that the SPM and XPM can be interpreted as degenerate FWM processes that are self-phase matched. For example, f_{1122} corresponds to an intermodal XPM process that is phase matched because $\Delta\beta_{1122} = 0$.

One more change is often made before solving Eq. (5.4.9) numerically. The exponential term in this equation, containing the phase-mismatch $\Delta\beta_{mnpq}$, oscillates rapidly along the fiber. It can be eliminated by defining

$$\mathbf{B}_m = \mathbf{A}_m\exp[i(\beta_{0m} - \beta_{01})z], \qquad T = t - \beta_{11}z, \quad (5.4.12)$$

where β_{01} and β_{11} serve as reference values. In terms of $\mathbf{B}_m$, Eq. (5.4.9) takes the form

$$\frac{\partial \mathbf{B}_m}{\partial z} + d_m\frac{\partial \mathbf{B}_m}{\partial T} + \frac{i\beta_{2m}}{2}\frac{\partial^2 \mathbf{B}_m}{\partial T^2} = i(\beta_{0m} - \beta_{01})\mathbf{B}_m + i\sum_n \kappa_{mn}\mathbf{B}_n e^{i(\beta_{0n}-\beta_{0m})z}$$

$$+ \frac{i\gamma}{3}\sum_n\sum_p\sum_q f_{mnpq}[2(\mathbf{B}_p^H\mathbf{B}_n)\mathbf{B}_q + (\mathbf{B}_n^T\mathbf{B}_q)\mathbf{B}_p^*], \quad (5.4.13)$$

where $d_m = \beta_{1m} - \beta_{11}$ is the DGD parameter. Note that the fundamental mode is used as a reference for DGDs associated with other modes. Considerable computational resources are needed to solve the preceding set of coupled nonlinear equations. The random nature of κ_{mn} dictates that the entire set should be solved many times before the average behavior of the system can be predicted. For this reason, attention has been paid to developing numerical techniques that solve Eq. (5.4.13) in an efficient manner [15, 52, 53].

It is clear from Eq. (5.4.9) that the nonlinear evolution of pulses in GRIN fibers is much more complex compared to single-mode fibers. The simplest situation occurs for a GRIN fiber designed to support just the LP_{01} and LP_{11} mode groups. When we include the even and odd versions of the LP_{11} mode, such a fiber support three spatial modes ($m = 1, 2, 3$), the mode 1 being the LP_{01} mode. The first step is to find 81 overlap factors f_{mnpq} resulting from three possible values of the four subscripts. Their magnitudes can be calculated using the mode profiles and carrying out the spatial integrals in Eq. (5.4.10). It turns out that 60 out of 81 coefficients vanish. The remaining 21 elements take five distinct values. Three SPM coefficients are $f_{1111} = 1$ and $f_{2222} = f_{3333} = 3/4$. The 12 intermodal XPM coefficients break into three groups:

$$f_{1122} = f_{2211} = f_{1133} = f_{3311} = 1/2,$$
$$f_{1212} = f_{2121} = f_{1313} = f_{3131} = 1/2,$$
$$f_{2233} = f_{3311} = f_{2323} = f_{3232} = 1/4. \tag{5.4.14}$$

The remaining six coefficients govern intermodal FWM and have values

$$f_{1221} = f_{2112} = f_{1331} = f_{3113} = 1/2, \quad f_{2332} = f_{3223} = 1/4. \tag{5.4.15}$$

As a simple example, assume that the three modes remain linearly polarized without any coupling to their orthogonally polarized modes. If we ignore linear coupling among these modes, Eq. (5.4.13) leads to the following coupled NLS equations:

$$\frac{\partial B_1}{\partial z} + \frac{i\beta_{21}}{2}\frac{\partial^2 B_1}{\partial T^2} = i\gamma\left[\left(|B_1|^2 + |B_2|^2 + |B_3|^2\right)B_1 + \frac{1}{2}\left(B_2^2 + B_3^2\right)B_1^*\right], \tag{5.4.16}$$

$$\frac{\partial B_2}{\partial z} + d_2\frac{\partial B_2}{\partial T} + \frac{i\beta_{22}}{2}\frac{\partial^2 B_2}{\partial T^2} = i(\beta_{02} - \beta_{01})B_2 + i\gamma\left[\left(\frac{3}{4}|B_2|^2 + |B_1|^2 + |B_3|^2\right)B_2\right.$$
$$\left. + \frac{1}{4}\left(2B_1^2 + B_3^2\right)B_2^*\right]. \tag{5.4.17}$$

The equation for B_3 is obtained by interchanging the subscripts 2 and 3 in Eq. (5.4.17). Five nonlinear terms in these equations include the effects of SPM, XPM, and FWM among the three modes. For example, $|B_1|^2 B_1$ in Eq. (5.4.16) is a SPM term, $|B_2|^2 B_1$ is a XPM term, and $B_2^2 B_1^*$ is a FWM term.

5.4.2 Intermodal Cross-Phase Modulation

As mentioned in Section 1.3.3, XPM refers to the nonlinear phase shift of an optical wave induced by another wave differing from the first one in some way. In the case of intermodal XPM, the two waves propagate in different modes of the same waveguide. To understand such nonlinear phenomena as simply as possible, we focus on two specific modes of a GRIN fiber with amplitudes A_p and A_q, neglect linear coupling between them, and assume no other mode of the fiber is excited. From Eq. (5.4.9), we obtain the following two coupled NLS equations:

$$\frac{\partial A_p}{\partial z} + \frac{i\beta_{2p}}{2}\frac{\partial^2 A_p}{\partial T^2} = i\gamma\left[\left(f_{pppp}|A_p|^2 + 2f_{ppqq}|A_q|^2\right)A_p + f_{pqqp}A_q^2 A_p^* e^{2i\Delta\beta z}\right], \quad (5.4.18)$$

$$\frac{\partial A_q}{\partial z} + d_{qp}\frac{\partial A_q}{\partial T} + \frac{i\beta_{2q}}{2}\frac{\partial^2 A_q}{\partial T^2} = i\gamma\left[\left(f_{qqqq}|A_q|^2 + 2f_{qqpp}|A_p|^2\right)A_q + f_{qppq}A_p^2 A_q^* e^{-2i\Delta\beta z}\right], \quad (5.4.19)$$

where $\Delta\beta = \beta_{0q} - \beta_{0p}$ and $d_{qp} = \beta_{1q} - \beta_{1p}$ is the DGD between the two modes.

If the two modes are not degenerate ($\Delta\beta \neq 0$) and the fiber's length L is long enough that $\Delta\beta L \gg 1$, the last term in Eqs. (5.4.18) and (5.4.19) can be neglected because of its oscillating nature. In this specific case, we obtain

$$\frac{\partial A_p}{\partial z} + \frac{i\beta_{2p}}{2}\frac{\partial^2 A_p}{\partial T^2} = i\gamma\left(\overline{f}_{pp}|A_p|^2 + 2\overline{f}_{pq}|A_q|^2\right)A_p, \quad (5.4.20)$$

$$\frac{\partial A_q}{\partial z} + d_{qp}\frac{\partial A_q}{\partial T} + \frac{i\beta_{2q}}{2}\frac{\partial^2 A_q}{\partial T^2} = i\gamma\left(\overline{f}_{qq}|A_q|^2 + 2\overline{f}_{pq}|A_p|^2\right)A_q, \quad (5.4.21)$$

where we used the contracted notation $\overline{f}_{pq} \equiv f_{ppqq}$. These equations appear similar to the coupled NLS equations obtained for two pulses of different wavelengths propagating inside a single-mode fiber [12]. In fact, they reduce to them if all overlap factors are set to 1.

The last term in Eqs. (5.4.20) and (5.4.21) is the XPM term. Its presence shows that the two modes are coupled nonlinearly and affect each other. The physical implication of the intermodal XPM is that the phase of each mode is affected by the other mode. Such phase changes affect the spectra of two pulses but do not lead to transfer of energy between them. They also affect the nature of modulation instability. In the absence of XPM, this instability requires GVD to be anomalous. It was found in 1987 that modulation instability can occur even in the normal-GVD region of a fiber when the XPM term is present [54]. A linear stability analysis

based on Eqs. (5.4.20) and (5.4.21) shows [12] that such an instability can occur
only when the overlap factors satisfy the condition:

$$2\overline{f}_{pq} > \sqrt{\overline{f}_{pp}\overline{f}_{qq}}. \tag{5.4.22}$$

In a 2017 study, performed using a 40-cm-long GRIN fiber with 11-μm core radius
[55], spectral sidebands were observed for the mode pair involving the LP_{01} and
LP_{02} modes because the condition (5.4.22) was satisfied for this pair ($\overline{f}_{pp} = 1$,
$\overline{f}_{qq} = \overline{f}_{pq} = 1/2$). The observed frequency shift was close to 10 THz at power levels
of 50 kW for 400-ps input pulses used in the experiment. In contrast, the condition
(5.4.22) was not satisfied for the (LP_{01}, LP_{21}) mode pair, and no sidebands were
observed.

5.4.3 Intermodal Four-Wave Mixing

FWM occurs when two photons, taken from ore or two pump beams, are converted
into two new photons whose frequencies are shifted such that their energy is conserved. If the two pump photons come from the same pump beam at the frequency
ω_p, energy conservation requires $\omega_s + \omega_i = 2\omega_p$, where ω_s and ω_i are frequencies
of the signal and idler photons. In practice, the signal and pump beams are launched
into an optical fiber, and the Kerr nonlinearity of the fiber generates the idler beam
through the FWM process.

Significant FWM occurs only when the total momentum of two input photons is also conserved. Momentum conservation requires the phase-matching
condition [12],

$$\Delta\beta = \beta(\omega_s) + \beta(\omega_i) - 2\beta(\omega_p) = 0, \tag{5.4.23}$$

where $\beta(\omega)$ is the propagation constant at the frequency ω. This condition can be
satisfied when the pump, signal, and idler beams propagate in different modes of
a multimode fiber. Intermodal FWM was observed as early as 1974 by launching
pump pulses at 532 nm into a 9-cm-long fiber, together with a CW signal that was
tunable from 565 to 640 nm [56]; FWM created an idler wave in the blue region at
the fiber's output. A GRIN fiber was first used in 1981 for FWM [57].

The phase-matching condition (5.4.23) can be used to find the signal and idler
frequencies by writing the propagation constant, given in Eq. (2.2.23) for a GRIN
fiber, in the form $\beta_g = n_0 k_0 - gb$, where the integer $g = \ell + 2m - 1$ identifies a mode
group. The phase mismatch in Eq. (5.4.23) comes from two sources, representing the
material and waveguide contributions. The material part comes from the frequency
dependence of n_0. Expanding $\beta(\omega_s)$ and $\beta(\omega_i)$ in a Taylor series around ω_p with
$\omega_{s,i} = \omega_p \pm \Omega$, the material contribution is found to be

$$\Delta\beta_M = \beta_2\Omega^2 + \beta_4\Omega^4/12 + \cdots. \tag{5.4.24}$$

The waveguide contribution depends on the GRIN parameter b and is found to be [58]

$$\Delta\beta_W = -m_g b, \qquad m_g = g_s + g_i - 2g_p, \tag{5.4.25}$$

where g_p, g_s, and g_i are the mode groups for the pump, signal, and idler waves, respectively.

Phase matching requires $\Delta\beta = \Delta\beta_M + \Delta\beta_W = 0$. Keeping only the first two terms in Eq. (5.4.24), phase matching occurs when

$$\beta_2\Omega^2 + \beta_4\Omega^4/12 = m_g b. \tag{5.4.26}$$

This condition can be satisfied for specific vales of Ω when the integer m_g is positive. As Ω^2 satisfies a quadratic equation, its solution is given by [59]

$$\Omega^2 = \frac{6}{\beta_4}\left(-\beta_2 + \sqrt{\beta_2^2 + \frac{m_g b \beta_4}{3}}\right). \tag{5.4.27}$$

When β_4 is negligible, Ω is given by a remarkably simple expression: $\Omega = \pm\sqrt{m_g b/\beta_2}$. The integer m_g can take only even values by a constraint set by the conservation of angular momentum [58]. Even with this restriction, FWM can occur for several mode combinations when the pump beam propagates in the LP_{01} mode of a GRIN fiber.

In a 2017 experiment, a 1-m-long GRIN fiber was used to avoid random mode coupling [60]. Its 22-μm core diameter supported only a few mode groups. The pump was at 1064 nm (in the form of a 400-ps pulse train) and its spatial size was controlled to ensure that it excited mostly the LP_{01} mode. Even though no signal beam was launched with the pump, spontaneous intermodal FWM generated many signal-idler pairs over a wide spectral range. In a later experiment, a GRIN fiber with 50-μm core was used to observe intermodal FWM [59]. The 1064-nm pump was launched with another beam at 1555 nm that seeded the FWM process. Seeding amplified the 1555-nm signal and also created new spectral peaks in the visible and infrared regions. In essence, the GRIN fiber was used as a parametric amplifier.

5.4.4 Intermodal Raman Scattering

Stimulated Raman Scattering (SRS) is a well-known nonlinear process. When an intense pump wave passes through a molecular medium, SRS creates a Stokes wave at a frequency shorter than that of the pump. The frequency shift depends on the vibrational frequency of molecules of the medium and is near 13.2 THz for silica fibers, a value that corresponds to the peak of the Raman gain spectrum in Figure 5.5. In the case of a pulsed pump beam, the Stokes beam also contains pulses whose width is comparable to that of pump pulses. No Stokes beam is generated for femtosecond pump pulses. For such short pulses, intrapulse Raman Scattering encountered in Section 5.3.3 leads to a continuous red-shift of the pump's wavelength.

In the case of single-mode fibers, Stokes pulses appear in the same mode as the pump but travel at different speeds because their spectra are shifted by 13.2 THz [12]. The situation becomes different for a multimode GRIN fiber because the Stokes pulses can be generated in a different mode of the fiber when pump pulses are intense enough to exceed the Raman threshold. This process is called intermodal Raman Scattering and is a special case of the SRS process. For sufficiently intense pump pulses, each Stokes pulse can itself act as a pump and create another pulse that is shifted again in frequency by the same 13.2 THz. This can occur multiple times, and such a process is referred to as the cascaded SRS. It results in the formation of multiple Stokes pulses at longer and longer wavelengths that may propagate in different modes of a multimode fiber.

In a 1988 experiment, 150-ps pump pulses at 1064 nm were injected into a 50-m-long GRIN fiber with a core diameter of 38 μm [61]. The output spectrum exhibited multiple spectral bands extending up to 1700 nm, each belonging to a different Stokes pulse and red-shifted by 13.2 THz from the preceding band. Pulses whose wavelengths exceeded 1300 nm were narrower than the pump pulse because they experienced anomalous GVD and formed solitons. An investigation of cascaded SRS was carried out in 1992 by launching 3-ns pulses (at 585 nm) into a 30-m-long GRIN fiber with a core diameter of 50 μm [62]. A grating was employed to separate the pump and Stokes pulses and to record their spatial patterns at the fiber output. With careful adjustment of the launch conditions, four different Stokes pulses were found to propagate in different low-order modes of the fiber.

A much longer GRIN fiber was used in a 2013 experiment [63]. Cascaded SRS over a wide wavelength range was observed by launching 8-ns pulses at 523 nm into a 1-km-long GRIN fiber with 50-μm core diameter. Figure 5.12 shows the spectrum recorded at the fiber's output end in the visible (a) and infrared (b) regions. It contains many peaks that cover a large wavelength range extending up to 1750 nm. All peaks are separated by about 13.2 THz when plotted on a frequency scale. The intermodal nature of SRS was verified in the experiment through spatial profiles of different Stokes peaks. It was also found that Stokes beams were narrower at the output end of the GRIN fiber, compared to the pump beam. Such beam narrowing is due to a phenomenon known as the Raman-induced beam cleanup [64]. Its origin lies in the modal dependence of the Raman gain, occurring when the pump pulse excites several modes at the input end of a GRIN fiber. This gain is typically larger for the LP_{01} mode of the fiber. Because of the exponential nature of the Stokes

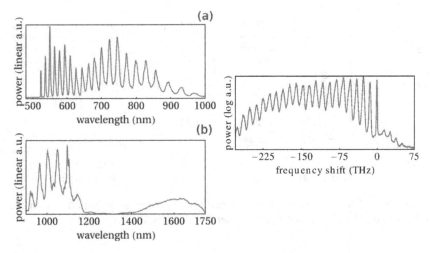

Figure 5.12 (a) Cascaded SRS peaks observed at the end of a 1-km-long GRIN fiber by launching 8-ns pulses at 523 nm with 22 kW peak power; (b) spectrum in the infrared region at higher pulse energies. Spectral data is plotted on a log scale (right) as a function of the frequency shift. (After Ref. [63]; ©2013 AIP.)

growth inside the fiber, most of the energy of any Stokes pulse ends up in the LP_{01} mode at the output end of the GRIN fiber.

5.4.5 *Intermodal Brillouin Scattering*

A backward propagating Stokes wave is generated though stimulated Brillouin Scattering (SBS) in single-mode fibers when the injected pump beam is intense enough to exceed the Brillouin threshold [12]. Its frequency is down-shifted from that of the pump by about 11 GHz for pump wavelengths near 1550 nm. This shift corresponds to the frequency of an acoustic wave that accompanies the SBS process.

In the case of a multimode GRIN fiber, the pump's power is typically divided into several modes, all of which participate in generating the Stokes wave through excitation of multiple acoustic modes of the fiber. In this situation, the net Brillouin gain for the Stokes wave propagating in a specific mode of the fiber has a spectrum composed of multiple Lorentzian peaks of different widths centered at slightly different frequencies [65–67]. Each peak corresponds to a different acoustic mode of the GRIN fiber.

The multi-peak spectrum of the Brillouin gain was measured in a 2021 experiment for a 1-km-long GRIN fiber that was designed to support only the LP_{01} and LP_{11} mode groups. Figure 5.13 shows the measured SBS spectra at several power levels when the pump beam was launched (a) on-axis or (b) off-axis with respect to the fiber's central axis [68]. Only the LP_{01} mode was excited in the on-axis case, but

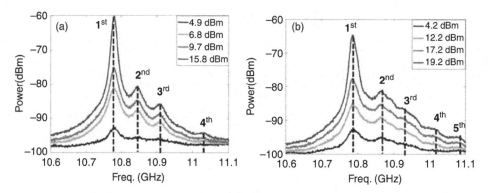

Figure 5.13 Measured SBS spectra at several power levels for (a) on-axis or (b) off-axis launch of the pump beam. (After Ref. [68]; ©2021 Elsevier.)

two mode groups were excited by the pump in the off-axis case. In both cases, the SBS spectrum exhibited several peaks that originated from different acoustic modes of the GRIN fiber.

Similar to the case of intermodal Raman scattering, SBS can be used to cleanup the spatial profile at the fiber's output end [69]. The reason is again related to the Brillouin gain experienced by the Stokes wave propagating in a specific mode, when the pump excites several modes of a GRIN fiber. As the Stokes wave builds up from noise, modal overlap factors play an important role in selecting the mode in which the Stokes wave ends up at the end of the fiber. In practice, most of the Stokes power appears in the LP_{01} mode of the fiber, resulting in a cleaner spatial profile than that of the pump.

5.5 Applications

In this section we focus on the applications that make use of the nonlinear effects in GRIN fibers. Some of these applications make use of the spatiotemporal nature of the nonlinear evolution. This is so because the spatial and temporal features of a pulsed optical beam become coupled inside a GRIN medium through multiple processes such as self-imaging, self-focusing, SPM, SRS, and dispersion. Indeed, spatiotemporal dynamics has been observed in GRIN fibers [16].

5.5.1 *Supercontinuum Generation*

Supercontinuum generation refers to the extreme spectral broadening of optical pulses induced by the nonlinear effects [70]. This process has been studied extensively in the context of single-mode fibers [12] and is used to make commercial supercontinuum sources, useful for biomedical imaging among other things.

Supercontinuum generation in multimode fibers requires an extension of the multimode nonlinear equations given in Eq. (5.4.13) to include the effects of intrapulse Raman scattering [14].

If a fiber is short enough that linear mode coupling is not a major factor, one can set $\kappa_{mn} = 0$ in Eq. (5.4.13) to simplify it. The resulting set is often solved numerically to study the nonlinear intermodal effects in few-mode fibers. Typical simulations consider femtosecond pulses launched in the anomalous-GVD region of the fiber such that their energy is carried by the first few LP_{0m} modes. During propagation inside the fiber, other LP_{0m} modes may also be excited, but the fraction of energy transferred to them remains relatively small [71]. Experiments have shown that when short optical pulses propagate inside a GRIN fiber, some of the spectral bands within the supercontinuum belong to higher-order modes of the fiber.

Starting in 2013, standard GRIN fibers (core diameter 50 μm) were used in several experiments with pulse widths ranging from 100 fs to 1 ns [72–75]. The size of the input beam was kept large enough that multiple modes of the GRIN fiber were excited simultaneously. In one experiment, 0.5-ps pulses at 1550 nm were launched into a 1-m-long GRIN fiber [72]. Figure 5.14 shows the observed spectrum (bottom trace) together with the result of numerical simulations (top trace). Experimental spectrum does not extend into the mid-infrared region because of limitations imposed by the optical spectrum analyzer, but it still covers a wide wavelength range (400 to 1700 nm). The predicted spectrum extends into the mid-infrared region (up to 5 μm). An important feature is that the spectrum exhibits multiple peaks in the visible region that are not equally spaced. These peaks correspond to multiple dispersive waves that have the same origin as the peaks seen in Figure 5.6. They originate from the self-imaging phenomenon that produces spatial oscillations of the beam's width inside a GRIN fiber.

Dispersive waves are produced when a soliton sheds energy in response to its perturbation by higher-order dispersive effects. This phenomenon also occurs in single-mode fibers, but it typically results in a single dispersive wave at a specific frequency for which the phase-matching condition, $\beta(\omega_d) = \beta_s(\omega_s)$, is satisfied [12]. Here ω_d is the frequency of the dispersive wave and β_s is the propagation constant of the soliton at its frequency ω_s. In the case of GRIN fibers, the Kerr-induced index grating, created by spatial oscillations of the peak intensity along the fiber's length, modifies the phase-matching condition as $\beta(\omega_d) - \beta_s(\omega_s) = 2\pi m/L_p$, where m is an integer and L_p is the self-imaging period. This modification is similar to the case of modulation instability in Section 5.3.1. Using $\omega_d = \omega_s + \Omega$ and expanding $\beta(\omega_d)$ in a Taylor series around ω_s (retaining the terms up to third-order), the frequency shift of the dispersive wave is found by solving the following cubic polynomial [27]:

$$\frac{\beta_2}{2}\Omega^2 + \frac{\beta_3}{6}\Omega^3 - \delta\beta_1\Omega = \frac{2\pi m}{L_p} + \frac{\gamma P_s}{2C_f}, \tag{5.5.1}$$

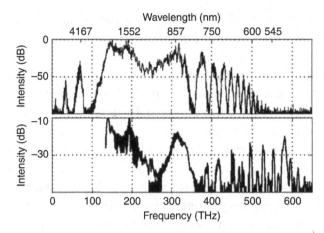

Figure 5.14 Numerical (top) and experimental (bottom) spectra at the end of a GRIN fiber for 0.5-ps input pulses launched at 1550 nm. (After Ref. [72] ©2015 APS.)

where we used $\beta_s(\omega_s) = \beta(\omega_s) + \gamma P_s/2C_f$ after averaging $\bar{\gamma}$ over one self-imaging period. In Eq. (5.5.1), P_s is the soliton's peak power and $\delta\beta_1$ is change in its velocity from that of the input pulse.

In the absence of the Kerr-induced nonlinear grating ($m = 0$), Eq. (5.5.1) predicts a single dispersive wave at a frequency close to $\Omega = -3\beta_2/\beta_3$ [12]. However, a large number of dispersive waves can form for different values of m when the nonlinear grating is present. Their frequencies differ from those of the sidebands generated through modulation instability because of the effects of third-order dispersion in Eq. (5.5.1). The predicted values are close to those observed experimentally or found numerically [27].

In a 2021 study, a modal approach was adopted for calculating the frequencies of dispersive waves, assuming that each femtosecond pulse forms a multimode soliton [76]. This model was also able to predict the frequencies observed in an experiment performed by launching 100-fs pulses at 1550 nm into a 10-m-long GRIN fiber. Figure 5.15 shows the visible part of the observed spectrum with multiple dispersive waves. Vertical lines show the predictions of their frequencies based on the modal approach. When a step-index fiber was used in place of the GRIN fiber, the comb-like structure was replaced with a single broad peak covering nearly the same spectral range.

It is not essential to employ femtosecond pulses for creating a supercontinuum. In two 2016 experiments [73, 74], longer pulses with widths close to 1 ns, obtained from a Nd:YAG laser operating at 1064 nm, were launched into GRIN fibers of up to 30 m length. The observed supercontinuum extended from 500 to 2000 nm for a 12-m-long fiber, with multiple peaks in the visible reason. When the length

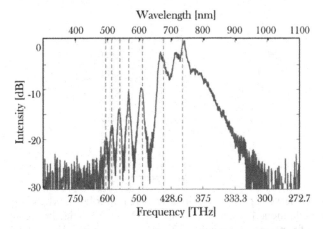

Figure 5.15 Visible part of the observed spectrum when 100-fs pulses at 1550 nm were launched into a 10-m-long GRIN fiber. Vertical lines mark the predicted frequencies of various dispersive waves. (After Ref. [76] ©2021 CC BY.)

was increased to 30 m, the supercontinuum extended to near 2500 nm and became smoother. The physical mechanism leading to the spectral broadening is different for long pulses propagating in the normal-GVD region of the GRIN fiber. The spectrum broadens initially around the pump wavelength through the SPM and SRS processes. Soon after that, the onset of modulation instability creates multiple widely separated sidebands whose origin has been discussed in Section 5.3.1. These bands interact with each other and with the pump through XPM and FWM, and they can also create their own Stokes peaks through SRS. Thus, more and more frequencies are generated nonlinearly, and a relatively smooth supercontinuum forms if the GRIN fiber is long enough.

Situation changes when pulses with intermediate widths ($\sim$ 100 ps) are employed for supercontinuum generation. In a 2018 experiment, 70-ps pulses at 1040 nm were launched into the normal-GVD region of a 20-m-long GRIN fiber [77]. Figure 5.16 shows the observed spectra at different power levels on the (a) logarithmic (dB) and (b) linear scales. The curve for 0.82 W of power shows multiple peaks generated through a cascaded Raman process. This is more clearly seen in part (c) where the spectra are shown on a magnified scale at several power levels. The important point is that more and more peaks are created as the average power is increased beyond 1 W. These peaks eventually broaden and merge together to create a supercontinuum that extends mostly on the long-wavelength side of the input wavelength. Spatial profiles of the output beam in different wavelength ranges were similar to that of a Gaussian beam, indicating the presence of Raman-induced beam cleanup.

Considerable attention has been paid in recent years to extend the supercontinuum into the mid-infrared region because of practical applications of such sources in

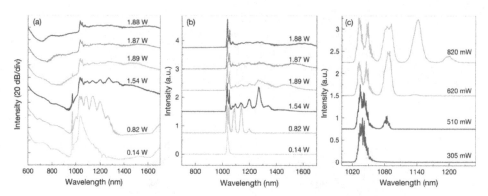

Figure 5.16 Observed spectra at various average power levels when 70-ps pulses at 1040 nm were launched into a 20-m-long GRIN fiber; (a) decibel scale, (b) linear scale, and (c) formation of cascaded SRS peaks. (After Ref. [77] ©2018 CC BY.)

a variety of fields [78]. As silica-based fibers have higher losses at wavelengths above 1.7 μm, fluoride or chalcogenide fibers are often used for extending the output spectrum beyond 2 μm. Although step-index fibers are commonly employed, GRIN fibers have also been used recently. In a 2021 experiment [79], a silica-based GRIN fiber was used to generate a spectral sideband at 1870 nm through modulation instability by launching 65-ps pulses at 1064 nm into the fiber. This radiation was amplified inside a thulium-doped fiber amplifier and launched into a 10-m-long fluoride (InF$_3$) fiber, where its spectrum broadened through SPM and covered a spectral range from 1.7 to 3.4 μm.

In a novel approach, a non-silica GRIN fiber was fabricated with a nanostructured core, consisting of two types of lead-bismuth-gallate glasses with different refractive indices [80]. A 20-cm-long piece of this fiber was pumped with 350-fs pulses (wavelength 1.7 μm) to generate a two-octave-wide supercontinuum extending from 0.7 to 2.8 μm. Figure 5.17 shows the observed spectra when the beam was launched (a) normal to the fiber's axis, (b) with a small tilt, and (c) with a large tilt. The output spatial profile is also shown in each case. In case (a), the intensity profile is Gaussian-like because the input beam excites only a few radially symmetric LP$_{0m}$ modes of the fiber. In case (b), multiple low-order LP$_{lm}$ modes of the fiber become excited, and the spatial profile exhibits two distinct lobes. Many more modes are excited in case (c), resulting in a speckled spatial pattern. Three supercontinua exhibit multiple peaks that can only be interpreted as the spectral sidebands generated through modulation instability. The reason is that the GRIN fiber exhibited normal dispersion at wavelengths up to 2 μm. As a result, the input pulse did not propagate as a soliton. Indeed, the vertical lines in Figure 5.17 show the frequencies of the sidebands predicted by Eq. (5.3.9). Their reasonable match with the peaks' locations indicates that this interpretation is justified.

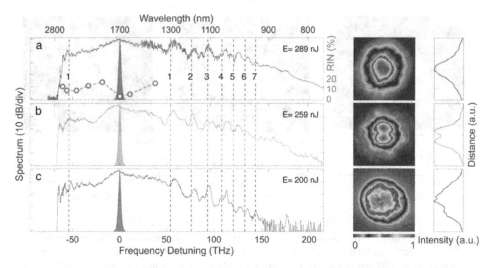

Figure 5.17 Observed spectra in three launch conditions when 350-fs pulses at 1700 nm were launched into a 20-cm-long non-silica GRIN fiber. Spatial profile and intensity distribution are also shown in each case. Vertical lines show the frequencies predicted by Eq. (5.3.9). (After Ref. [80] ©2022 CC BY.)

5.5.2 Spatial Beam Cleanup

As described in Section 2.4.2, the spatial profile of an optical beam at the end of a GRIN fiber often exhibits a speckle pattern because of random mode coupling and modal interference occurring inside the fiber. The poor quality of such beams hinders their use in many applications such as medical imaging (see Section 3.4.2). In this section we discuss how the nonlinear effects within GRIN fibers can be employed to improve the spatial quality of intense pulsed beams.

We have seen in Section 5.4.4 that the Raman-induced beam cleanup produces a spatial profile for the Stokes beam at the end of a GRIN fiber that appears much smoother than that of the pump beam. In this case, a cleaner beam is obtained at a wavelength different from that of the pump. Several experiments have shown that the Kerr nonlinearity itself can produce a cleaner beam at the original wavelength. This effect is called the Kerr-induced beam cleanup [81–86]. The term "beam cleanup" generally implies that the optical beam at the end of the fiber has a smooth spatial intensity profile with a single peak.

In a 2016 experiment, relatively wide 900 ps pulses (at the 1064-nm wavelength) were launched into a 6-m-long standard GRIN fiber in the form of a Gaussian beam with a diameter of 35 μm [35]. Figure 5.18 shows the observed spatial intensity profiles in three cases. The linear case is shown in part (a), where the peak power of input pulses was reduced to 60 W. As expected, the output beam exhibits a speckle

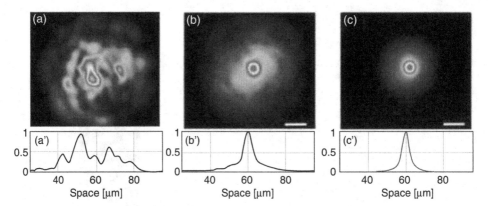

Figure 5.18 Spatial distribution at the output of a 6-m-long GRIN fiber when a pulsed (width 500 ps) Gaussian beam at 1064 nm was launched at the input end. Peak power was 0.06 kW in part (a) and 50 kW in part (b). Part (c) shows the output near 750 nm in case (b). Bottom panels show the intensity profiles. (After Ref. [35] ©2016 APS.)

pattern resulting from modal interference. Part (b) shows the nonlinear case, where the input power was increased so much that the peak power of each pulse was close to 50 kW. The spatial profile is much smoother with a central bright spot because of the Kerr-induced beam cleanup. Part (c) shows the beam shape in the nonlinear case at a wavelength of 750 nm that corresponds to the first sideband generated through modulation instability. Recall from Section 5.3.1 that modulation instability can produce multiple spectral sidebands for intense input beams (see Figure 5.6). Figure 5.18 indicates that the Kerr-induced beam cleanup occurs over a broad wavelength range.

Kerr-induced beam cleanup has been observed over a wide range of pulse widths. A 20-cm-long GRIN fiber was used in one experiment with 80-fs input pulses at 1030 nm [82]. Near-field images of the fiber's output indicated that the spatial pattern changed considerably as pulse energy was increased from 0.5 to 46 nJ. At the highest energy level, the Kerr nonlinearity led to a smooth intensity profile with a nearly circular spot. In the time-domain, the pulse at this energy level was only 34 fs wide and had a broad spectrum. The soliton effects did not play a significant role in this experiment because the GVD of the GRIN fiber was normal ($\beta_2 > 0$) at the input wavelength.

Several experiments have shown that the extent of Kerr-induced cleanup depends on many factors. The parabolic nature of the index profile of the GRIN fiber appears to be important because self-imaging occurs only for this profile. As the index profile becomes flatter, much larger peak powers are needed [16] for the beam cleanup to occur. Also, the effectiveness of the Kerr-induced cleanup decreases

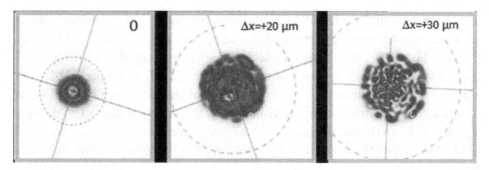

Figure 5.19 Spatial distribution at the output of a 12-m-long GRIN fiber when a pulsed (width 500 ps) Gaussian beam at 1064 nm was launched with its center shifted from the core's center by 0, 20, and 30 μm. Peak power of input pulses was 10 kW in each case. (After Ref. [16]; ©2019 CC BY.)

as the number of modes excited by the input beam increases. This feature can be observed with the off-center launch of an input Gaussian beam. As discussed in Section 2.4.1, many more modes of a GRIN fiber are excited by the optical beam under such conditions. Figure 5.19 shows the observed spatial intensity distributions at the output of a 12-m-long GRIN fiber when 500-ps-wide pulses with 10-kW peak power were launched into the GRIN fiber such the Gaussian beam was shifted from the core's center by 0, 20, and 30 μm [16]. A single smooth spot was observed only in the first case (no shift from the center) for which the beam excites only low-order radially symmetric LP_{0m} modes. Almost no Kerr-induced cleanup occurred for the 30-μshift of the Gaussian beam from the core's center.

In a 2020 experiment, Kerr-induced cleanup was observed to occur at lower input powers when 160-ps pulses were launched such that only a few low-order modes of the GRIN fiber were excited [86]. The input wavelength was 1562 nm, at which the 12-m-long GRIN fiber exhibited anomalous GVD ($\beta_2 < 0$). Input pulses were chirped with a relatively large bandwidth of 6 nm. At low peak powers, spatial profile of the beam was relatively broad and exhibited a double-lobe shape. At power levels close to 1 kW, the beam became narrower and had a nearly circular shape indicative of the LP_{01} mode of the fiber. By adjusting the launch conditions at the input end, it was possible to produce Kerr-induced cleaning such that the output spatial shape was close to that of the LP_{11} mode. The nature of GVD did not play a significant role because the input pulses were relatively wide and their peak powers were low enough to avoid the onset of modulation instability.

The physical mechanism behind the spatial beam cleanup in multimode fibers has two origins. First, self-focusing must play a role because it reduces a beam's size even in a bulk medium and leads to its collapse when the input power exceeds a critical level. In most experiments, power levels have remained below this critical

level but some self-focusing of the beam did occur. Numerical results also show a
nonlinear transfer of energy from higher-order modes to the fundamental mode of
the same fiber [87]. Second, periodic self-imaging plays a significant role in GRIN
fibers, as it results in spatial oscillations of a beam's width with a period of about
1 mm. The Kerr nonlinearity converts these oscillations into periodic variations of
the refractive index, resulting in a Kerr-induced grating along the fiber's length. Such
a grating helps in transferring energy from higher-order modes to the fundamental
mode. This is the reason why the threshold of Kerr self-cleaning is lower for GRIN
fibers.

Two approaches have been developed in recent years that shed some light into the
process of Kerr-induced cleanup of optical beams in GRIN fibers. In one approach,
theory of hydrodynamic turbulence is employed to reveal nonlinear redistribution of
energy among modes above a certain power threshold [88]. In the second approach,
thermodynamics is evoked to show how the modal distribution of input power is
affected by the nonlinear effects inside a multimode fiber [89–92]. A discrete kinetic
equation has been used to show that structural disorder can accelerate the process of
thermalization and condensation by several orders of magnitude.

5.5.3 Second Harmonic Generation

Second harmonic generation (SHG) is a common nonlinear phenomenon that is
used to generate light in the visible and ultraviolet regions [2]. Strictly speaking,
SHG should not occur in silica-based GRIN fibers because the inversion symmetry
of SiO_2 molecules forbids all nonlinear effects resulting from the second-order
susceptibility $\chi^{(2)}$. However, this type nonlinearity can be induced in silica glass
by a process known as poling [93]. It requires exposing the fiber's core to a strong
electric field to provide quasi-phase matching (QPM) for SHG.

Another process for SHG in silica fibers makes use of optical poling. When an
intense pump wave is launched into a fiber, its absorption generates charge carriers
that create a modulated $\chi^{(2)}$ whose period satisfies the QPM condition. In a 1986
experiment [94], the SHG power grew exponentially with time when 100-ps pump
pulses were propagated through a 1-m-long fiber, before saturating after 10 hours.
The 530-nm pulses at the fiber's output had a width of about 55 ps and were intense
enough to pump a dye laser. In another experiment, a silica fiber could be prepared
in only a few minutes, provided a weak second-harmonic signal, acting as a seed,
was launched together with the pump pulses [95]. In this situation, a permanent $\chi^{(2)}$
grating was written to the fiber such that its period satisfied the QPM condition.

This technique was used in 2017 to write a $\chi^{(2)}$ grating inside a standard 6-m-long
GRIN fiber by using a 1064-nm pump with a 532-nm seed [96]. As the spatial width
of the pump beam varied periodically because of the self-imaging phenomenon, the

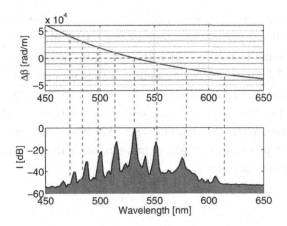

Figure 5.20 Observed SHG spectrum at the output of a 6-m-long GRIN fiber when 500-ps pulses at the wavelength of 1064 nm were launched with 37.5-kW peak power. (After Ref. [96]; ©2017 OSA.)

amplitude of the $\chi^{(2)}$ grating exhibited double periodicity, with the QPM period $\Lambda = 48$ μm and the self-imaging period $L_p = 600$ μm. Such a poled GRIN fiber was used to study the SHG at 532 nm when 500-ps pump pulses were launched into it with a peak power of 37.5 kW. As the second harmonic built up within the fiber at the 532-nm wavelength, several spectral sidebands also developed on its both sides. Figure 5.20 shows the observed multipeak SHG spectrum at the end of the GRIN fiber (bottom panel). A total of eight peaks can be seen at wavelengths 476, 488, 501, 515, 532, 552, 576, and 606 nm. The spectrum of the pump also broadened and covered a wide wavelength range from 1000 to 1500 nm.

The wavelengths of spectral peak seen in Figure 5.20 can be predicted using the phase-matching condition for SHG to occur while accounting for the double periodicity of the $\chi^{(2)}$ grating. For a pump at the frequency ω_p, this condition can be written as [96]

$$\Delta\beta = \beta(\omega) - 2\beta(\omega_p) - \frac{2\pi}{\Lambda} = \frac{2\pi m}{L_p}, \tag{5.5.2}$$

where m is an integer. Top panel in Figure 5.20 shows the phase mismatch $\Delta\beta$ as a function of wavelength. Horizontal lines show the value of $2\pi m/L_p$ for $m = -4$ to 4. Wavelengths of different peaks correspond to the intersection of these lines with the $\Delta\beta$ curve. In the absence of self-imaging, only one spectral peak will form at 532 nm corresponding to the choice $m = 0$. Remaining peaks have their origin in the self-imaging of an optical beam occurring for a parabolic index profile of a GRIN medium.

In another experiment, SHG with a conversion efficiency of 6.5% occurred even without a seed, when a germanium-doped silica GRIN fiber was pumped at 1064 nm [97]. At the same time, the GRIN fiber also produced infrared radiation at a wavelength of 2128 nm, which corresponds to a subharmonic at a frequency that is one half of the pump frequency. When the peak power of 400-ps pulses was increased beyond 100 kW, the observed spectrum showed not only multiple peaks centered near 532 nm, but also near 450 nm and near 2128 nm. These results indicate that multiple nonlinear processes can occur simultaneously inside a germanium-doped GRIN fiber made of silica glass.

5.5.4 Saturable Absorber for Mode-Locking

Passive mode-locking of lasers often requires the use of a saturable absorber, a device whose insertion loss is high at low input powers but is reduced substantially at high power levels. Early mode-locked lasers used materials such as a dye solution, whose relatively long response time limited the width of emitted pulses. Since then, many nonlinear schemes have been developed for mode-locked fiber lasers that make use of the Kerr nonlinearity and act as an "effective saturable absorber" with an ultrashort ($\sim$ 1 fs) response time [98]. It has been found recently that GRIN fibers can also be used for making such an effective saturable absorber [99–103].

In its simplest form, such a saturable absorber is made by connecting the two ends of a GRIN fiber of suitable length to two single-mode fibers (SMFs), as shown schematically in Figure 5.21. The beam entering from the input SMF excites multiple modes of the GRIN fiber that interfere with each other because of their different phase shifts. At low input power levels, power entering the output SMF is reduced by a larger factor, resulting in high losses, when the GRIN fiber's length is chosen suitably to utilize multimode interference. As larger input powers, one must consider several intermodal nonlinear effects discussed in Section 5.4. It was found in a 2013 study [99] that these nonlinear effects modify the modal distribution within the GRIN fiber such that much more power enters the output SMF when the input power exceeds a certain level. This power-dependent coupling loss converts the GRIN fiber into an effective saturable absorber.

We can understand the operation of such a device in terms of the self-imaging phenomenon. In the low-power case, the transmission through the device varies periodically with a period equal to the self-imaging period. Losses are relatively large when the fiber's length is chosen to be $L = (m + \frac{1}{2})L_p$, where m is an integer. In the high-power regime, the SPM and XPM effects introduce additional phase shifts such that a much larger fraction of input power is transmitted by such a device.

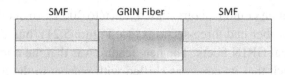

Figure 5.21 Saturable absorber made by connecting two ends of a GRIN fiber of specific length to two single-mode fibers (SMFs).

A GRIN-fiber-based saturable absorber was used in 2015 for passive Q-switching of an erbium-doped fiber laser [100]. Mode locking of a thulium-doped fiber laser was demonstrated [101] in 2017 by using a modified form of the design in Figure 5.21. A small piece of a step-index multimode fiber (length about 0.3 mm) was inserted at the input end of the 7.8-cm-long GRIN fiber. This additional piece relaxed restriction on the precise length of the GRIN fiber and improved the performance of the resulting saturable absorber. Stable mode locking produced pulses of 1.4-ps width at wavelengths near 1850 nm, when a thulium-doped fiber laser was pumped continuously at 1570 nm. The laser's wavelength could also be tuned over 50 nm. A similar device was used for mode-locking of a ytterbium-doped fiber laser [102]. The resulting pulses were 4-ps wide because of the normal GVD of the laser's cavity at the laser's wavelength. However, they were linearly chirped and could be compressed down to 276 fs.

Mode-locking of an erbium-doped fiber laser was realized in 2018 [103] using a modified form of the design shown in Figure 5.21. The input end of the GRIN fiber was etched before splicing with the single-mode fiber. The resulting hole formed a nonuniform cavity whose maximum dimension was 45 μm. This cavity modified the coupling of light into the GRIN fiber and resulted in the excitation of many more modes. The GRIN fiber was also bent to produce pulses shorter than 1 ps by mode-locking the laser.

The design of Figure 5.21 without any modification was used in a 2019 experiment for mode-locking an erbium-doped fiber laser [104]. The measured transmission for the device containing a 37-cm-long GRIN fiber increased from 44% to 75% when the energy fluence was close to 0.1 μJ/cm^2. The laser produced 2-ps wide pulses with energies as high as 13.6 nJ. In another experiment, a thulium-doped fiber laser was mode-locked with a similar GRIN-based device that acted both as a saturable absorber and as an optical filter [105]. The laser produced 1.2-ps-wide pulses at the wavelength 1.93 μm, and this wavelength could be tuned over 5 nm.

GRIN-fiber-based saturable absorbers have remained an attractive option. In a 2020 experiment, the design was similar to that used in Ref. [101], but the section added to the saturable absorber before the GRIN fiber had no core [106]. Such a device was used to mode-lock a Yb-doped fiber laser that produced 2.4-ps pulses at 1034.2 nm, which could be compressed down to 300 fs. In another experiment, the saturable absorber had three sections (see Figure 5.21), but all three sections employed identical GRIN fibers. The new feature was that the middle section was offset-spliced such that its core was shifted with respect to the input and output fibers [107]. This shift breaks the radial symmetry when light enters the middle section, resulting in the excitation of many more modes (see Section 2.4). Such a device was used to mode-lock a Yb-doped fiber laser. It was found that even a two-section device, consisting of two offset-spliced GRIN fibers, acts as a saturable absorber and produces mode-locked pulses.

References

[1] R. Y. Chiao, E. Garmire, and C. H. Townes, *Phys. Rev. Lett.* **13**, 479 (1964); https://doi.org/10.1103/PhysRevLett.13.479.

[2] R. W. Boyd, *Nonlinear Optics*, 4th ed. (Academic Press, 2020).

[3] J.-M. Liu, *Nonlinear Photonics* (Cambridge University Press, 2022).

[4] P. L. Baldeck, F. Raccah, and R. R. Alfano, *Opt. Lett.* **12**, 588 (1987); https://doi.org/10.1364/OL.12.000588.

[5] B. Bendow, P. D. Gianino, and N. Tzoaro, *J. Opt. Soc. Am.* **71**, 656 (1981); https://doi.org/10.1364/OSA.71.000656.

[6] J. T. Manassah, P. L. Baldeck, and R. R. Alfano, *Opt. Lett.* **13**, 589 (1988); https://doi.org/10.1364/OL.13.000589.

[7] M. Karlsson, D. Anderson, and M. Desaix, *Opt. Lett.* **17**, 22 (1992); https://doi.org/10.1364/OL.17.000022.

[8] I. S. Gradshteyn and I. M. Ryzhik, Eds., *Table of Integrals, Series, and Products*, 6th ed. (Academic Press, 2000).

[9] S. Longhi and D. Janner, *J. Opt. B: Quantum Semiclass. Opt.* **6**, S303 (2004); https://doi.org/10.1088/1464-4266/6/5/019.

[10] T. Hansson, A. Tonello, T. Mansuryan et al., *Opt. Express* **28**, 24005 (2020); https://doi.org/10.1364/OE.398531.

[11] B. Bendow, P. D. Gianino, N. Tzoar, and M. Jain, *J. Opt. Soc. Am.* **70**, 539 (1980); https://doi.org/10.1364/JOSA.70.000539.

[12] G. P. Agrawal, *Nonlinear Fiber Optics*, 6th ed. (Academic Press, 2019).

[13] B. Crosignani, A. Cutolo, and P. Di Porto, *J. Opt. Soc. Am.* **72**, 1136 (1982); https://doi.org/10.1364/JOSA.72.001136.

[14] F. Poletti and P. Horak, *J. Opt. Soc. Am. B* **25**, 1645 (2008); https://doi.org/10.1364/JOSAB.25.001645.

[15] L. G. Wright, Z. M. Ziegler, P. M. Lushnikov et al., *IEEE J. Sel. Topics Quantum Electron.* **24**, 5100516 (2018); https://doi.org/10.1109/JSTQE.2017.2779749.

[16] K. Krupa, A. Tonello, A. Barthélémy et al., *APL Photon.* **4**, 110901 (2019); https://doi.org/10.1063/1.5119434.

[17] L. G. Wright, W. H. Renninger, D. N. Christodoulides, and F. W. Wise, *Optica* **9**, 824 (2022); https://doi.org/10.1364/OPTICA.461981.

[18] M. J. Ablowitz and P. A. Clarkson, *Solitons, Nonlinear Evolution Equations, and Inverse Scattering* (Cambridge University Press, 1991).

[19] H. Hasegawa and Y. Kodama, *Solitons in Optical Communications* (Oxford University Press, 1995).

[20] Y. S. Kivshar and G. P. Agrawal, *Optical Solitons: From Fibers to Photonic Crystals* (Academic Press, 2003).

[21] Y. Silberberg, *Opt. Lett.* **15**, 1282 (1990); https://doi.org/10.1364/OL.15.001282.

[22] S.-S. Yu, C.-H. Chien, Y. Lai, and J. Wang, *Opt. Commun.* **119**, 167 (1995); https://doi.org/10.1016/0030-4018(95)00377-K.

[23] S. Raghavan and G. P. Agrawal, *Opt. Commun.* **180**, 377 (2000); https://doi.org/10.1016/S0030-4018(00)00727-6.

[24] O. V. Shtyrina, M. P. Fedoruk, Y. S. Kivshar, and S. K. Turitsyn, *Phys. Rev. A* **97**, 013841 (2018); https://doi.org/10.1103/PhysRevA.97.013841.

[25] S. V. Sazonov, *Phys. Rev. A* **100**, 043828 (2019); https://doi.org/10.1103/PhysRevA.100.043828.

[26] Y. V. Kartashov, G. E. Astrakharchik, B. A. Malomed, and L. Torner, *Nat. Rev. Phys.* **1**, 185 (2019); https://doi.org/10.1038/s42254-019-0025-7.

[27] M. Conforti, C. M. Arabi, A. Mussot, and A. Kudlinski, *Opt. Lett.* **42**, 4004 (2017); https://doi.org/10.1364/OL.42.004004.

[28] R. H. Stolen, J. P. Gordon, W. J. Tomlinson, and H. A. Haus, *J. Opt. Soc. Am. B* **6**, 1159 (1989); https://doi.org/10.1364/JOSAB.6.001159.

[29] K. J. Blow and D. Wood, *IEEE J. Quantum Electron.* **25**, 2665 (1989); https://doi.org/10.1109/3.40655.

[30] Q. Lin and G. P. Agrawal, *Opt. Lett.* **31**, 3086 (2006); https://doi.org/10.1364/OL.31.003086.

[31] A. Hasegawa, *Opt. Lett.* **5**, 416 (1980); https://doi.org/10.1364/OL.5.000416.

[32] B. Crosignani and P. Di Porto, *Opt. Lett.* **6**, 329 (1981); https://doi.org/10.1364/OL.6.000329.

[33] S. Longhi, *Opt. Lett.* **28**, 2363 (2003); https://doi.org/10.1364/OL.28.002363.

[34] F. Matera, A. Mecozzi, M. Romagnoli, and M. Settembre, *Opt. Lett.* **18**, 1499 (1993); https://doi.org/10.1364/OL.18.001499.

[35] K. Krupa, A. Tonello, A. Barthélémy et al., *Phys. Rev. Lett.* **116**, 183901 (2016); https://doi.org/10.1103/PhysRevLett.116.183901.

[36] L. G. Wright, Z. I. Liu, D. A. Nolan et al., *Nat. Photon.* **10**, 771 (2016); https://doi.org/10.1038/nphoton.2016.227.

[37] C. M. Arabi, A. Kudlinski, A. Mussot, and M. Conforti, *Phys. Rev. A* **97**, 023803 (2018); https://doi.org/10.1103/PhysRevA.97.023803.

[38] A. Hasegawa and Y. Kodama, *Opt. Lett.* **15**, 1443 (1990); https://doi.org/10.1364/OL.15.001443.

[39] G. P. Agrawal, *Fiber-Optic Communication Systems*, 5th ed. (Wiley, 2021).

[40] W. H. Renninger and F. W. Wise, *Nat. Commun.* **4**, 1719 (2013); https://doi.org/10.1038/ncomms2739.

[41] L. G. Wright, W. H. Renninger, D. N. Christodoulides, and F. W. Wise, *Opt. Express* **23**, 3492 (2015); https://doi.org/10.1364/OE.23.003492.

[42] A. S. Ahsan and G. P. Agrawal, *Opt. Lett.* **43**, 3345 (2018); https://doi.org/10.1364/OL.43.003345.

[43] M. Zitelli, F. Mangini, M. Ferraro et al., *Opt. Express* **28**, 20473 (2020); https://doi.org/10.1364/OE.394896.

[44] M. Zitelli, M. Ferraro, F. Mangini, and S. Wabnitz, *Photon. Res.* **9**, 741 (2021); https://doi.org/10.1364/PRJ.419235.

[45] M. Zitelli, M. Ferraro, F. Mangini, O. Sidelnikov, and S. Wabnitz, *Commun. Phys.* **4**, 182 (2021); https://doi.org/10.1038/s42005-021-00687-0.

[46] A. S. Ahsan and G. P. Agrawal, *Opt. Lett.* **44**, 2637 (2019); https://doi.org/10.1364/OL.44.002637.

[47] A. Mafi, *J. Lightwave Technol.* **30**, 2803 (2012); https://doi.org/10.1109/JLT.2012
.2208215.

[48] A. Mecozzi, C. Antonelli, and M. Shtaif, *Opt. Express* **20**, 23436 (2012); https:
//doi.org/10.1364/OE.20.023436.

[49] S. Mumtaz, R.-J. Essiambre, and G. P. Agrawal, *J. Lightwave Technol.* **31**, 398 (2013);
https://doi.org/10.1109/JLT.2012.2231401.

[50] C. Antonelli, M. Shtaif, and A. Mecozzi, *J. Lightwave Technol.* **34**, 36 (2016);
https://doi.org/10.1109/JLT.2016.2519240.

[51] S. Buch, S. Mumtaz, R.-J. Essiambre, A. M. Tulino, and G. P. Agrawal, *Opt. Fiber
Technol.* **47**, 123 (2019); https://doi.org/10.1016/j.yofte.2018.12.020.

[52] J. Laegsgaard, *J. Opt. Soc. Am. B* **34**, 2266 (2017); https://doi.org/10.1364/JOSAB.34
.002266.

[53] J. Laegsgaard, *J. Opt. Soc. Am. B* **36**, 2235 (2019); https://doi.org/10.1364/JOSAB.36
.002235.

[54] G. P. Agrawal, *Phys. Rev. Lett.* **59**, 880 (1987); https://doi.org/10.1103/PhysRevLett
.59.880.

[55] R. Dupiol, A. Bendahmane, K. Krupa et al., *Opt. Lett.* **42**, 3419 (2017); https://
doi.org/10.1364/OL.43.003419.

[56] R. H. Stolen, *IEEE J. Quantum Electron.* **11**, 100 (1975); https://doi.org/10.1109/JQE
.1975.1068571.

[57] K. O. Hill, D. C. Johnson, and B. S. Kawasaki, *Appl. Opt.* **20**, 1075 (1981); https:
//doi.org/10.1364/AO.20.001075.

[58] E. Nazemosadat, H. Pourbeyram, and A. Mafi, *J. Opt. Soc. Am. B* **33**, 144 (2016);
https://doi.org/10.1364/JOSAB.33.000144.

[59] A. Bendahmane, K. Krupa, A. Tonello, et al., *J. Opt. Soc. Am. B* **35**, 295 (2018);
https://doi.org/10.1364/JOSAB.35.000295.

[60] R. Dupiol, A. Bendahmane, K. Krupa et al., *Opt. Lett.* **42**, 1293 (2017); https://
doi.org/10.1364/OL.43.001293.

[61] A. B. Grudinin, E. M. Dianov, D. V. Korbkin, A. M. Prokhorov, and D. V. Khaidarov,
JETP Lett. **47**, 356 (1988).

[62] K. S. Chiang, *Opt. Lett.* **17**, 352 (1992); https://doi.org/10.1364/OL.17.000352.

[63] H. Pourbeyram, G. P. Agrawal, and A. Mafi, *Appl. Phys. Lett.* **102**, 201107 (2013);
https://doi.org/10.1063/1.4807620.

[64] N. B. Terry, T. G. Alley, and T. H. Russell, *Opt. Express* **15**, 17509 (2007); https:
//doi.org/10.1364/OE.15.017509.

[65] A. Minardo, R. Bernini, and L. Zeni, *Opt. Express* **22**, 17480 (2014); https://doi.org/
10.1364/OE.22.017480.

[66] W.-W. Ke, X.-J. Wang, and X. Tang, *IEEE J. Sel. Topics Quantum Electron.* **20**,
0901610 (2014); https://doi.org/10.1109/JSTQE.2014.2303256.

[67] H. Lu, P. Zhou, and X. Wang, *J. Lightwave Technol.* **33**, 4464 (2015); https://doi.org/
10.1109/JLT.2015.2476364.

[68] Suchita, B. Srinivasan, G. P. Agrawal, and D. Venkitesh, *Opt. Commun.* **494**, 127052
(2021); https://doi.org/10.1016/j.optcom.2021.127052.

[69] Q. Gao, Z. Lu, C. Zhu, and J. Zhang, *Appl. Phys. Express* **8**, 052501 (2015);
https://doi.org/10.7567/APEX.8.052501.

[70] R. R. Alfano, Ed., *The Supercontinuum Laser Source*, 4th ed. (Springer, 2022).

[71] F. Poletti and P. Horak, *Opt. Express* **17**, 6134 (2009); https://doi.org/10.1364/OE.17
.006134.

[72] L. G. Wright, S. Wabnitz, D. N. Christodoulides, and F. W. Wise, *Phys. Rev. Lett.* **115**,
223902 (2015); https://doi.org/10.1103/PhysRevLett.115.223902.

[73] G. Lopez-Galmiche, Z. S. Eznaveh, M. A. Eftekhar et al., *Opt. Lett.* **41**, 2553 (2016); https://doi.org/10.1364/OL.41.002553.

[74] K. Krupa, C. Louot, V. Couderc et al., *Opt. Lett.* **41**, 5785 (2016); https://doi.org/10.1364/OL.41.005785.

[75] M. A. Eftekhar, L. G. Wright, M. S. Mills et al., *Opt. Express* **25**, 9078 (2017); https://doi.org/10.1364/OE.25.009078.

[76] M. A. Eftekhar, H. Lopez-Aviles, F. W. Wise, R. Amezcua-Correa, and D. N. Christodoulides, *Commun. Phys.* **4**, 137 (2021); https://doi.org/10.1038/s42005-021-00640-1.

[77] U. Teğin and B. Ortaç, *Sci. Rep.* **8**, 12470 (2018); https://doi.org/10.1038/s41598-018-30252-9.

[78] K. L. Vodopyanov, *Laser-based Mid-infrared Sources and Applications* (Wiley, 2020).

[79] Y. Leventoux, G. Granger, K. Krupa et al., *Opt. Lett.* **46**, 3717 (2021); https://doi.org/10.1364/OL.428623.

[80] Z. Eslami, L. Salmela, A. Filipkowski et al., *Nat. Commun.* **13**, 2126 (2022); https://doi.org/10.1038/s41467-022-29776-6.

[81] L. G. Wright, D. N. Christodoulides, and F. W. Wise, *Nat. Photon.* **9**, 306 (2015); https://doi.org/10.1038/nphoton.2015.61.

[82] Z. Liu, L. G. Wright, D. N. Christodoulides, and F. W. Wise, *Opt. Lett.* **41**, 3675 (2016); https://doi.org/10.1364/OL.41.003675.

[83] K. Krupa, A. Tonello, B. M. Shalaby et al., *Nat. Photon.* **11**, 237 (2017); https://doi.org/10.1038/nphoton.2017.32.

[84] R. Dupiol, K. Krupa, A. Tonello et al., *Opt. Lett.* **43**, 587 (2018); https://doi.org/10.1364/OL.43.000587.

[85] K. Krupa, G. G. Castañeda, A. Tonello et al., *Opt. Lett.* **44**, 171 (2019); https://doi.org/10.1364/OL.44.000171.

[86] Y. Leventoux, A. Parriaux, O. Sidelnikov et al., *Opt. Express* **28**, 14333 (2020); https://doi.org/10.1364/OE.392081.

[87] S. Buch and G. P. Agrawal, *Opt. Lett.* **40**, 225 (2015); https://doi.org/10.1364/OL.40.000225.

[88] E. V. Podivilov, D. S. Kharenko, V. A. Gonta et al., *Phys. Rev. Lett.* **122**, 103902 (2019); https://doi.org/10.1103/PhysRevLett.122.103902.

[89] A. Fusaro, J. Garnier, K. Krupa, G. Millot, and A. Picozzi, *Phys. Rev. Lett.* **122**, 123902 (2019); https://doi.org/10.1103/PhysRevLett.122.123902.

[90] F. O. Wu, A. U. Hassan, and D. N. Christodoulides, *Nat. Photon.* **13**, 776 (2019); https://doi.org/10.1038/s41566-019-0501-8.

[91] H. Pourbeyram, P. Sidorenko, F. Wu et al., *Nat. Phys.* **18**, 685 (2022); https://doi.org/10.1038/s41567-022-01579-y.

[92] F. Mangini, M. Gervaziev, M. Ferraro et al., *Opt. Express* **30**, 10850 (2022); https://doi.org/10.1364/OE.449187.

[93] A. Canagasabey, C. Corbari, A. V. Gladyshev et al., *Opt. Lett.* **34**, 2483 (2009); https://doi.org/10.1364/OL.34.002483.

[94] U. Österberg and W. Margulis, *Opt. Lett.* **11**, 516 (1986); https://doi.org/10.1364/OL.11.000516.

[95] R. H. Stolen and H. W. K. Tom, *Opt. Lett.* **12**, 585 (1987); https://doi.org/10.1364/OL.12.000585.

[96] D. Ceoldo, K. Krupa, A. Tonello et al., *Opt. Lett.* **42**, 971 (2017); https://doi.org/10.1364/OL.42.000971.

[97] M. A. Eftekhar, Z. Sanjabi-Eznaveh, J. E. Antonio-Lopez et al., *Opt. Lett.* **42**, 3478 (2017); https://doi.org/10.1364/OL.42.003478.

[98]   G. P. Agrawal, *Applications of Nonlinear Fiber Optics*, 3rd ed. (Academic Press, 2020).

[99]   E. Nazemosadat and A. Mafi, *J. Opt. Soc. Am. B* **30**, 1357 (2013); https://doi.org/10.1364/JOSAB.30.001357.

[100] S. Fu, Q. Sheng, X. Zhu et al., *Opt. Express* **23**, 17255 (2015); https://doi.org/10.1364/OE.23.017255.

[101] H. Li, Z. Wang, C. Li, J. Zhang, and S. Xu, *Opt. Express* **25**, 26546 (2017); https://doi.org/10.1364/OE.25.026546.

[102] U. Teğin and B. Ortaç, *Opt. Lett.* **43**, 1611 (2018); https://doi.org/10.1364/OL.43.001611.

[103] F. Yang, D. N. Wang, Z. Wang et al., *Opt. Express* **26**, 927 (2018); https://doi.org/10.1364/OE.26.000927.

[104] Z. Wang, J. Chen, T. Zhu, D. N. Wang, and F. Gao, *Photon. Res.* **7**, 1214 (2019); https://doi.org/10.1364/PRJ.7.001214.

[105] H. Li, F. Hu, Y. Tian et al., *Opt. Express* **27**, 14437 (2019); https://doi.org/10.1364/OE.27.014437.

[106] S. Thulasi and S. Sivabalan, *Appl. Opt.* **59**, 7357 (2020); https://doi.org/10.1364/AO.395356.

[107] W. Pan, L. Jin, J. Wang et al., *Appl. Opt.* **60**, 923 (2021); https://doi.org/10.1364/AO.413601.

6

Effects of Loss or Gain

As we saw in Section 1.3.1, all materials exhibit losses that reduce the power of an optical beam as it propagates through a material. We have ignored such losses so far in this book, assuming short propagation distances. Some applications require long GRIN fibers made with plastic materials, for which losses cannot be ignored. Losses can be compensated using a suitable gain mechanism that amplifies the optical beam inside a GRIN medium pumped optically. In Section 6.1, we discuss the impact of losses and focus in Section 6.2 on the mechanisms used for providing optical gain inside a GRIN medium. Section 6.3 is devoted to Raman amplifiers and lasers, built with GRIN fibers and pumped suitably to provide optical gain. Parametric amplifiers are discussed in Section 6.4, together with the phase matching required for four-wave mixing to occur. The focus of Section 6.5 is on amplifiers and lasers made by doping a GRIN fiber with rare-earth ions. Section 6.6 includes the nonlinear effects and describes the formation of spatial solitons and similaritons inside an active GRIN medium under suitable conditions.

6.1 Impact of Optical Losses

Optical losses are negligible for GRIN lenses and other devices because of their short lengths (< 0.1 m). However, losses can become relevant for GRIN fibers, depending on the application. The impact of losses in GRIN fibers was investigated during the 1970s, motivated mainly by their use in optical communication systems [1–3]. Losses became even more relevant for GRIN fibers made with plastic materials and used routinely for data transfer over short distances [4, 5]. Figure 1.4 of Section 1.3 compares the wavelength dependence of measured loss in silica fibers to losses in two types of plastic. As seen there, plastic fibers exhibit much larger losses compared to those of silica fibers, whose loss is typically below 0.2 dB/km at wavelengths near 1550 nm.

Losses resulting from absorption can be included in the formalism of Section 2.1 by treating the refractive index $n(\mathbf{r})$ in the Helmholtz equation (2.1.13) as a complex quantity. More specifically, n_0 in the parabolic index profile given in Eq. (1.2.5) becomes complex such that $n_0 = n_r + in_i$, where the imaginary part n_i accounts for material's absorption at the frequency of incident radiation. In practice, n_i is relatively small, and the condition $n_i \ll n_r$ is satisfied. Its use allows us to write Eq. (2.1.13) as

$$\nabla^2 \tilde{\mathbf{E}} + k(k + i\alpha)(1 - b^2\rho^2)\tilde{\mathbf{E}} = 0, \tag{6.1.1}$$

where $k = n_r k_0$ and $\alpha = 2k_0 n_i$ is the absorption coefficient of the GRIN material. It is common to add other types of losses, such as those resulting from Rayleigh scattering, to α so that it accounts for all losses. Also, optical gain within a GRIN medium can be included using a negative value for α.

Equation (6.1.1) shows that the loss affects a GRIN medium in two ways. It not only reduces the optical power inside such a medium but also replaces the parameter b with the complex quantity

$$b' = b(1 + i\alpha/k)^{1/2} \approx b(1 + i\alpha/2k), \tag{6.1.2}$$

where we used the approximation $\alpha \ll k$ that holds in practice. This change has important consequences. For example, one may ask how the Hermite–Gauss or the Laguerre–Gauss modes found in Section 2.2 are affected by losses [6]. As the analysis there did not depend on b being real, spatial profiles of the Laguerre–Gauss modes are still given by Eq. (2.2.20) with $q^2 = kb'$. The complex values of q affect the spatial form of the modes of a GRIN medium. A fundamental change is that the phase front of any mode does not remain planar and acquires a curvature. This can be seen by noting that the Gaussian factor in Eq. (2.2.20) can be written as

$$\exp\left(-\frac{kb'}{2}\rho^2\right) = \exp\left(-\frac{kb}{2}\rho^2\right)\exp\left(-\frac{i\alpha b}{4}\rho^2\right). \tag{6.1.3}$$

It follows that the curvature of the phase front scales linearly with α and becomes planar for $\alpha = 0$. The propagation constant β_{lm} of any LP$_{lm}$ mode also becomes complex and can be written from Eq. (2.2.23) as

$$\beta_{lm}(\omega) = k[1 - 2(\ell + 2m - 1)b'/k]^{1/2} \approx k - (\ell + 2m - 1)(b + ib\alpha/2k). \tag{6.1.4}$$

The imaginary part of β_{lm} indicates that losses become different for different modes.

In the non-modal approach discussed in Section 2.5, we can still introduce the slowly varying amplitude as $\tilde{E} = Ae^{ikz}$ and make the paraxial approximation to obtain

$$2ik\frac{\partial A}{\partial z} + \nabla_T^2 A + ik\alpha A - k^2 b'^2(x^2 + y^2)A = 0. \tag{6.1.5}$$

Again, losses impact the evolution of an optical beam in two ways. The third term reduces the beam's power as it propagates inside a GRIN medium, and b' in the fourth term becomes complex as indicated in Eq. (6.1.2). We can remove the third term using $A = A' \exp(-\alpha z/2)$, resulting in

$$2ik\frac{\partial A'}{\partial z} + \nabla_T^2 A' - k^2 b'^2 (x^2 + y^2)A' = 0. \tag{6.1.6}$$

This equation is formally identical to Eq. (2.5.5) of Section 2.5 obtained for a lossless GRIN fiber, with the only difference that b' has an imaginary part. As losses are typically such that $\alpha \ll k$, often one can replace b' with b. In that case, losses reduce an optical beam's power, but have negligible effect on the periodic self-imaging of this beam inside a GRIN medium. In particular, the results obtained in Section 2.5 for variations in the beam's width remain valid.

In the case of a lossy, nonlinear GRIN medium, we should add the Kerr nonlinearity to Eq. (6.1.5), as indicated in Eq. (5.1.6). The resulting equation becomes

$$2ik\frac{\partial A}{\partial z} + \nabla_T^2 A + ik\alpha A - k^2 b'^2 (x^2 + y^2)A + 2kk_0 n_2 |A|^2 A = 0. \tag{6.1.7}$$

Introducing again $A = A' \exp(-\alpha z/2)$, the equation for A' takes the form

$$2ik\frac{\partial A'}{\partial z} + \nabla_T^2 A' - k^2 b'^2 (x^2 + y^2)A' + 2kk_0 n_2 e^{-\alpha z}|A'|^2 A' = 0. \tag{6.1.8}$$

The factor of $e^{-\alpha z}$ in the last term indicates that the nonlinear effects become weaker as the beam propagates down the GRIN medium. This is not surprising if we note that the nonlinear effects depend on the local intensity, which is reduced because of losses.

Equation (6.1.7) can also be used when the GRIN medium is pumped suitably to provide a net optical gain. In this case, α is replaced with $-g$, where g is the gain coefficient, defined in a way similar to α. Even though α can be treated as a constant, the gain g may vary spatially depending on the pumping mechanism employed. We discuss in the next section several schemes for realizing optical gain inside a GRIN medium.

6.2 Mechanisms for Providing Gain

Three mechanisms can be used for the amplification of optical signals inside an optical fiber [7–9]. One of them requires doping of the fiber with a suitable material during its fabrication. The other two make use of the nonlinear effects such as stimulated Raman scattering (SRS) and four-wave mixing (FWM). All three require a pump laser at a suitable wavelength that transfers some of its energy to the signal being amplified.

6.2.1 Doping with Rare-Earth Materials

Amplifiers made by doping a single-mode fiber with rare-earth materials (such as erbium and ytterbium) were developed during the 1980s. The erbium-doped fiber amplifiers are used routinely for compensating losses of single-mode silica fibers in optical communication systems [10–13]. After 2010, such amplifiers were developed for telecom systems designed with step-index multimode fibers [14]. Around 2018, the doping of GRIN fibers was carried out successfully for making amplifiers and lasers [15, 16].

A GRIN fiber was fabricated in 2019 by doping it with ytterbium during the preform stage [16]. Its core had a diameter of 30 μm with a nearly parabolic refractive-index profile. A 3-m-long piece of such a doped GRIN fiber was used to make an amplifier that was pumped at 976 nm using a semiconductor laser. The input signal was in the form of 70-ns pulses at 1074 nm with an average power of < 0.1 W. The amplified signal at the fiber's output had an average power of 10 W at a pump power of 17 W. Thus, amplification by a factor of more than 100 (20 dB) was realized with an efficiency of 60%. The beam's quality did not degrade in this experiment because the density of dopants was graded such that the gain was larger in the central region of the fiber's core.

In a novel approach, nano-engineering was used to produce an index gradient within the core of a fiber [15]. At the prefrom stage, two types of glass rods were used, made of either pure silica or germanium-doped silica. More than 2000 of such rods were stacked such that the density of germanium-doped rods decreased radially from the central region of the preform. Figure 6.1 shows an example of the distribution of the two types of rods that produced a nearly parabolic index profile when the preform was used to make an optical fiber. This method can be used to make GRIN fibers containing cores of different shapes and sizes with an arbitrary index distribution. It can also be used to dope the core of a GRIN fiber with erbium or ytterbium by using tubes containing such materials.

To understand the physics behind the signal's amplification inside a doped GRIN fiber, details of pumping are not important. Pumping creates the population inversion between two energy states that is required for the optical gain. For a homogeneously broadened gain medium, the gain varies with frequency as [8]

$$g(\omega) = \frac{g_0}{1 + (\omega - \omega_a)^2/(\delta\omega)^2 + P/P_s}, \tag{6.2.1}$$

where g_0 is the peak value of gain, ω is the frequency of the incident signal, ω_a is the resonance frequency of the dopant material, and P is the optical power of the signal being amplified. The saturation power P_s depends on parameters such as the fluorescence time of dopants. The parameter $\delta\omega$ in Eq. (6.2.1) sets the bandwidth

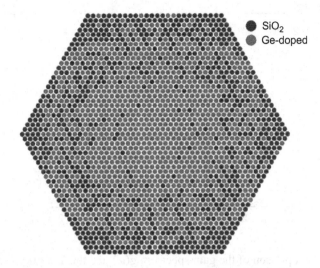

SiO$_2$
Ge-doped

Figure 6.1 Distribution of glass rods in the core region of a nano-engineered GRIN fiber. The core contains more than 2000 glass rods of pure and Ge-doped silica. (After Ref. [15]; ©2018 CC BY.)

over which amplification can occur. It is inversely related to the dipole relaxation time of dopants.

Equation (6.2.1) can be used to discuss several important characteristics of optical amplifiers. The power dependence of the gain leads to its saturation when amplified signal's power becomes large enough to approach the saturation power P_s. Saturation of the gain can be neglected for $P \ll P_s$, resulting in

$$g(\omega) = g_0[1 + (\omega - \omega_a)^2/(\delta\omega)^2]^{-1}. \tag{6.2.2}$$

The frequency dependence of the gain shows that maximum amplification occurs at resonance ($\omega = \omega_a$), and the gain is reduced as the input frequency is detuned from this value. The gain bandwidth, defined as the full width at half maximum (FWHM) of the gain spectrum $g(\omega)$, is found to be $\Delta\omega_g = 2(\delta\omega)$.

A related concept of the amplifier bandwidth is also used in place of the gain bandwidth. The difference becomes clear when one considers the amplification factor, defined as $G = P_{out}/P_{in}$, where P_{in} and P_{out} are the input and output powers of the CW signal being amplified. The amplification factor is obtained by solving the simple equation,

$$\frac{dP}{dz} = g(\omega)P(z), \tag{6.2.3}$$

where $P(z)$ is the optical power at a distance z from the input end of the amplifier. Integrating this equation with the conditions $P(0) = P_{in}$ and $P(L) = P_{out}$, the amplification factor for an amplifier of length L is found to be

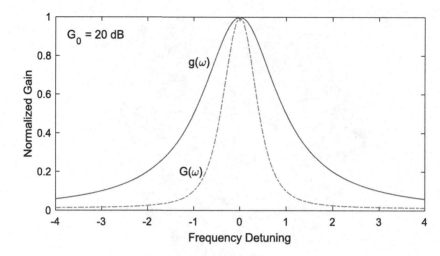

Figure 6.2 Comparison of the gain spectrum $g(\omega)$ and the amplifier's spectrum $G(\omega)$ for an amplifier with a peak gain of 20 dB.

$$G(\omega) = \exp\left[\int_0^L g(\omega)\,dz\right] = \exp[g(\omega)L], \qquad (6.2.4)$$

where $g(\omega)$ was assumed to remain constant along the amplifier's length.

Figure 6.2 shows the gain spectrum $g(\omega)$ and the amplification factor $G(\omega)$ by plotting g/g_0 and $G/G(\omega_0)$ as a function of $(\omega - \omega_a)/\delta\omega$. Both $G(\omega)$ and $g(\omega)$ peak at $\omega = \omega_a$ and decrease when $\omega \neq \omega_a$. However, $G(\omega)$ decreases much faster than $g(\omega)$ because of the exponential dependence seen in Eq. (6.2.4). The amplifier's bandwidth is defined as the FWHM of $G(\omega)$ and is related to the gain bandwidth as

$$\Delta\omega_A = \Delta\omega_g \left(\frac{\ln 2}{\ln G_0 - \ln 2}\right)^{1/2}, \qquad (6.2.5)$$

where $G_0 = G(\omega_0)$ is the peak value of $G(\omega)$.

The origin of gain saturation lies in the power dependence of the gain coefficient in Eq. (6.2.1). To simplify the following discussion, we consider the on-resonance case $\omega = \omega_a$. By substituting g from Eq. (6.2.1) into Eq. (6.2.3), we obtain

$$\frac{dP}{dz} = \frac{g_0 P}{1 + P/P_s}. \qquad (6.2.6)$$

This equation can be integrated over the amplifier length to obtain the implicit relation [8]

$$G = G_0 \exp\left(-\frac{G-1}{G}\frac{P_{out}}{P_s}\right). \qquad (6.2.7)$$

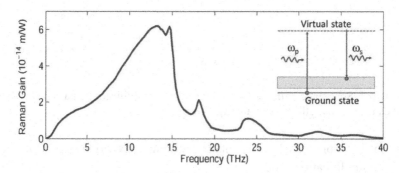

Figure 6.3 Raman gain spectrum of silica glass. The inset shows energy levels involved in the Raman scattering process.

Equation (6.2.7) can be used to calculate the saturated gain G. The results show that G remains close to its unsaturated value G_0 as long as the amplified signal's power at the output end remains well below the saturation power P_s. When that is not the case, the ratio G/G_0 decreases, and P_{out} becomes nearly independent of G_0 for $G_0 > 20$ dB. A quantity of practical interest is the output saturation power P_{out}^s, defined as the output power at which the amplifier gain G is reduced by a factor of 2 from its unsaturated value G_0. Using $G = G_0/2$ in Eq. (6.2.7), P_{out}^s is found to be related to P_s as

$$P_{out}^s = \frac{G_0 \ln 2}{G_0 - 2} P_s. \tag{6.2.8}$$

As $G_0 \gg 2$ in practice, $P_{out}^s \approx (\ln 2)P_s \approx 0.69 P_s$.

6.2.2 Stimulated Raman Scattering

It has been known since 1973 that SRS can be used to amplify optical beams propagating inside an optical fiber [17]. This technique requires a pump laser whose wavelength is shorter than that of the signal being amplified such that the pump-signal frequency difference, $\omega_p - \omega_s$, is close to 13.2 THz. This value corresponds to the peak of the Raman gain spectrum of silica fibers shown in Figure 6.3.

The inset in Figure 6.3 indicates how the pump's energy is transferred to the signal through SRS [18]. The energy $\hbar\omega_p$ of each pump photon is used to transfer a silica molecule to a transient virtual state with an ultrashort lifetime. Stimulated by signal photons, the molecule ends up in a vibrational state, after emitting a photon of energy $\hbar\omega_s$. The net result is that the number of photons at the frequency ω_s increases exponentially as the signal propagates inside the fiber. The virtual state can also decay spontaneously, resulting in photons whose phase and frequencies

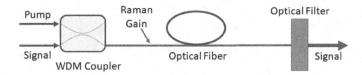

Figure 6.4 Schematic of the setup used for Raman amplification of a signal. WDM
stands for wavelength-division multiplexing.

differ from signal photons. Such spontaneous Raman scattering acts as noise during
the Raman-amplification process.

Figure 6.4 shows schematically how the Raman amplification scheme can be
implemented in practice. A device known as the fiber coupler is used to send both
the signal and pump beams through the same optical fiber. At the output end, an
optical filter blocks the pump but lets the amplified signal pass through it. The
signal's intensity grows inside the fiber as the pump beam transfers power to the
signal. Although a detailed description should consider the multimode nature of the
fiber, a simple model can be used if we assume that both the pump and the signal are
propagating in the LP_{01} mode of the GRIN fiber. In this case, the signal's intensity
I_s grows as [18]

$$\frac{dI_s}{dz} = g_R I_p I_s, \tag{6.2.9}$$

where I_p is the pump intensity and g_R is the Raman-gain coefficient, whose value
depends on the frequency difference $\Omega = \omega_p - \omega_s$ (see Figure 6.3). If we neglect
the pump's depletion over short lengths and treat I_p as a constant, we can integrate
Eq. (6.2.9) to obtain $I_s(z) = I_s(0)\exp(g_R I_p z)$. This relation shows that the signal's
intensity grows exponentially inside the fiber.

The situation is more complicated in the case of long fibers because we should
include fiber's losses as well as the pump's depletion. When these effects are
included, the Raman amplification process is governed by the following two coupled
equations [18]:

$$\frac{dI_s}{dz} = g_R I_p I_s - \alpha_s I_s, \tag{6.2.10}$$

$$\frac{dI_p}{dz} = -\frac{\omega_p}{\omega_s} g_R I_p I_s - \alpha_p I_p, \tag{6.2.11}$$

where α_s and α_p account for fiber losses at the signal and pump frequencies, respectively. These equations can be written by considering the processes through which
photons appear in or disappear from each beam. The frequency ratio ω_p/ω_s has its
origin in the fact that the total number of photons is conserved during the amplification process in the absence of losses. When $\alpha_s = \alpha_p = 0$, one can readily verify
that

$$\frac{d}{dz}\left(\frac{I_s}{\hbar\omega_s} + \frac{I_p}{\hbar\omega_p}\right) = \frac{d}{dz}(F_p + F_s) = 0, \qquad (6.2.12)$$

where F_p and F_s represent the flux of photons.

The solution of Eqs. (6.2.10) and (6.2.11) is difficult because they are coupled in a nonlinear fashion. They can be solved if we neglect the first term on the right side of Eq. (6.2.11) that leads to pump's depletion. Substituting its solution, $I_p = I_0 \exp(-\alpha_p z)$, into Eq. (6.2.10), we obtain

$$\frac{dI_s}{dz} = g_R I_0 \exp(-\alpha_p z) I_s - \alpha_s I_s, \qquad (6.2.13)$$

where I_0 is the input pump intensity at $z = 0$. This equation is easily solved to obtain

$$I_s(L) = I_s(0) \exp(g_R I_0 L_{\text{eff}} - \alpha_s L), \qquad L_{\text{eff}} = [1 - \exp(-\alpha_p L)]/\alpha_p, \qquad (6.2.14)$$

where L is the fiber's length. The solution shows that the effective length is reduced from L to L_{eff} because of fiber's losses.

Fiber-based Raman amplifiers typically require long lengths ($L > 1$ km). They can be pumped in the backward direction by launching the pump from the output end of the fiber. A bidirectional pumping scheme can also be used in which two pump lasers are used at two ends of the fiber. As the gain is distributed over long lengths, the device is said to provide distributed amplification. Such a scheme is useful for optical communication systems because it can compensate for losses and maintain the signal's power over long lengths of fibers.

All amplifiers exhibit gain saturation when input power of the signal is so large that it becomes comparable to the pump's power inside the amplifier. In this situation, the pump is depleted and the gain of the amplifier saturates. Clearly, we need to solve Eqs. (6.2.10) and (6.2.11) as a coupled nonlinear system to account for pump depletion and gain saturation. It turns out they can be solved when losses are nearly the same for the pump and the signal ($\alpha_s = \alpha_p \equiv \alpha$). Making the transformation $I_j = \omega_j F_j \exp(-\alpha z)$ with $j = s$ or p, we obtain two simpler equations:

$$\frac{dF_s}{dz} = \omega_p g_R F_p F_s e^{-\alpha z}, \qquad \frac{dF_p}{dz} = -\omega_p g_R F_p F_s e^{-\alpha z}. \qquad (6.2.15)$$

Adding them, we deduce $F_p(z) + F_s(z) = C$, where C is a constant (conservation of the photon flux). Using $F_p = C - F_s$ in the F_s equation, we can integrate it to obtain

$$G_s = \frac{F_s(L)}{F_s(0)} = \left(\frac{C - F_s(L)}{C - F_s(0)}\right) \exp(\omega_p g_R C L_{\text{eff}} - \alpha L). \qquad (6.2.16)$$

Using $C = F_p(0) + F_s(0)$ in this equation, the saturated gain can be written as

$$\frac{G_s}{G_u} = \frac{(1 + r_0)e^{-\alpha L}}{r_0 G_u + G_u^{-r_0}}, \qquad (6.2.17)$$

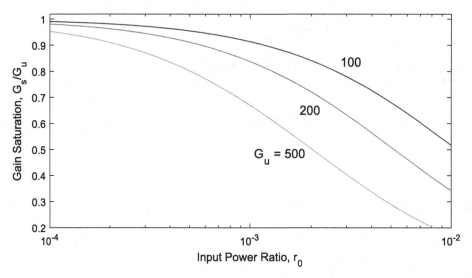

Figure 6.5 Gain-saturation characteristics of Raman amplifiers for several values
of the unsaturated amplifier gain G_u.

where $G_u = \exp(g_R I_0 L_{\text{eff}})$ is the unsaturated gain of the amplifier and r_0 is defined as

$$r_0 = \frac{F_s(0)}{F_p(0)} = \frac{\omega_p}{\omega_s} \frac{I_s(0)}{I_0}. \tag{6.2.18}$$

Figure 6.5 shows the saturation characteristics of a Raman amplifier by plotting G_s/G_u as a function of r_0 for several values of G_u using $\alpha = 0$. Physically, the ratio G_s/G_u represents reduction in the signal's amplification owing to gain saturation. Its value depends on the factor r_0 by which the signal's power is smaller than that of the pump at the input end. In practice, r_0 should be kept below 10^{-3} to avoid gain saturation. Using $\alpha = 0$ and $r_0 \ll 1$ in Eq. (6.2.17), we can write it as

$$\frac{G_s}{G_u} \approx \frac{1}{r_0 G_u + 1}, \tag{6.2.19}$$

It follows that G_s is reduced by a factor of 2 when $r_0 G_u = 1$. From Eq. (6.2.18), gain saturation sets in when the intensity of the amplified signal inside the Raman amplifier becomes a few percent of the pump intensity I_0.

6.2.3 Four-Wave Mixing

FWM is a nonlinear process involving four optical waves. Its origin lies in the third-order nonlinear polarization of the form [18]

$$\mathbf{P}_{\text{NL}} = \epsilon_0 \chi^{(3)} \mathbf{E}\mathbf{E}\mathbf{E}, \tag{6.2.20}$$

where $\chi^{(3)}$ is the third-order susceptibility of the medium. If we consider four CW waves at frequencies ω_1, ω_2, ω_3, and ω_4 polarized linearly along the x axis and treat them as plane waves propagating along the z axis, the total electric field can be written as

$$\mathbf{E} = \frac{1}{2}\hat{x} \sum_{j=1}^{4} E_j \exp[i(\beta_j z - \omega_j t)] + \text{c.c.}, \qquad (6.2.21)$$

where $\beta_j(\omega_j)$ is the propagation constant at the frequency ω_j.

We use Eq. (6.2.21) in Eq. (6.2.20) and express $\mathbf{P}_{\text{NL}}$ in the same form as $\mathbf{E}$ using

$$\mathbf{P}_{\text{NL}} = \frac{1}{2}\hat{x} \sum_{j=1}^{4} P_j \exp[i(\beta_j z - \omega_j t)] + \text{c.c.} \qquad (6.2.22)$$

The P_4 term responsible for generating the fourth wave is given by

$$\begin{aligned} P_4 = \frac{3\epsilon_0}{4}\chi^{(3)}_{xxxx} \big[|E_4|^2 E_4 &+ 2(|E_1|^2 + |E_2|^2 + |E_3|^2)\, E_4 \\ &+ 2E_1 E_2 E_3 \exp(i\theta_+) + 2E_1 E_2 E_3^* \exp(i\theta_-) + \cdots \big], \end{aligned} \qquad (6.2.23)$$

where θ_+ and θ_- are defined as

$$\theta_+ = (\beta_1 + \beta_2 + \beta_3 - \beta_4)z - (\omega_1 + \omega_2 + \omega_3 - \omega_4)t, \qquad (6.2.24)$$
$$\theta_- = (\beta_1 + \beta_2 - \beta_3 - \beta_4)z - (\omega_1 + \omega_2 - \omega_3 - \omega_4)t. \qquad (6.2.25)$$

Consider the practical situation where $E_4 = 0$ initially. The three input waves mix together nonlinearly to produce the last two terms in P_4, which are the FWM terms that create the fourth wave. Which FWM term dominates depends on the phase mismatch governed by θ_+ and θ_-. The θ_+ term creates a wave at the sum frequency $\omega_4 = \omega_1 + \omega_2 + \omega_3$ provided θ_+ nearly vanishes through phase matching. This is hard to realize in optical fibers. It is easier to make θ_- small or zero. This step requires matching of the frequencies as well as of the propagation constants. The frequency matching occurs when

$$\omega_3 + \omega_4 = \omega_1 + \omega_2. \qquad (6.2.26)$$

Physically, two pump photons at frequencies ω_1 and ω_2 are annihilated, while two photons at frequencies ω_3 and ω_4 are created simultaneously. The increase in photon numbers at the signal's frequency represents amplification of the signal. The phase-matching requirement for this process is $\Delta\beta = 0$, where

$$\Delta\beta = \beta(\omega_3) + \beta(\omega_4) - \beta(\omega_1) - \beta(\omega_2). \qquad (6.2.27)$$

The terms containing E_4 in Eq. (6.2.23) are responsible for the SPM and XPM effects (see Section 5.4).

The special case $\omega_1 = \omega_2$ is of practical interest because a single pump beam can be used to amplify the signal at the frequency ω_3, while creating a new wave (called the idler) at the frequency ω_4. In this case, only two optical beams are launched at the input end of the fiber, and the setup shown in Figure 6.4 can be used to amplify the signal. Such amplifiers are called parametric amplifiers. The generation of the idler at a new frequency is also useful for several applications.

Theoretical modeling of the FWM process is complicated because we must deal with the complex amplitudes of waves, rather than their intensities, to account for phase matching. In the general case, one needs to solve a set of four coupled nonlinear equations satisfied by these amplitudes [18]. If we assume that a single pump is employed, which is so intense that it remains undepleted, this set is reduced to the following two linear equations for the signal and idler fields:

$$\frac{dA_3}{dz} = 2i\gamma P_0 e^{-i\kappa z} A_4^*, \tag{6.2.28}$$

$$\frac{dA_4}{dz} = 2i\gamma P_0 e^{-i\kappa z} A_3^*, \tag{6.2.29}$$

where γ is the nonlinear parameter introduced in Section 5.2.3, P_0 is the pump power, and the phase mismatch is given by

$$\kappa = \Delta\beta + 2\gamma P_0. \tag{6.2.30}$$

Notice that A_3 couples to A_4^* and A_4 to A_3^*, that is, the idler wave generated through FWM is a phase-conjugated version of the signal when pump's phase remains relatively constant during the FWM process. This feature makes the idler useful for a variety of applications [8].

Equations (6.2.28) and (6.2.29) can be solved because of their linear nature. If we differentiate A_3 a second time and eliminate A_4^*, we obtain

$$\frac{d^2 A_3}{dz^2} + i\kappa \frac{dA_3}{dz} - (2\gamma P_0)^2 A_3 = 0. \tag{6.2.31}$$

The solution to this equation is given by

$$A_3(z) = [a_3 e^{gz} + b_3 e^{-gz}] \exp(-i\kappa z/2), \tag{6.2.32}$$

where a_3, b_3 are determined from the initial conditions at $z = 0$ and the parametric gain g is defined as

$$g = \sqrt{(2\gamma P_0)^2 - \kappa^2/4}. \tag{6.2.33}$$

In practice, $A_4 = 0$ at the input end at $z = 0$. In this case, a_3 and b_3 can be found in terms of $A_3(0)$ and the solution becomes

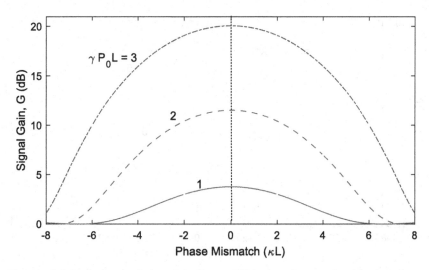

Figure 6.6 Variations of the signal's amplification factor G with the phase-mismatch parameter κL for several values $\gamma P_0 L$. Maximum amplification occurs for $\kappa L = 0$.

$$A_3(z) = A_3(0)[\cosh(gz) + (i\kappa/2g)\sinh(gz)]e^{-i\kappa z/2}. \qquad (6.2.34)$$

The signal power, $P_3 = |A_3|^2$, grows with z as

$$P_3(z) = P_3(0)[1 + (1 + \kappa^2/4g^2)\sinh^2(gz)]. \qquad (6.2.35)$$

For an amplifier of length L, the power gain, $G = P_3(L)/P_3(0)$, is given by

$$G = 1 + (1 + \kappa^2/4g^2)\sinh^2(gL). \qquad (6.2.36)$$

The signal's amplification factor G depends on the parametric gain g, which in turn depends on the phase mismatch κ. Figure 6.6 shows how G varies with κL for several values of the nonlinear phase shift $\gamma P_0 L$. Amplification by a factor of 100 (or 20 dB) occurs for $\gamma P_0 L = 3$ provided κ is close to zero. Maximum amplification occurs for $\kappa = 0$. In this case, the signal's power grows exponentially with z:

$$P_3(z) = P_3(0)\cosh^2(\gamma P_0 z) \approx \frac{1}{4}P_3(0)\exp(2\gamma P_0 z). \qquad (6.2.37)$$

The phase-matching condition in Eq. (6.2.30) requires $\Delta\beta = -2\gamma P_0$, that is, $\Delta\beta$ should be negative. In practice, the pump's wavelength is adjusted to realize phase matching.

We can calculate $\Delta\beta$ from Eq. (6.2.27) by using $\omega_1 = \omega_2 = \omega_p$, $\omega_3 = \omega_p + \Omega$ and $\omega_4 = \omega_p - \Omega$, where Ω is the shift of the signal's frequency from that of the

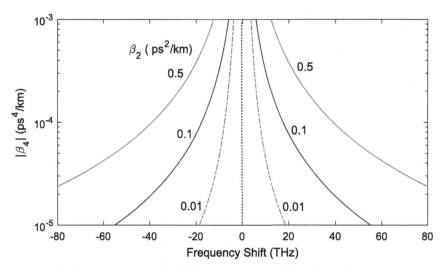

Figure 6.7 Phase matching of the FWM process in a GRIN fiber with $\beta_2 > 0$, while $\beta_4 < 0$ at the pump's wavelength. Frequency shifts from the pump's frequency (dotted vertical line) are plotted as a function of $|\beta_4|$ for several values of β_2.

pump. Expanding $\beta(\omega_3)$ and $\beta(\omega_4)$ around the pump's frequency in a Taylor series, we obtain

$$\Delta\beta = \beta_2\Omega^2 + (\beta_4/12)\Omega^4 + \cdots, \tag{6.2.38}$$

where all odd order terms cancel out. If we keep the dominant first term in this expansion, phase matching requires β_2 at the pump's frequency to be negative (anomalous GVD), and it occurs for values of Ω such that $\Omega = \pm(2\gamma P_0/|\beta_2|)^{1/2}$. These two values correspond to the frequencies of signal and idlers waves involved in the FWM process. As an example, if $\beta_2 = -0.1$ ps^2/km and $\gamma = 0.1$ W^{-1}/km for a GRIN fiber at a pump wavelength near 1.3 μm, the signal can be detuned from the pump by 1 THz when P_0 is close to 20 W.

If β_2 is positive at the pump's wavelength for a GRIN fiber, but β_4 is negative at this wavelength, the phase-matching condition in Eq. (6.2.30) can be satisfied for frequencies shifts such that

$$\Omega^2 = \frac{6}{|\beta_4|}\left(\sqrt{\beta_2^2 + 2|\beta_4|\gamma P_0/3} + \beta_2\right). \tag{6.2.39}$$

Figure 6.7 shows how the frequency shift, $\Omega/(2\pi)$, varies with β_2 and $|\beta_4|$ for $\gamma P_0 = 10$ km^{-1}. The positive and negative values represent the signal and idler sidebands on opposite sides of the pump. Clearly, the fourth-order dispersion affects phase matching considerably. For values of $\beta_2 > 0.1$ ps^2/km, frequency shift can exceed 30 THz, indicating that the signal can be more than 200 nm away from the pump's wavelength.

6.3 Raman Amplifiers and Lasers

Single-mode fibers were used for Raman amplifiers during the 1980s, mainly motivated by their applications in optical communication systems [19, 20]. Such amplifiers are called distributed Raman amplifiers because amplification of a signal occurs over long lengths (exceeding 10 km). When the fiber's length in a Raman amplifier is 1 km or less, it is called a lumped amplifier. We focus in this section on lumped Raman amplifiers made with GRIN fibers.

6.3.1 Raman-Induced Beam Cleanup

An important feature of GRIN Raman amplifiers is a better quality of the amplified beam compared to that of the pump (see Section 5.4.4). To understand this feature, we consider the case in which the CW pump and signal beams excite several modes of the GRIN fiber at the input end (see Section 2.4). A mode-based description of the amplification process provides the most physical insight in such a situation.

Equation (5.4.9) of Section 5.4.1 shows how each mode exited by the pump or the signal evolves inside a nonlinear GRIN medium. This equation can be extended to include the Raman gain that transfers energy from the pump to the signal [21]. If we use the notation A_{js} and A_{kp}, where j and k denote specific modes and p and s stand for the pump and signal waves respectively, the intermodal Raman scattering transfers energy from A_{kp} to A_{js} through the modal gain g_{jk} as the two waves propagate inside the GRIN medium. The important question is how g_{jk} varies from one mode pair to another. It is expected that the modal gain g_{jk} will depend on the spatial overlap of the two modes involved.

The modal approach requires a numerical solution of $2M$ coupled equation of the form given in Eq. (5.4.9), if we assume that both the pump and signal waves excite M spatial modes of a GRIN fiber. Some insight can be gained through Eq. (6.2.14), showing how the signal's power in any mode increases exponentially through the Raman gain. If we neglect fiber's losses over the amplifier's length, the amplification factor for the jth mode can be written as [23]

$$G_j = \exp\left[g_R(P_0/A_{\text{eff}})L \sum_k \overline{f}_{jk} p_k \right], \tag{6.3.1}$$

where A_{eff} is the effective mode area of the fundamental mode, p_k is the fraction of the pump power P_0 in the kth mode, and the overlap factor $\overline{f}_{jk}$ is found from Eq. (5.4.10) to be:

$$\overline{f}_{jk} = f_{jjkk} = A_{\text{eff}} \iint |F_{js}(x,y)|^2 |F_{kp}(x,y)|^2 dx \, dy. \tag{6.3.2}$$

Here $|F_{kp}|^2$ is the spatial profile for the pump in the kth mode. As expected, the Raman gain for any mode pair depends on the overlap between the spatial profiles of the two modes involved.

Even though the signal in all modes is amplified by the pump, the mode for which G_j is the largest would become dominant at the fiber's output end because of the exponential nature of the amplification process. We know from the mode-excitation analysis of Section 2.4.1, relatively large fractions of the pump and signal's powers appear in the fundamental mode. As the overlap factor $\bar{f}_{11}$ is the largest for the LP$_{01}$ mode, G_j in Eq. (6.3.1) has the largest magnitude for $j = 1$. It follows that the signal in the LP$_{01}$ mode will be amplified the most. This is the reason why a large fraction of the signal's power ends up in the fundamental mode of the GRIN fiber at its output end. As a result, spatial quality of the signal beam, quantified in practice through the M^2 factor [24], is better compared to that of the pump beam because of the Raman-induced beam cleanup.

Even when an intense pump beam is launched into a GRIN fiber without any input signal, spontaneous Raman scattering acts as the seed that is amplified by the pump in all modes excited by the pump beam. Again, as G_1 has the largest magnitude in Eq. (6.3.1), the Stokes beam, red-shifted in frequency by 13.2 THz, ends up in the fundamental mode at the end of the GRIN fiber because of the exponential nature of the amplification process. The Raman-induced beam cleanup leads to a Stokes beam with a narrower and smoother spot size compared to that of the pump [22].

When the pump is in the form of a short optical pulse, the effects of dispersion begin to play an important role. As the group velocity in a fiber depends on the wavelength of the pulse, the Raman and Stokes pulses do not travel at the same speed and separate from each other after a distance that depends on the width of pump pulses. In practice, energy transfer between the pump and Stokes pulses is reduced considerably for pump pulses shorter than 10 ps. In fact, no Stokes pulse is generated through SRS when femtosecond pump pulses propagate inside single-mode fibers [18]. Rather, the spectrum of pump pulse undergoes a continuous red-shift through intrapulse Raman scattering (see Section 5.3.3).

6.3.2 Evolution of Pump and Signal Beams

When a large number of a GRIN fiber's modes are excited by the pump and signal beams, it is better to use a non-modal approach and solve the Helmholtz equation (2.1.13) with n^2 given in Eq. (5.1.5). The total electric field inside a Raman amplifier can be written as

$$\tilde{\mathbf{E}} = \hat{\mathbf{p}}E = \hat{\mathbf{p}}[A_p e^{i\beta_p z} + A_s e^{i\beta_s z} e^{i\Omega t}], \qquad (6.3.3)$$

where $\beta_j = n_0(\omega_j)\omega_j/c$ with $j = p, s$ and $\Omega = \omega_p - \omega_s$ is the frequency shift of the signal from the pump (about 13.2 THz for maximum Raman gain). Both waves are assumed to remain linearly polarized along the $\hat{p}$ direction. The pump frequency ω_p is the reference frequency in Eq. (6.3.3).

For Raman amplification to occur, the nonlinear part, $n_2|E|^2$, of the refractive index is modified to include the Raman contribution as shown in Eq. (5.2.30):

$$n_2|E|^2 = (1 - f_R)n_2|E|^2 + f_R n_2 \int_0^\infty h_R(t')|E(t - t')|^2 dt', \qquad (6.3.4)$$

where f_R is the fractional Raman contribution (about 18% for silica glass) and $h_R(t)$ is the Raman response function of the material used to make the GRIN device; it is normalized such that $\int h_R(t) \, dt = 1$. Using Eq. (6.3.3) within the integral of Eq. (6.3.4), we obtain

$$\int_0^\infty h_R(t')|E(t - t')|^2 dt' = |A_p|^2 + |A_s|^2 + A_p^* A_s e^{i\delta\beta z} \tilde{h}_R(\Omega) + A_p A_s^* e^{-i\delta\beta z} \tilde{h}_R^*(\Omega),$$

$$(6.3.5)$$

where $\delta\beta = \beta_s - \beta_p$ and $\tilde{h}_R(\Omega)$ is the Fourier transform of $h_R(t)$. The imaginary part of $\tilde{h}_R(\Omega)$ is related to the Raman gain at the signal's frequency as $g_R = 2f_R(\omega_s/c)\text{Im}(h_R)$.

We use Eqs. (6.3.3) through (6.3.5) in Eq. (2.1.13), make the paraxial approximation, retain only the phase-matched terms, and separate the pump and signal terms to obtain the following two propagation equations:

$$\frac{\partial A_p}{\partial z} + \frac{\nabla_T^2 A_p}{2i\beta_p} + \frac{i}{2}\beta_p b^2 \rho^2 A_p = \frac{i\omega_p}{c} n_2(|A_p|^2 + 2|A_s|^2)A_p - \frac{\omega_p}{2\omega_s} g_R |A_s|^2 A_p - \alpha_p A_p,$$

$$(6.3.6)$$

$$\frac{\partial A_s}{\partial z} + \frac{\nabla_T^2 A_s}{2i\beta_s} + \frac{i}{2}\beta_s b^2 \rho^2 A_s = \frac{i\omega_s}{c} n_2(|A_s|^2 + 2|A_p|^2)A_s + \frac{1}{2} g_R |A_p|^2 A_s - \alpha_s A_s, \quad (6.3.7)$$

where the loss of the GRIN device has also been added through the last term. These equations include all linear and nonlinear effects for a GRIN Raman amplifier operating under the CW conditions. In general, they must be solved numerically if one wants to include the diffractive and self-imaging effects together with the depletion of the pump and the resulting gain saturation.

An analytic approach can be used if we focus on the intensities of the pump and signal beams, $I_p = |A_p|^2$ and $I_s = |A_s|^2$. It is easy to show that they satisfy the following simpler equations:

$$\frac{\partial I_p}{\partial z} = 2\text{Re}\left(A_p^* \frac{\partial A_p}{\partial z}\right) = -\frac{\omega_p}{\omega_s} g_R I_p I_s - \alpha_p I_p, \qquad (6.3.8)$$

$$\frac{\partial I_s}{\partial z} = 2\text{Re}\left(A_s^* \frac{\partial A_s}{\partial z}\right) = g_R I_p I_s - \alpha_s I_s. \qquad (6.3.9)$$

These equations appear identical to Eqs. (6.2.10) and (6.2.11) with one major difference: They contain partial derivatives because both I_p and I_s depend on the transverse coordinate ρ, in addition to being z dependent. As before, they can be simplified considerably when the pump remains much more intense than the signal over the entire length L of the Raman amplifier. In that situation, depletion of the pump can be ignored in Eq. (6.3.8).

If we further assume that both beams are initially in the form of Gaussian beams whose intensities depend on the radial coordinate ρ, the solution of Eq. (6.3.8) can be written as $I_p(\rho, z) = I_0 S_p(\rho, z) e^{-\alpha_p z}$, where I_0 is the input peak intensity at $\rho = 0$ of the pump beam, whose spatial profile evolves with z as

$$S_p(\rho, z) = \frac{1}{f_p} \exp\left[-\frac{\rho^2}{w_p^2 f_p}\right], \qquad f_p(z) = 1 - (1 - C_p^2)\sin^2(2\pi z/L_p), \quad (6.3.10)$$

where $C_p = (w_g/w_p)^2$, w_p is the pump's width at $z = 0$, L_p is the self-imaging period of the GRIN medium, and w_g is the spot size of the fundamental mode. This form of S_p corresponds to the solution obtained in Eq. (2.5.42) in the absence of the nonlinear terms, with $f_p(z)$ governing periodic self-imaging of the pump inside a GRIN medium. Using it in Eq. (6.3.9) and integrating over z, we obtain

$$I_s(\rho, z) = I_s(\rho, 0) e^{-\alpha_s z} G_s(\rho, z), \qquad G_s(\rho, z) = \exp\left[g_R I_0 \int_0^z S_p(\rho, z) e^{-\alpha_p z} \, dz\right],$$

$$(6.3.11)$$

where $I_s(\rho, 0)$ is the initial profile the signal beam at $z = 0$.

The preceding equation has a simple physical interpretation. The signal's input intensity is amplified by a Raman-amplification factor $G_s(\rho, z)$, whose magnitude is different for different radial distances across the transverse intensity profile of the signal beam. In practice, the central part of the signal beam near $\rho = 0$ is amplified more than the outer parts in its wings. As a result, a Raman amplifier reduces spatial width of a signal beam as it amplifies it, resulting in the Raman-induced beam cleanup. In the modal picture, a narrower beam contains a larger fraction of its power in the fundamental mode of the GRIN fiber. For a sufficiently long GRIN Raman amplifier, the amplified signal would be in the fundamental mode at the output end of the GRIN fiber.

6.3.3 Raman-Induced Narrowing of Signal Beam

It is possible to obtain an approximate equation governing changes in the signal beam's width. For this purpose, we solve the amplitude equations (6.3.6) and (6.3.7) with the assumption that the pump beam remains much more intense than the signal over the entire length L of a Raman amplifier. If we also neglect losses for simplicity, we obtain

$$\frac{\partial A_p}{\partial z} + \frac{\nabla_T^2 A_p}{2i\beta_p} + \frac{i}{2}\beta_p b^2 \rho^2 A_p = \frac{i\omega_p}{c} n_2 |A_p|^2 A_p, \tag{6.3.12}$$

$$\frac{\partial A_s}{\partial z} + \frac{\nabla_T^2 A_s}{2i\beta_s} + \frac{i}{2}\beta_s b^2 \rho^2 A_s = \frac{2i\omega_s}{c} n_2 |A_p|^2 A_s + \tfrac{1}{2} g_R |A_p|^2 A_s, \tag{6.3.13}$$

For a CW pump in the form of a Gaussian beam, the solution of Eq. (6.3.12) has been discussed in Section 5.1.2. When the pump power is below the self-focusing threshold, the solution takes the form

$$A_p(\rho, z) = \sqrt{I_0} F_p(\rho, z), \quad F_p(\rho, z) = \sqrt{\frac{1}{f_p}} \exp\left[-\frac{\rho^2}{2w_p^2 f_p} + i\phi_p(\rho, z)\right], \tag{6.3.14}$$

where I_0 is the peak intensity and w_p is the width of the pump beam at $z = 0$. If we ignore the XPM-induced phase shift in Eq. (6.3.13), we need to solve the following linear equation for the signal field:

$$\frac{\partial A_s}{\partial z} + \frac{\nabla_T^2 A_s}{2i\beta_s} + \frac{i}{2}\beta_s b^2 \rho^2 A_s = \tfrac{1}{2} g_R |A_p|^2 A_s. \tag{6.3.15}$$

We can solve Eq. (6.3.15) approximately using the techniques discussed in Section 2.5. We write its solution in a self-similar form as

$$A_s(\rho, z) = A_0 \exp\left(-\frac{\rho^2}{2w^2} + \frac{ikh}{2}\rho^2 + i\psi\right), \tag{6.3.16}$$

where the four parameters (A_0, w, h, and ψ) vary with z. We use this form in Eq. (6.3.15) and approximate $|A_p|^2$ as

$$|A_p(\rho, z)|^2 \approx \frac{I_0}{f_p}\left[1 - \frac{\rho^2}{w_p^2 f_p}\right]. \tag{6.3.17}$$

Equating the real and imaginary parts on the two sides of terms containing different powers of ρ, we obtain the following four first-order differential equations:

$$\frac{dA_0}{dz} = \frac{g_R I_0}{2f_p} A_0 - hA_0, \qquad \frac{d\psi}{dz} = -\frac{1}{\beta_s w^2}, \tag{6.3.18}$$

$$\frac{dw}{dz} = hw - \frac{g_R I_0}{2w_p^2 f_p^2} w^3, \qquad \frac{dh}{dz} = \frac{1}{\beta_s^2 w^4} - h^2 - b^2. \tag{6.3.19}$$

In the absence of the Raman gain, the preceding equations reduce to those obtained in Section 2.5.4. The presence of the Raman gain modifies equations for both A_0 and w. The amplitude in Eq. (6.3.18) is expected to increase with z because of Raman gain. However, as seen from Eq. (6.3.19), the Raman gain also affects the width of the signal beam. We combine the two equations in Eq. (6.3.19) to obtain a single equation for the width w. Using $s = w/w_s$ and $\xi = bz$, the normalized form of this equation is

$$\frac{d^2s}{d\xi^2} + s = \frac{C_s^2}{s^3} - \mu \frac{d}{d\xi}\left(\frac{s^3}{f_p^2}\right), \qquad \mu = \frac{g_R I_0 w_s^2}{2bw_p^2}, \qquad (6.3.20)$$

where $C_s = (w_g/w_s)^2$ and w_s is the initial width of the signal being amplified. The last term represents the effects of the Raman gain and depends on both the intensity and the width of the pump beam inside the GRIN medium.

The solution of Eq. (6.3.20) is known when $\mu = 0$ and is given by

$$s(\xi) = \sqrt{f_s(\xi)}, \qquad f_s(\xi) = 1 - (1 - C_s^2)\sin^2(\xi). \qquad (6.3.21)$$

The solution is oscillatory because of the GRIN-induced periodic focusing of the signal beam. The Raman term affects the solution over a much longer length scale compared to the period of self-imaging. However, this term depends on $f_p^2(\xi)$ and varies rapidly with ξ. To find its impact approximately, we replace $f_p^2(\xi)$ with its average value and write Eq. (6.3.20) as

$$\frac{d^2s}{d\xi^2} + \frac{3\mu s^2}{\langle f_p^2 \rangle}\frac{ds}{d\xi} + s = \frac{C_s^2}{s^3}, \qquad (6.3.22)$$

where the angle brackets denote average over one self-imaging period. To simplify it further, we replace s^2 with $\langle f_s \rangle$ and neglect the diffraction term containing C_s^2. This results in a linear equation corresponding to a damped harmonic oscillator:

$$\frac{d^2s}{d\xi^2} + 2\gamma_d\frac{ds}{d\xi} + s = 0, \qquad \gamma_d = \frac{3\mu\langle f_s \rangle}{2\langle f_p^2 \rangle}. \qquad (6.3.23)$$

The solution of Eq. (6.3.23) for an initially collimated beam is given by $s(\xi) = \exp(-\gamma_d\xi)\cos\xi$. We can include the diffraction effects by writing the approximate solution of Eq. (6.3.22) in terms of the original variables as

$$w(z) = w_s\sqrt{f_s(z)}\exp(-\gamma_d bz). \qquad (6.3.24)$$

It shows that the amplitude of width oscillations decreases with distance inside a GRIN Raman amplifier on a length scale governed by the decay rate γ_d. This scale depends on the input parameters as

$$L_d = \frac{1}{\gamma_d b} = \frac{2w_p^2\langle f_p^2 \rangle}{3g_R I_0 w_s^2\langle f_s \rangle}. \qquad (6.3.25)$$

The two averages in this equation can be done analytically and are found to be

$$\langle f_s \rangle = \frac{1}{2}(1 + C_s^2), \qquad \langle f_p^2 \rangle = C_p^2 + \frac{3}{8}(1 - C_p^2)^2. \qquad (6.3.26)$$

The length L_d depends on the peak intensity of the pump beam as well as on the widths of the pump and signal beams at the input end. As expected, L_d is relatively long (> 100 m) compared to the self-imaging period (about 1 mm) for typical values of input parameters.

6.3.4 Experimental Results

GRIN fibers have been used in several experiments as Raman amplifiers and lasers. In a 1988 experiment, a 50-m-long GRIN fiber with a core diameter of 38 μm was pumped with 150-ps pulses at 1064 nm [25]. No signal beam was launched at the input end of the fiber in this experiment. Rather, cascaded SRS generated multiple Stokes from noise that seeded the Raman amplifier. The output spectrum exhibited multiple spectral bands extending up to 1700 nm, each belonging to a different Stokes pulse and red-shifted by about 13.2 THz from the preceding band. Pulses whose wavelengths exceeded 1300 nm were narrower than the pump pulse because they experienced anomalous GVD and formed solitons.

In a 1992 experiment, a 30-m-long standard GRIN fiber (core diameter 50 μm) was pumped at 590 nm using 3-ns pulses, without any signal at the input end [23]. The new feature of this experiment was that the four Stokes pulses, generated through cascaded SRS, had most of their energy in the fundamental mode at the output of the fiber, even though energy of the input pump pulse was distributed among several low-order modes of the GRIN fiber. This feature is attributed to the Raman-induced beam cleanup [22] and has been observed in several other experiments [26–28].

In a 2013 experiment, multiple Stokes beams were generated through cascaded SRS covering a wide wavelength range by launching 8-ns pulses at 523 nm into a 1-km-long GRIN fiber with 50-μm core diameter [28]. Energy of pump pulses was distributed over several modes of the fiber at the input end. The Stokes beams generated within the fiber were much narrower in size compared to the pump beam at the output end of the fiber. Figure 6.8 shows three spatial profiles obtained with a beam profiler. The speckle pattern in part (a) was obtained without any optical filter. It is dominated by the pump beam and its shape results from interference and coupling among the modes excited by the pump (see Section 2.4). The spatial patterns in parts (b) and (c) were obtained when a single Stokes beam was selected using optical filters centered at 610 and 700 nm, respectively. A narrow spot size in part (b) resembles the shape of the fundamental mode of the GRIN fiber and shows a clear evidence of Raman-induced beam cleanup.

In a 2004 experiment, Raman-induced beam cleanup was used to advantage for making a Raman laser using a 40-m-long GRIN fiber with 50-μm core diameter [29]. The laser was pumped at 1064 nm. Two fiber Bragg gratings at two ends of the

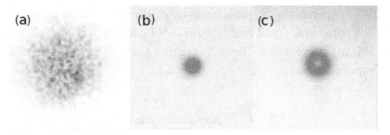

Figure 6.8 Spatial profiles measured using a CCD-based beam profiler. Image (a) was obtained without any optical filter. The other two images were obtained using optical filters centered at (b) 610 nm and (c) 700 nm. (After Ref. [28]; ©2013 AIP.)

GRIN fiber reflected radiation at 1116 nm and served as mirrors of the laser's cavity. The Raman laser reached its threshold when the pump power was about 3 W. The measured spot size of the beam emitted by the Raman laser was much smaller than that of the pump beam because the Stokes wave was confined to the fundamental mode of the GRIN fiber. In effect, the laser operated in a single mode, even though it was pumped in a multimode fashion. More recently, GRIN Raman lasers have been developed that are pumped with semiconductor or fiber lasers and provide large output powers with high beam quality in an efficient manner [30–33].

In a 2013 experiment, a 4.5-km-long GRIN fiber was pumped with a high-power semiconductor laser operating at 938 nm [30]. It emitted more than 2 W of power at the 980 nm wavelength when pumped with 45 W of pump power. A much higher efficiency was realized in a 2015 experiment in which pumping employed two semiconductor lasers operating at different wavelengths near 975 nm. Figure 6.9 shows schematically how the outputs of two lasers were combined before launching the pump beam into a GRIN fiber [31]. The laser's cavity is formed by two dichroic mirrors, designed to reflect only the Stokes beam at the wavelength of 1019 nm. This wavelength corresponds to a pump-Stokes frequency difference of 13.2 THz.

Figure 6.10 shows the output power as a function of pump power for two GRIN fibers of different lengths. In both cases, the laser reaches its threshold at a specific pump power. Beyond this threshold, emitted power grows linearly with increasing pump power. The slope of the resulting line is a measure of the laser's efficiency. The maximum output power for the 3-km-long Raman laser was close to 15 W at a pump power of 40 W. As the laser reached its threshold at 16 W, slope efficiency of this laser is 62%, in agreement with theoretical predictions. The slope efficiency could be increased to 81% by reducing the GRIN fiber's length to 1.5 km. Even though the threshold power increased to 20 W for the shorter laser, its output power reached 20 W at a pump power near 40 W. The reason for more efficient operation is related to lower cavity losses for the shorter GRIN fiber. Even more efficient

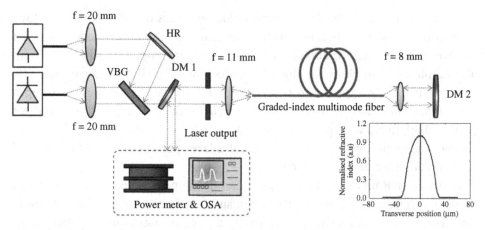

Figure 6.9 Experimental setup of a GRIN-fiber Raman laser pumped using two semiconductor lasers. VBG, volume Bragg grating; HR, high-reflectivity mirror; DM, dichroic mirror; OSA, optical spectrum analyzer. (After Ref. [31]; ©2015 CCBY.)

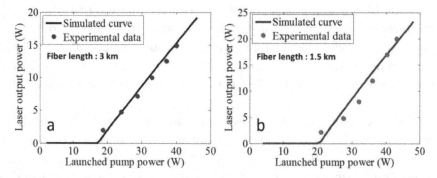

Figure 6.10 Output power emitted by a Raman laser as its pump power was increased. Length of the GRIN fiber was (a) 3 km or (b) 1.5 km. In both cases, experimental data (circles) is compared to theoretical predictions. (After Ref. [31]; ©2015 CCBY.)

operation of a Raman laser was realized in 2018 by using just 200 m of GRIN fiber [34]. The laser emitted 135 W of CW power at a wavelength of 1081 nm, when pumped at 1030 nm with 200 W of power obtained from four fiber lasers. By 2021, the highest output power of 443 W was obtained when a Raman laser was pumped with 665 W of pump power [35]. The beam quality, judged by the M^2 factor of 3.5, was somewhat degraded at such high power levels ($M^2 = 1$ for a perfect Gaussian beam).

Fiber-based Raman lasers can also generate optical pulses through mode locking. An active mode-locking technique was used in 2015 for a laser with a 500-m-long cavity containing a single-mode fiber, producing pulses of about 2-ns width [36]. Shorter pulses of 50 ps duration were generated in a later experiment in which a passive technique was combined with active mode-locking [37]. This experiment also employed a single-mode fiber. Mode locking of a GRIN Raman laser was realized in a 2022 experiment in which the laser's cavity was formed using a fiber grating at each end of a 3.7-km-long GRIN fiber. The use of a modulator produced relatively wide pulses (5-7 ns) at a repetition rate of 27.2 kHz [38]. As a semiconductor laser was used for pumping this laser, pulses could also be generated by modulating the pump laser's current.

GRIN-fiber Raman amplifiers have been developed in recent years that can provide even more output power than a Raman laser, when pumped with high-power fiber lasers. Raman laser requires a cavity and often makes use of fiber Bragg gratings, which can degrade at high power levels. This is not an issue for amplifiers as no cavity is involved. In a 2018 experiment, a 100-m-long, double-clad, GRIN fiber with the core diameter of 62.5 μm was pumped at 1018-nm using five Yb-doped fiber lasers capable of providing up to 800 W of CW power [39]. The input power of the 1060-nm signal was 25.6 W. The output power was 654 W at the input pump power of 766 W (conversion efficiency of 85.3%). The quality of the beam was quantified through the M^2 factor, whose measured value was 4.2. Within a year, the output power of the Raman amplifier could be increased to near 1 kW by optimizing length of the GRIN fiber that was used to make the Raman amplifier [40].

Further development of GRIN-fiber Raman amplifiers has led to considerable advances [41–45]. Figure 6.11 shows the setup of a 2020 experiment [41], in which a 20-m-long piece of GRIN fiber was employed to ensure suppression of the second-order Stokes inside the amplifier (amplified signal acing as a pump). The input signal with 135 W power was provided by a seed laser operating at 1130 nm. Six Yb-doped fiber lasers, operating at 1080 nm, were used for pumping the amplifier and were capable of providing up to 2400 W of power. The outputs of pump lasers were combined with the signal using a beam combiner, before injecting the combination into the GRIN fiber.

Figure 6.12 shows the output power of the Raman amplifier as the pump power is increased from 0 to 2375 W. As seen there, it exceeds 2 kW at the highest pump power, resulting in a conversion efficiency of almost 80%. The residual pump power increases initially before decreasing to 341 W at the maximum power output. The second-order Stokes created by the amplified signal was negligible up to the signal's power level of 1980 W. Such a high threshold was possible because of the relatively short length of the GRIN fiber (20 m) used to make the amplifier. The quality of the amplified beam was quantified by measuring its M^2 factor with a beam-quality analyzer. The measured value of M^2 varied with the pump power and exceeded 3 at

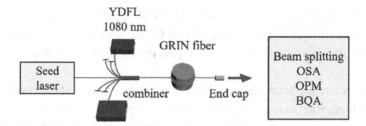

Figure 6.11 Setup of a GRIN-fiber Raman amplifier providing more than 2 kW output power. Outputs from the seed laser and six Yb-doped fiber lasers (YDFL) are combined before injecting into the 20-m-long GRIN fiber. OSA, optical spectrum analyzer; OPM, optical power meter; BQA, beam quality analyzer. (After Ref. [41]; ©2020 CCBY.)

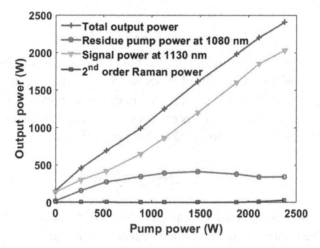

Figure 6.12 Measured output powers of the signal, the residual pump, and the second-order Stokes as a function of the input pump power launched into the GRIN fiber. Total output power is also shown. (After Ref. [41]; ©2020 CC BY.)

pump-power levels below 800 W. However, its value was close to 2.8 at higher power levels, indicating an acceptable beam quality. By 2021, the maximum output power of a Raman amplifier exceeded 3 kW, with a conversion efficiency of 78.7% [42]. The measured M^2 factor was 5.7 at this power level but became 3.1 when the power was reduced to near 2.4 W.

6.4 Parametric Amplifiers

As we discussed in Section 6.2.2, FWM inside a single-mode fiber has been used for making parametric amplifiers that amplify an input signal, while also generating a new wave known as the idler. Such amplifiers were pursued in the 1970s [46] and have remained of interest since then [47]. A GRIN fiber was first used in 1981 for FWM [48]. In this section, we focus on parametric amplifiers made with GRIN fibers.

A major difference between the Raman and parametric amplifiers is related to the phase matching. Raman amplifiers have no such requirement. In contrast, FWM can occur efficiently only if the phase-matching condition in Eq. (6.2.30) is satisfied, where $\Delta\beta$ is given in Eq. (6.2.27). In the single-pump case, two photons at the frequency ω_p create two photons at the signal and idler frequencies (ω_s and ω_i). If we ignore the nonlinear contribution, the phase-matching condition takes the form

$$\Delta\beta = \beta(\omega_s) + \beta(\omega_i) - 2(\omega_p) = 0. \tag{6.4.1}$$

6.4.1 Intermodal Four-Wave Mixing

There are several ways that the phase-matching condition in Eq. (6.4.1) can be satisfied. Intermodal FWM, discussed in Section 5.4.3, is the special case in which the pump, signal, and idler waves propagate in different modes of the same fiber. The phase-matching condition in this case is given in Eq. (5.4.26). Recalling that the self-imaging period is related to the GRIN parameter b as $L_p = 2\pi/b$, it can be written as

$$\beta_2\Omega^2 + \frac{\beta_4}{12}\Omega^4 = \frac{2\pi m_g}{L_p}, \tag{6.4.2}$$

where $\Omega = \omega_s - \omega_p$ is the relative frequency shift and β_2 and β_4 are the dispersion parameters. The solution of the preceding quadratic equation is

$$\Omega^2 = \frac{6}{\beta_4}\left[-\beta_2 + \left(\beta_2^2 + m_g\frac{2\pi\beta_4}{3L_p}\right)^{1/2}\right], \tag{6.4.3}$$

where the integer $m_g = g_s + g_i - 2g_p$ with g_j denoting the mode group for jth wave. Recall that m_g is restricted to take only even integer values because of a constraint set by the conservation of angular momentum [49]. Even with this restriction, FWM can occur for several mode combinations when the pump propagates in the LP_{01} mode ($g_p = 1$). For example, $m_g = 2$ for the two (g_s, g_i) combinations (1,3) and (3,1). As an example, the idler is generated in the LP_{02} mode when the pump and signal are launched into the LP_{01} mode of a GRIN fiber.

The sideband frequencies found in the context of modulation instability (see Section 5.3) coincide with the preceding Ω values when we use $m_g = 2m$ because the phenomenon of self-imaging is behind both processes. However, the analysis in Section 5.3.1 assumed that the pump beam excites many modes of the GRIN fiber simultaneously. The intermodal FWM requires that the pump propagate in a

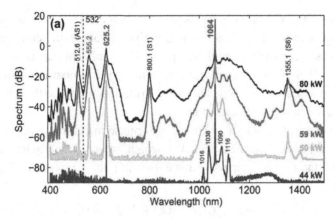

Figure 6.13 Measured spectra at the end of a 1-m-long GRIN fiber at four peak power levels of 1064-nm input pulses. The 625-nm peak acts as a pump at high power levels and creates its own FWM sidebands that extend in the UV region. (After Ref. [51]; ©2017 OSA.)

single mode. In a 2017 experiment [51], a short, 1-m-long, GRIN fiber was used to avoid random mode coupling. Its core diameter of 22 μm was chosen to ensure that it supported only a few mode groups. The pump was at 1064 nm in the form of a 400-ps pulse train, and its spatial size was controlled to ensure that it excited mostly the LP_{01} mode. As no signal was launched, noise from spontaneous FWM acted as the seed. Figure 6.13 shows the observed output spectra at four peak power levels ranging from 44 to 80 kW. As seen there, intermodal FWM generates many sidebands, covering a wide spectral range from 400 to 1600 nm. The four sidebands close to the pump result from nondegenerate FWM (pump in two different modes) and are affected by SRS.

An interesting feature of Figure 6.13 is the narrow spectral line at 625 nm at the 44-kW pump level. This line is due to intermodal FWM, and its frequency shift of 181 THz corresponds to the choice $m_g = 2$ in Eq. (6.4.3); its partner sideband lies in the mid-infrared region beyond 2.5 μm. As the pump power is increased, the 625-nm feature broadens, becomes more intense, and eventually acts as a pump to generate its own set of FWM sidebands. Their frequencies can also be predicted from Eq. (6.4.3) by using $g_p = 1$ because this wave was generated in the LP_{01} mode. The beam profiles recorded at 433, 473, and 625 nm agreed with this interpretation [51].

6.4.2 Phase Matching by the Kerr Nonlinearity

As discussed in Section 5.3.1, the nonlinear phenomenon of SPM, induced by the Kerr nonlinearity, leads to modulation instability in GRIN fibers. This instability can be interpreted as a FWM process that is phase matched by SPM and seeded by noise [18]. This view can be understood by noting that the phase-matching condition in Eq. (6.2.30) has the linear and nonlinear parts. The nonlinear part is always positive for silica and other glasses because the parameter γ has positive sign for $n_2 > 0$. but the linear part $\Delta\beta$ can be positive or negative, depending on the nature of GVD of the fiber at the pump's wavelength.

In the case of single-mode fibers, nonlinear phase matching requires anomalous GVD. This can be seen from Eq. (6.2.38), which provides the Taylor expansion of $\Delta\beta$. Retaining the first two terms in this expansion, the phase-matching condition becomes

$$\beta_2\Omega^2 + (\beta_4/12)\Omega^4 + 2\gamma P_0 = 0, \qquad (6.4.4)$$

where $\Omega = \omega_s - \omega_p$. If we neglect the β_4 term in this equation, the phase-matching condition can be satisfied only when the GVD is anomalous at the pump's wavelength ($\beta_2 < 0$). Assuming that to be the case, phase matching occurs when the signal's frequency is shifted from the pump's frequency such that

$$\Omega = \pm\sqrt{2\gamma P_0/|\beta_2|}. \qquad (6.4.5)$$

The situation changes considerably for multimode GRIN fibers because phase matching in such a fiber can occur even when the pump propagates in the normal-GVD region of the fiber. It was found in Section 5.3.1 that this feature results from the phenomenon of self-imaging. More specifically, periodic focusing of the pump beam creates a nonlinear index grating along the length of the GRIN fiber with the period $L_p/2$ that modifies the phase-matching condition in Eq. (6.4.4) as

$$\beta_2\Omega^2 + (\beta_4/12)\Omega^4 + 2\langle\gamma\rangle P_0 = mK_g, \qquad (6.4.6)$$

where $K_g = 4\pi/L_p$ and m can be any integer including zero. As γ varies in a periodic fashion because of self-imaging, the average value of the nonlinear parameter appears in this equation. Recall from Section 5.3.1 that periodic focusing of a beam enhances local value of γ by reducing its effective cross-section area. Averaging γ over the self-imaging period, we obtain $\langle\gamma\rangle = \gamma/C_f$, where C_f is the factor by which the pump beam is compressed during each cycle. If we neglect the small β_4 term in Eq. (6.4.6), its solution for each value of m is given by

$$\Omega_m = \pm\left[\frac{4\pi m}{\beta_2 L_p} - \frac{2\gamma P_0}{\beta_2 C_f}\right]^{1/2}. \qquad (6.4.7)$$

These frequencies are identical to those in Eq. (5.3.7), obtained in the context of modulation instability.

For each value of the integer m, Eq. (6.4.7) provides two frequencies that correspond to the frequencies of the signal and idler waves. It is common to refer them as

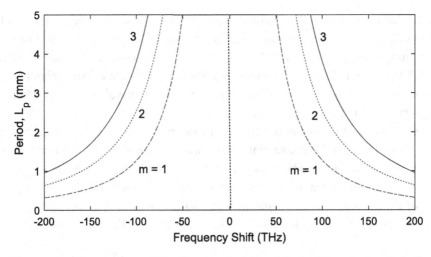

Figure 6.14 Dependence of the frequency shift on the self-imaging period of a GRIN fiber L_p when FWM occurs in its normal-GVD region ($\beta_2 = 25$ ps^2/km). For each value of the integer m, phase matching occurs for two frequencies located on opposite sides of the pump's frequency (dotted vertical line).

a pair of sidebands located symmetrically around the pump's frequency. The closest pair occurs for $m = 0$, and it exists only when the GVD is anomalous at the pump wavelength ($\beta_2 < 0$). However, other pairs can form for $m > 0$ even when the GVD is normal ($\beta_2 > 0$). Recalling that L_p is <1 mm for most GRIN fibers, the first term in Eq. (6.4.7) dominates over the second term. As a result, the frequencies can be estimated for $\beta_2 > 0$ using the simple relation given in Eq. (5.3.9). It follows that phase-matched FWM can be realized in the normal-GVD region of a GRIN medium by launching a signal at frequencies such that $\omega_s = \omega_p + \Omega_m$. There are two frequencies for each value of m, located on opposite sides of the pump, whose values depend mainly on the GVD parameter β_2 and the self-imaging period L_p.

Figure 6.14 shows how the frequency shift varies with the self-imaging period L_p for several values of the integer m using $\beta_2 = 25$ ps^2/km, a value suitable for pump wavelengths near 1 μm. The contribution of the second term in Eq. (6.4.7) was included using $C_f = 0.2$ and $L_{NL} = (\gamma P_0)^{-1} = 1$ m. For $m = 1$, the shift from the pump frequency is about 100 THz for $L_p = 1$ mm, and this shift increases with m by a factor of $\sqrt{m}$. If a signal is launched at any of the phase-matched frequencies together with the pump, it will be amplified inside the GRIN fiber, and an idler will be generated at a frequency set by Eq. (6.4.7) on the opposite side of the pump's wavelength. The phase-matched signal and idler frequencies depend on the pump power so weakly that they remain nearly unchanged.

In contrast, if the pump wavelength is close to 1550 nm, where $\beta_2 \approx -20$ ps^2/km, the $m = 0$ sidebands also form; the nonlinear index grating plays a minor role in their formation. For $m = 0$, the phase-matched signal and idler frequencies in Eq. (6.4.7) scale with pump power as $\sqrt{P_0}$, and the signal's frequency is closer to that of the pump because Ω_0 obtained from Eq. (6.4.7) is considerably smaller. Using $\gamma = 0.1$ W^{-1}/km and $C_f = 0.2$ as typical values, the frequency shift is about 1 THz (or 8 nm in wavelength) at 1 kW pump-power level. This value doubles when the pump's power is increased by a factor of four.

As we have seen earlier in the context of intermodal FWM, the small β_4 term in Eq. (6.4.6) affects the phase-matching condition considerably and can lead to qualitative changes [50]. The reason is that it converts Eq. (6.4.6) into a quartic equation for Ω, which can provide up to four phase-matched frequencies for negative value of β_4, in sharp contrast with only two values indicated in Eq. (6.4.7). If we neglect the small P_0 term, the four values are found from Eq. (6.4.6) to be

$$\Omega^2 = \frac{6}{\beta_4}\left[-\beta_2 \pm \left(\beta_2^2 + m\frac{2\pi\beta_4}{3L_p}\right)^{1/2}\right]. \tag{6.4.8}$$

This equation is identical to Eq. (6.4.3) with m playing the role of m_g. Figure 6.15 shows how the frequency shift varies with L_p for several values of the integer m using $\beta_2 = 25$ ps^2/km and $\beta_4 = -5 \times 10^{-5}$ ps^2/km. It should be compared with Figure 6.14 obtained under identical conditions but with β_4 set to zero. The qualitative behavior is quite different for $\beta_4 < 0$. Only two solutions exist for small values of L_p. A bifurcation occurs as L_p increases, and the number of phase-matched solutions changes from two to four.

In a 2018 experiment, a 1-m-long, GRIN fiber, whose 50-μm-radius core supported a large number of modes, was used to observe seeded FWM [52]. The index-gradient parameter, $b = \sqrt{2\Delta/a}$, was 14 mm^{-1} for this fiber with $\Delta = 0.024$, resulting in the self-imaging period $L_p = 0.4$ mm. The 1064-nm pump beam was in the form of a 400-ps pulse train with peak powers as high as 70 kW, and its spatial width was controlled to ensure that pump's power coupled mostly into the LP$_{01}$ mode. The signal was at 1555 nm, a wavelength that corresponds to the choice $m = 1$ in Eq. (6.4.7), and its power was much smaller than that of the pump (about 0.1 W). The measured spectra for different pump-power levels extended from 400 to beyond 1700 nm with many peaks and were similar to those seen in Figure 6.13. The signal peak is at 1555 nm and the idler peak located near 800 nm became stronger as the pump power was increased. The other peaks represented spontaneous FWM at phase-matched frequencies with $m > 1$ in Eq. (6.4.8). At high pump-power levels, cascaded FWM was observed for which the signal itself acted as the pump and produced additional spectral bands.

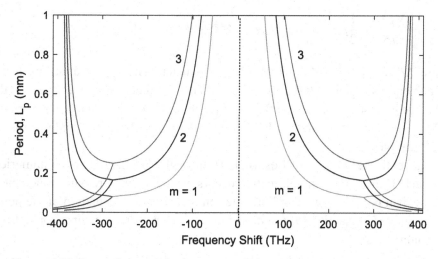

Figure 6.15 Dependence of the frequency shift on the self-imaging period of a GRIN fiber L_p for FWM to occur when the β_4 term is included in the phase-matching condition. For each value of m, phase matching occurs for up to four frequencies.

6.4.3 FWM Theory for GRIN Fibers

The phase-matching condition can only be used to find the appropriate signal and pump wavelengths for which FWM occurs in an efficient manner. To find how the two waves evolve inside a GRIN fiber, we must solve the Helmholtz equation (2.1.13) with n^2 given in Eq. (5.1.5). The total electric field inside a parametric amplifier consists of the pump, signal, and idler waves oscillating at different frequencies:

$$\tilde{E} = \hat{p}[A_p e^{i\beta_p z} + A_s e^{i\beta_s z - i\Omega t} + A_i e^{i\beta_i z + i\Omega t}], \tag{6.4.9}$$

where $\beta_j = n_0(\omega_j)\omega_j/c$ with $j = p, s, i$ and $\Omega = \omega_s - \omega_p = \omega_p - \omega_i$. All three waves are assumed to be identically polarized along the $\hat{p}$ direction. The pump frequency ω_p is the reference frequency in Eq. (6.4.9).

When we substitute $\tilde{E}$ from Eq. (6.4.9) into Eq. (2.1.13), the nonlinear term, $n_2|E|^2 E$, produces many terms oscillating at different frequencies. Retaining only the terms oscillating at the original frequencies, we obtain the following three coupled propagation equations for the pump, signal, and idler waves [18]:

$$\frac{\partial A_p}{\partial z} + \frac{\nabla_T^2 A_p}{2i\beta_p} + \frac{i}{2}\beta_p b^2 \rho^2 A_p = i k_0 n_2[(|A_p|^2 + 2|A_s|^2 + 2|A_i|^2)A_p + 2A_s A_i A_p^* e^{i\Delta\beta z}], \tag{6.4.10}$$

$$\frac{\partial A_s}{\partial z} + \frac{\nabla_T^2 A_s}{2i\beta_s} + \frac{i}{2}\beta_s b^2 \rho^2 A_s = i k_0 n_2[(|A_s|^2 + 2|A_p|^2 + 2|A_i|^2)A_s + 2A_p^2 A_i^* e^{-i\Delta\beta z}], \tag{6.4.11}$$

$$\frac{\partial A_i}{\partial z} + \frac{\nabla_T^2 A_i}{2i\beta_i} + \frac{i}{2}\beta_i b^2 \rho^2 A_s = ik_0 n_2 [(|A_i|^2 + 2|A_p|^2 + 2|A_s|^2)A_i + 2A_p^2 A_s^* e^{-i\Delta\beta z}],$$

$$(6.4.12)$$

where $\Delta\beta = \beta_s + \beta_i - 2\beta_p$ is the phase mismatch. The second and third terms in each equation govern the diffractive and GRIN effects, while the terms on the right side include all nonlinear effects. Three intensity-dependent terms govern the SPM and XPM effects, and the last term is responsible for FWM and parametric amplification.

In general, the solution of Eqs. (6.4.10) through (6.4.12) requires a numerical approach. Considerable simplification occurs when the pump remains much more intense than both the signal and the idler. In that situation, depletion of the pump can be ignored. Assuming $|A_s|^2$ and $|A_i|^2$ to be much smaller in comparison to $|A_p|^2$, we obtain

$$\frac{\partial A_p}{\partial z} + \frac{\nabla_T^2 A_p}{2i\beta_p} + \frac{i}{2}\beta_p b^2 \rho^2 A_p = ik_0 n_2 |A_p|^2 A_p, \qquad (6.4.13)$$

$$\frac{\partial A_s}{\partial z} + \frac{\nabla_T^2 A_s}{2i\beta_s} + \frac{i}{2}\beta_s b^2 \rho^2 A_s = 2ik_0 n_2 \left(|A_p|^2 A_s + A_p^2 A_i^* e^{-i\Delta\beta z}\right), \qquad (6.4.14)$$

$$\frac{\partial A_i}{\partial z} + \frac{\nabla_T^2 A_i}{2i\beta_i} + \frac{i}{2}\beta_i b^2 \rho^2 A_i = 2ik_0 n_2 \left(|A_p|^2 A_i + A_p^2 A_s^* e^{-i\Delta\beta z}\right), \qquad (6.4.15)$$

We can solve the pump equation (6.4.13) with the variational technique of Section 5.1. When the pump power is considerably below the self-focusing threshold, the solution for a Gaussian pump beam takes the form

$$A_p(\rho, z) = P_0 F_p(\rho, z), \qquad F_p(\rho, z) = \sqrt{\frac{\pi w_p^2}{f_p(z)}} \exp\left[-\frac{\rho^2}{2w_p^2 f_p(z)} + i\phi_p(\rho, z)\right].$$

$$(6.4.16)$$

where P_0 is the peak power of the pump beam, w_p is its width at $z = 0$, and its periodic self-imaging is governed by

$$f_p(z) = \cos^2(2\pi z/L_p) + C_p^2 \sin^2(2\pi z/L_p). \qquad (6.4.17)$$

Recall L_p is the self-imaging period and $C_p = (w_g/w_p)^2$, w_g being the spot size of the fundamental mode. The phase of the pump also varies with z (see Section 5.1.2).

The signal and idler beams also experience periodic focusing as they are amplified by the pump. If we assume that self-imaging is not affected much by the nonlinear terms in Eqs. (6.4.14) and (6.4.15), we can solve these equations approximately using the technique discussed in Section 5.2.3. We write the solution for A_s as $A_s(\rho, z) = B_s(z)F_s(\rho, z)$, where $F_s(\rho, z)$ has the form similar to Eq. (6.4.16):

$$F_s(\rho, z) = \sqrt{\frac{\pi w_s^2}{f_s(z)}} \exp\left[-\frac{\rho^2}{2w_s^2 f_s(z)} + i\phi_s(\rho, z)\right], \qquad (6.4.18)$$

and $f_s(z)$ has the form of Eq. (6.4.17) with $C_s = (w_g/w_s)^2$. The idler beam is assumed to have the same spatial profile as the signal, that is, $A_i(\rho, z) = B_i(z)F_s(\rho, z)$. If the initial widths are the same for the pump and signal, all three beams evolve in a self-similar fashion with different amplitudes.

Following the procedure of Section 5.2.3, B_s and B_i are found to satisfy

$$\frac{dB_s}{dz} = 2i\gamma_s P_0\left[f_1 B_s + f_2 B_i^* \exp(-i\Delta\beta z)\right], \qquad (6.4.19)$$

$$\frac{dB_i}{dz} = 2i\gamma_i P_0\left[f_1 B_i + f_2 B_s^* \exp(-i\Delta\beta z)\right], \qquad (6.4.20)$$

where P_0 is the pump's peak power, $\gamma_j = n_2\omega_j/(c\pi w_j^2)$, and the overlap factors f_1 and f_2 are defined as

$$f_1(z) = (\pi w_0^2)\iint |F_p|^2|F_s|^2\, dx\, dy, \qquad f_2(z) = (\pi w_0^2)\iint |F_p|^2|F_s|^2 e^{i\delta\phi}\, dx\, dy, \qquad (6.4.21)$$

with $\delta\phi = 2\phi_p - \phi_s - \phi_i$. Both f_1 and f_2 vary with z because of self-imaging inside a GRIN medium. The first nonlinear term in Eqs. (6.4.19) and (6.4.20) results from the pump-induced XPM, while the second governs the FWM-induced parametric amplification. These equations can be solved numerically to study how the signal beam is amplified inside a GRIN fiber through the FWM process.

A further simplification is possible if we assume that B_s and B_i don't change much over one self-imaging period L_p. As this length is typically small ($L_p < 1$ cm), this assumption is justified in practice, and we can replace the overlap factors f_1 and f_2 in Eqs. (6.4.19) and (6.4.20) with their average values $\bar{f}_1$ and $\bar{f}_2$, respectively. The XPM term can be removed using $B_j = D_j \exp(2i\gamma_j\bar{f}_1 P_0)$ for $j = s, i$ to obtain the simpler equations:

$$\frac{dD_s}{dz} = 2i\gamma_s\bar{f}_2 P_0 e^{-i\kappa z}D_i^*, \qquad (6.4.22)$$

$$\frac{dD_i}{dz} = 2i\gamma_i\bar{f}_2 P_0 e^{-i\kappa z}D_s^*, \qquad (6.4.23)$$

where $\kappa = \Delta\beta + 2\gamma_p\bar{f}_1 P_0$ is the net phase mismatch with the XPM contribution included. These equations are similar to Eqs. (6.2.28) and (6.2.29) and differ only by the overlap factors that reduce the FWM efficiency. They can be solved analytically following the procedure of Section 6.2.2.

6.5 Doped-Fiber Amplifiers and Lasers

Step-index fibers with ytterbium (Yb) doping have been available since the 1990s and are used routinely for making high-power amplifiers and lasers [12, 13]. As we saw in Section 6.2.1, Yb-doped GRIN fibers have also been fabricated in recent years [16] and used as amplifiers [53–55]. In this section, we discuss such GRIN-fiber amplifiers and lasers with attention to nonlinear phenomena such as spatial beam cleaning and supercontinuum generation.

6.5.1 *Amplification of CW Beams*

Consider the situation in which a Yb-doped, GRIN-fiber amplifier is pumped continuously to provide optical gain that amplifies a signal beam launched at the input end of the GRIN fiber. A modal approach is too complicated for such amplifiers as many modes of the GRIN fiber are excited by both the pump and signal beams. A non-modal approach has recently been used for studying spatial narrowing of the signal beam being amplified [56]. The resulting numerical model is comprehensive and is capable of including the nonlinear effects such as Kerr-induced beam cleaning. It requires a solution of the coupled partial differential equations satisfied by the pump and signal beams, while also accounting for the radial and axial variations of the optical gain through the atomic rate equations.

Here we focus on a simple model of the amplification process in GRIN-fiber amplifiers based on reasonable assumptions. Its use permits an analytic approach that provides considerable physical insight. The model is based on the observation that the pumping of the amplifier inverts population of the dopants in the doped region of the GRIN fiber. The resulting optical gain varies in the radial direction, even when the density of dopants is uniform, because it depends on the spatial intensity profile of the pump beam. If we neglect gain saturation and assume that the frequency of CW signal is close to the resonance frequency ω_a in Eq. (6.2.1), we can approximate the spatially varying gain $g(\rho)$ with the parabolic profile,

$$g(\rho) \approx g_0 - g_2 \rho^2, \tag{6.5.1}$$

where g_0 depends on the peak power of the pump and g_2 depends on details of the pumping scheme such as shape and width of the pump beam in the doped region of the GRIN fiber. A parabolic form is justified in practice when the width of the doped region is narrower compared to the width of the pump beam.

In general, both g_0 and g_2 vary with distance along the GRIN-fiber amplifier. Their z dependence has its origin in power depletion and periodic self-imaging of the pump beam [56]. Here we neglect this z dependence assuming (i) amplifier's length is short enough that the pump remains nearly undepleted and (ii) the self-imaging period, $L_p = 2\pi/b$, of the GRIN fiber (about 1 mm) is so short compared to its length (> 1 m) that we can replace g_0 and g_2 with their average values over one period. With these simplifications, we only need to solve a single equation for the signal being amplified. If we neglect the self-focusing effects, assuming that the signal power remains well below the critical power at which beam collapse occurs, we only need to modify the last term in Eq. (6.3.15) as

$$\frac{\partial A_s}{\partial z} + \frac{\nabla_T^2 A_s}{2i\beta_s} + \frac{i}{2}\beta_s b^2 \rho^2 A_s = \tfrac{1}{2}(g_0 - g_2\rho^2)A_s. \tag{6.5.2}$$

We solve Eq. (6.5.2) with the technique used in Section 6.3.3. For an initially Gaussian-shape signal beam, the solution can be written in the form

$$A_s(\rho,z) = A_0 \exp\left(-\frac{\rho^2}{2w^2} + \frac{i}{2}\beta_s h\rho^2 + i\psi\right), \tag{6.5.3}$$

where the four parameters (A_0, w, h, and ψ) vary with z. Recall that the parameter h governs curvature of the beam's wavefront. Using Eq. (6.5.3) into Eq. (6.5.2) and equating the terms with and without ρ^2 on the two sides of the resulting equation, we obtain the following equations for the four parameters:

$$\frac{dA_0}{dz} = (\tfrac{1}{2}g_0 - h)A_0, \qquad \frac{d\psi}{dz} = -\frac{1}{\beta_s w^2}, \tag{6.5.4}$$

$$\frac{dw}{dz} = hw - \tfrac{1}{2}g_2 w^3, \qquad \frac{dh}{dz} = \frac{1}{\beta_s^2 w^4} - h^2 - b^2. \tag{6.5.5}$$

These equations should be solved numerically with the known initial conditions at $z = 0$. For example, for an initially collimated signal beam with a width w_s, one can use $w(0) = w_s$, $h(0) = 0$, $A_0(0) = 1$, and $\psi(0) = 0$.

The width equation can be solved in an analytic form in the specific case of the radially uniform gain ($g_2 = 0$). Equation (6.5.5) then becomes identical to the case of an undoped GRIN fiber (discussed in Section 2.5.3), and the solution for the beam's width becomes

$$w(z) = w_s\sqrt{f(z)}, \qquad f(z) = 1 - (1 - C_s^2)\sin^2(bz), \tag{6.5.6}$$

where $C_s = (w_g/w_s)^2$ and w_s is the input width of the beam being amplified. This result shows that the beam's width inside the amplifier varies in the same periodic fashion as in an undoped GRIN fiber (because of self-imaging in both cases). However, the beam's power increases exponentially owing to its amplification. We can show this feature by using $dw/dz = hw$ in Eq. (6.5.4) and writing it as

$$\frac{d}{dz}(w^2 A_0^2) = g_0(w^2 A_0^2). \tag{6.5.7}$$

Recalling that the signal's power at any distance z is given by

$$P(z) = \int_0^{2\pi} d\phi \int_0^{\infty} |A_s(\rho, z)|^2 \rho d\rho = (\pi w^2) A_0^2, \qquad (6.5.8)$$

we obtain $dP/dz = g_0 P$ or $P(z) = P(0) \exp(g_0 z)$. As expected, the beam's power increases exponentially inside the amplifier in the absence of gain saturation.

The situation is considerably different when the gain is radially nonuniform and g_2 is finite in Eq. (6.5.1). To the first order in g_2, $w(z)$ satisfies an equation similar to that in Eq. (6.3.20):

$$\frac{d^2 w}{dz^2} + b^2 w = \frac{1}{\beta_s^2 w^3} - 2g_2 w^2 \frac{dw}{dz}. \qquad (6.5.9)$$

The last term in this equation governs the effects of a spatially nonuniform gain in the transverse dimensions. Its presence modifies the periodic solution given in Eq. (6.5.6). On physical grounds, we expect the amplitude of periodic variations to decrease with increasing z because of a larger gain in the central region of the signal beam compared to its outer regions. In other words, the beam should become narrower as the signal is amplified inside the fiber amplifier.

Before solving Eq. (6.5.9) numerically, we normalize it using $s = w/w_s$ and $\xi = bz$. The resulting equation for the normalized width s is

$$\frac{d^2 s}{d\xi^2} + s = \frac{C_s^2}{s^3} - 2\mu s^2 \frac{ds}{dz}, \qquad (6.5.10)$$

where $\mu = (g_2/b)w_s^2$ is a dimensionless parameter such that $\mu \ll 1$. Figure 6.16 shows the plots obtained by solving Eq. (6.5.10) with $C_s = 0.2$ for $\mu = 0.02$ (top) and $\mu = 0.04$ (bottom). The beam's width oscillates in both cases because of self-imaging, but the amplitude of oscillations decreases width distance. The reduction in amplitude does not occur for $\mu = 0$, and the damping rate of oscillations increases with μ.

The important question is what values of μ are realistic for a GRIN-fiber amplifier. Assuming that the gain in Eq. (6.5.1) is reduced by a factor of 4 at $\rho = a$, we find $g_2 = 3g_0/(4a^2)$. Its use allows us to write $\mu = (g_2/b)w_s^2$ as

$$\mu = (3g_0 L_p/8\pi)(w_s/a)^2. \qquad (6.5.11)$$

where $L_p = 2\pi/b$ is the self-imaging period. For an amplifier providing 20 dB amplification over its 10 m length, g_0 is found to be 0.46 m^{-1} from $e^{gL} = 100$. The self-imaging period of a GRIN fiber is close to 1 mm. If we use $w_s = 15$ μm with $a = 25$ μm, the estimated value of μ is $\sim 10^{-5}$. Such low values of μ indicate that decrease in the amplitude of oscillations will become apparent only after thousands of oscillations in Figure 6.16.

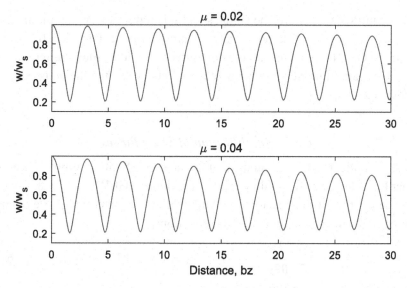

Figure 6.16 Oscillations in the width of the signal beam for two values of the parameter μ. The amplitude of oscillations does not change when $\mu = 0$.

We solve Eq. (6.5.9) approximately and follow the approach used for Eq. (6.3.20) to estimate the damping rate of oscillations. In the absence of the g_2 term, the solution of Eq. (6.5.9) is given in Eq. (6.5.6). We use this solution to replace w^2 in the g_2 term with $w_s^2 \langle f(z) \rangle$, where the average is over one self-imaging period. This allows us to write Eq. (6.5.9) as

$$\frac{d^2 w}{dz^2} + 2\gamma_d \frac{dw}{dz} + b^2 w = \frac{1}{\beta_s^2 w^3}, \qquad \gamma_d = g_2 w_s^2 \langle f(z) \rangle, \tag{6.5.12}$$

where the second term acts as a damping term for a harmonic oscillator. This form suggests that an approximate solution of Eq. (6.5.12) is given by

$$w(z) = w_s \sqrt{f(z)} \exp(-\gamma_d z). \tag{6.5.13}$$

The beam's width oscillates with the self-imaging period, as expected for a GRIN medium, but its amplitude decreases with z, resulting in spatial narrowing of the signal beam inside the amplifier. The length over which the width is reduced by a factor of $1/e$ is given by

$$L_d = 1/\gamma_d = [\tfrac{1}{2} g_2 w_s^2 (1 + C_s^2)]^{-1}, \tag{6.5.14}$$

where we calculated $\langle f(z) \rangle$ from Eq. (6.5.6).

The parameter C_s depends on the width of input beam and has a value of about 0.1 for $w_s = 15 \ \mu$m. Using $g_2 = 3g_0/(4a^2)$ with $g_0 = 0.46 \ \text{m}^{-1}$ and $a = 25 \ \mu$m, L_d is estimated to be 15 m. Thus, considerable narrowing of the signal beam can occur over the amplifier's length. If the signal's width is reduced from its initial value of 15 μm to 7 μm, most of its power at the output end will be in the fundamental mode of the GRIN fiber. This conclusion agrees with a detailed numerical model that includes gain saturation and many other effects by solving the pump and signal equations together [56].

6.5.2 Amplification of Short Pulses

When short optical pulses are amplified inside a GRIN-fiber amplifier, one must consider the dispersive and nonlinear effects within the GRIN fiber. In this case, we should use Eq. (5.2.10) of Section 5.2. After adding a gain term to it, we obtain the following four-dimensional propagation equation:

$$\frac{\partial A_s}{\partial z} + \frac{\nabla_T^2 A_s}{2i\beta_s} + \frac{i\beta_2}{2}\frac{\partial^2 A_s}{\partial T^2} + \frac{i\beta_s}{2}b^2\rho^2 A_s = ik_0 n_2 |A_s|^2 A_s + \frac{g}{2}A_s, \qquad (6.5.15)$$

where β_2 is the second-order dispersion parameter and the β_1 term has been removed with the transformation $T = t - \beta_1 z$. The gain g depends on the pumping scheme employed for the amplifier and depends, in general, on both ρ and z. Equation (6.5.15) governs the spatiotemporal evolution of a pulsed beam as it is amplified inside a GRIN fiber. It includes the effects of diffraction, dispersion, and the Kerr nonlinearity. The Raman contribution to the nonlinearity should also be included for short pulses (see Section 5.2.3).

Equation (6.5.15) should be solved numerically for an exact analysis. Some insight can be gained with the approximation indicated in Eq. (5.2.24) and using $A_s(\mathbf{r}, T) = A_t(z, T)F_s(\mathbf{r})$, where $F_s(\mathbf{r})$ governs periodic self-imaging of the signal. Using the method of Section 5.2.3, the following equation is obtained for the temporal part:

$$\frac{\partial A_t}{\partial z} + \frac{i\beta_2}{2}\frac{\partial^2 A_t}{\partial T^2} = i\bar{\gamma}(z)|A_t|^2 A_t + \frac{1}{2}\bar{g}(z)A_t, \qquad (6.5.16)$$

where $\bar{\gamma}(z)$ and $\bar{g}(z)$ are averaged over spatial profile of the signal being amplified:

$$\bar{\gamma}(z) = \frac{\iint_{-\infty}^{\infty} k_0 n_2 |F_s(\mathbf{r})|^4 dx\, dy}{\iint_{-\infty}^{\infty} |F_s(\mathbf{r})|^2 dx\, dy}, \qquad \bar{g}(z) = \frac{\iint_{-\infty}^{\infty} g(\mathbf{r})|F_s(\mathbf{r})|^2 dx\, dy}{\iint_{-\infty}^{\infty} |F_s(\mathbf{r})|^2 dx\, dy}. \qquad (6.5.17)$$

Equation (6.5.16) is much simpler to solve compared to Eq. (6.5.15). The question is under what condition its use can be justified. In the absence of the gain term, we were able to assume in Section 5.3.2 that $F_s(\mathbf{r})$ is a periodic function whose form for an initially Gaussian beam is known and is not affected by temporal changes dictated in Eq. (6.5.16). This assumption may not hold for GRIN-fiber amplifiers in which a nonuniform gain affects spatial width of the beam being amplified. One possible solution is to make use of the analysis in Section 6.5.1 and include the gain-induced changes in the beam's width before doing the averaging indicated in Eq. (6.5.17). An alternative is to solve Eq. (6.5.15) numerically that includes spatiotemporal coupling automatically. One can also employ the coupled multimode equations after expanding A_s in terms of the modes of the GRIN fiber being used for amplification.

6.5.3 Experimental Results

A Yb-doped multimode fiber amplifier was used in a 2017 experiment [53] for spatial beam narrowing, but index profile of the fiber used was closer to a step-index profile. By 2019, a GRIN fiber with a parabolic index profile and with ytterbium doping was used as an amplifier to study spatial cleanup of the signal beam being amplified [54]. The GRIN fiber used in this experiment was tapered such that the core's diameter decreased from 120 μm to 37 μm over its 10-m length. The fiber was pumped at 940 nm by injecting 10 W of CW power with a spot size of 200 μm. The 1064-nm, quasi-CW signal from a Q-switched laser had a spot size of about 20 μm at the input end of the fiber.

The beam narrowing observed in the 2019 experiment [54] was mostly due to the Kerr-induced beam cleanup, and amplification of the signal played a relatively minor role. The reason is related to the enhancement of the nonlinear effects inside the tapered fiber and the use of relatively large signal powers at the input end. This feature generated a supercontinuum extending from 520 to 2800 nm and reduced the extent of the signal's amplification through gain saturation. In fact, the factor by which the signal could be amplified was less than two in this experiment.

In a 2020 experiment, the same tapered GRIN fiber was used, but the input signal was launched at the narrow end of the fiber [55]. As the core's diameter increased from 47 to 96 μm over its 5-m-length, the nonlinear effects became weaker, resulting in little spectral broadening of the signal. As shown in Figure 6.17, the Yb-doped fiber was pumped backward at 940 nm using 10 W of CW power with a spot size of 105 μm. The 1064-nm quasi-CW signal had a spot size of 25 μm at the input end of the fiber. The impact of signal's amplification on the output beam's quality was quantified by measuring the M^2 factor for different values of the amplification factor G_s. Its value exceeded 15 for $G_s = 1$ (no amplification) but became close to 5 for $G_s > 3$. The Kerr effect also played a role in the process of spatial beam cleanup.

GRIN fibers have also been used for making mode-locked fiber lasers. In one approach, the gain is provided by a step-index Yb-doped fiber, and an undoped GRIN fiber is used within the cavity for mode locking the laser. Such a scheme is called spatiotemporal mode locking (STML) because it provides synchronization of the spatial as well as temporal (longitudinal) modes of a laser's cavity [57–61]. In a

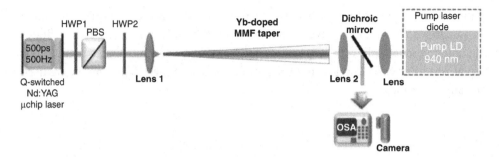

Figure 6.17 Experimental setup used for amplifying a 1064-nm signal inside a Yb-doped, tapered GRIN fiber by pumping it backward with a CW laser diode (LD) operating at 940 nm. HWP, half-wave plate; PBS, polarizing beam splitter. (After Ref. [55]; ©2020 CC BY.)

2017 experiment, a 10-m-long, step-index, Yb-doped fiber, supporting three spatial modes at the laser's wavelength (core diameter 10 μm), was spliced directly to a GRIN fiber supporting hundreds of spatial modes [57]. The cavity included spatial and spectral filters and employed a polarization-rotation scheme for passive mode locking. With suitable adjustments of the filters, STML led to short output pulses (widths $\sim$ 1 ps) composed of multiple spatial modes. The optical beam's spatial profile had a two-lobe structure and indicated the multimode nature of the laser's output.

A similar setup was used in a 2021 experiment to demonstrate STML, even when the cavity's modal dispersion was relatively large [59]. A 0.6-m-long piece of step-index Yb-doped fiber was combined with a 2.4 m-long undoped GRIN fiber within the laser's cavity, which included a spectral filter. The cavity employed a polarization-rotation scheme that not only provided passive mode locking but also helped in reducing the impact of intermodal dispersion. Figure 6.18 shows (a) spatial, (b) temporal, and (c) spectral features of the STML pulse train emitted by the laser. The 4-ps-wide pulses in part (b) are separated by 17.8 ns at the 56.1-MHz repetition rate set by the cavity's length. The spatial profile of the pulsed beam in part (a) shows multimode nature of the laser's output. With suitable adjustments, the laser could also be operated in the single LP_{11} mode such that the spatial profile exhibited a two-lobe pattern.

In a more direct approach to STML, the GRIN fiber itself is doped with Yb to provide the gain in the laser's cavity, and some other mechanism is used for mode locking. In a recent experiment, mode-locking was realized with the so-called Mamyshev mechanism in which a spectral filter, offset from the central frequency of the laser, is employed inside the cavity [62]. The ring cavity in this experiment used a 2-m-long Yb-doped GRIN fiber (core diameter 30-μm), together with a piece of Yb-doped single-mode fiber (length 1.5 m). Both fibers were pumped at 976 nm

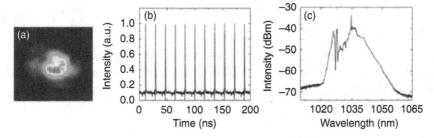

Figure 6.18 (a) Spatial, (b) temporal, and (c) spectral features of the STML pulse train emitted by a fiber laser whose cavity combined a 0.6-m-long Yb-doped fiber with a 2.4 m-long undoped GRIN fiber. (After Ref. [59]; ©2021 APS.)

using diode lasers. The laser was started by seeding with 1.3-ps pulses at 1040 nm, which were blocked after it began operating in the pulsed mode. The laser emitted short mode-locked pulses with energies ∼10 nJ at a repetition rate of 23 MHz. The spatial profile of the pulsed beam contained multiple spatial modes associated with the GRIN fiber. Self-starting of such a STML laser has been realized by optimizing the wavelength offset of the intracavity spectral filter [63].

A passive GRIN device, discussed in Section 5.5.4 and acting as an effective saturable absorber [64], has also been used for mode locking a fiber laser. In this case, a step-index Yb-doped fiber provides gain in the laser's cavity, and the GRIN device is used to mode-lock the laser. In a 2020 experiment, a Yb-doped fiber laser was mode-locked with such a GRIN device to emit 2.4-ps pulses that could be compressed down to 300 fs [65]. More recently, STML of an Er-doped fiber laser was realized by using a GRIN-based directional coupler within the laser's cavity [66]. This GRIN device acted as an effective saturable absorber, and it was also used as the output coupler through which the mode-locked pulses exited the cavity. The mode-locking threshold of the self-starting laser was as low 32 mW. Because of the anomalous GVD of the fiber at the laser's wavelength, the laser produced multimode solitons (width around 1 ps).

6.6 Spatial Solitons and Similaritons

In this section, we consider amplification of a CW beam inside a GRIN medium in which the refractive index increases radially from its value at core's center, rather than decreasing. Without the gain, the width of a low-power beam keeps increasing with distance in such a medium (see Section 2.5.3). This spreading is reduced for high-power beams because of the self-focusing effects induced by the Kerr nonlinearity. To model the beam's evolution in the presence of the gain, we modify Eq. (5.2.10) of Section 5.2 to include the local gain but neglect all time derivatives:

$$\frac{\partial A}{\partial z} + \frac{1}{2ik}\left(\frac{\partial^2 A}{\partial x^2} + \frac{\partial^2 A}{\partial y^2}\right) - \frac{ik}{2}b^2(x^2 + y^2)A = ik_0 n_2 |A|^2 A + \frac{g(z)}{2}A, \qquad (6.6.1)$$

where $k = n_0 k_0$ and the sign of the GRIN term has been reversed. The gain $g(z)$ is assumed to be spatially uniform in the radial direction but can change with z. Before considering the three-dimensional case, we focus on the case of a planar waveguide with index gradient in only the x direction.

6.6.1 Case of Planar Waveguides

Equation (6.6.1) is simplified for a planar waveguide, whose refractive index changes in a parabolic fashion along the x axis but remains constant in the y direction. In this situation, Eq. (6.6.1) becomes

$$\frac{\partial A}{\partial z} + \frac{1}{2ik}\frac{\partial^2 A}{\partial x^2} - \frac{ik}{2}b^2 x^2 A = ik_0 n_2 |A|^2 A + \frac{g}{2}A. \qquad (6.6.2)$$

When $b = 0$ (no index gradient) and $g = 0$ (no gain), this equation reduces to the nonlinear Schrödinger (NLS) equation,

$$i\frac{\partial A}{\partial z} + \frac{1}{2k}\frac{\partial^2 A}{\partial x^2} + k_0 n_2 |A|^2 A = 0, \qquad (6.6.3)$$

which supports spatial solitons – optical beams whose spatial shape does not change with propagation [67]. The question we ask is whether soliton-like solutions of Eq. (6.6.2) exist in a nonlinear GRIN-waveguide amplifier.

It is useful to introduce the dimensionless variables as

$$X = x/w_g, \quad \xi = bz, \quad G = g/b, \quad U = (k_0 n_2/b)^{1/2} A, \qquad (6.6.4)$$

where $w_g = (kb)^{-1/2}$. In terms of these variables, Eq. (6.6.2) takes the form

$$i\frac{\partial U}{\partial \xi} + \frac{1}{2}\frac{\partial^2 U}{\partial X^2} + \frac{1}{2}X^2 U + |U|^2 U - \frac{iG}{2}U = 0. \qquad (6.6.5)$$

This equation has attracted considerable attention in recent years [68–72]. Its solutions are referred to as *similaritons* because they evolve with distance in a self-similar fashion.

It was found in 2006 that self-similar solutions of Eq. (6.6.5) exist in the form [68]

$$U(X,\xi) = \frac{V(\chi)}{W(\xi)} \exp\left[\frac{i}{2}C(\xi)X^2 + iB(\xi)X + i\theta(\xi)\right], \qquad \chi = \frac{X - X_c(\xi)}{W(\xi)}, \qquad (6.6.6)$$

where all parameters (W, X_c, B, C,θ) depend on ξ and $V(\chi)$ is a function of the single variable χ that depends on both X and ξ, as indicated in Eq. (6.6.6). Physically, $V(\chi)$ governs the spatial shape of a beam whose center X_c and width W

change with ξ in a self-similar manner, maintaining overall shape of the beam. The phase front of such a beam is not planar and is curved in a parabolic fashion.

To find the beam's shape and its parameters, we substitute Eq. (6.6.6) into Eq. (6.6.5), collect similar terms, and set the real and imaginary parts of each term equal to zero. This procedure results in a set of first-order differential equations for the beam's parameters and requires a consistency condition for the solution to exist. In the case of a constant gain G, this condition is found to be

$$G = -C = 1. \tag{6.6.7}$$

This constraint requires C to be a constant whose value is set by the gain G. Further, the solution exists only for $G = 1$. It follows from Eq. (6.6.4) that amplifier's gain g should have a magnitude such that $g = b$. When the preceding condition is satisfied, the beam's parameters evolve with ξ as [68]

$$W(\xi) = W_0 e^{-\xi}, \quad B(\xi) = B_0 e^{\xi}, \quad X_c(\xi) = X_0 e^{-\xi} + B_0 \sinh(\xi), \tag{6.6.8}$$

where X_0, W_0, and B_0 are the initial values at $\xi = 0$. If both X_0 and B_0 vanish initially for a beam, it remains centered at $X = 0$, and only its width decreases with ξ in an exponential fashion. This decrease in width may appear surprising because a low-power beam would expand rapidly in a defocusing GRIN medium when $g = 0$. However, both the self-focusing and amplification counter this expansion. It is a combination of these two physical processes that leads to a decrease in the beam's width.

The preceding procedure also provides the following equations for the function $V(\chi)$ (governing shape of the optical beam) and the phase θ:

$$\frac{1}{2}\frac{d^2 V}{d\chi^2} + V^3 = KV, \qquad \frac{d\theta}{d\xi} = \frac{K}{W^2} - \frac{B^2}{2}, \tag{6.6.9}$$

where K plays the role of an eigenvalue. The equation for V is identical to that satisfied by a spatial soliton in a self-focusing nonlinear medium [67]. It can be solved by multiplying it with $dV/d\chi$ and integrating over χ. The result is found to be

$$(dV/d\chi)^2 = 2KV^2(1 - V^2) + C_1, \tag{6.6.10}$$

where C_1 is a constant of integration. Using the boundary condition that both V and $dV/d\chi$ vanish as $|\chi| \to \infty$ for a finite-size beam, we obtain $C_1 = 0$. The constant $K = \frac{1}{2}$ is found using $V = 1$ and $dV/d\chi = 0$ at the beam's center. Using these values, Eq. (6.6.10) can be integrated to obtain the well-known solution

$$V(\chi) = \text{sech}(\chi). \tag{6.6.11}$$

The solution in Eq. (6.6.6) is called a similariton because the beam's shape remains similar to a soliton even though its width changes with distance. It depends

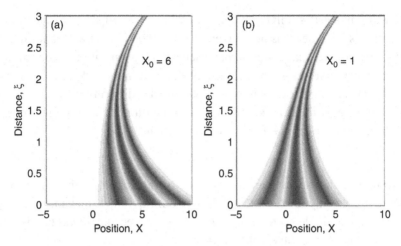

Figure 6.19 Evolution of a spatial similariton inside a GRIN amplifier for (a) $X_0 = 6$ and (b) $X_0 = 1$ for the initial values $W_0 = 2$ and $B_0 = 0.5$.

on three free parameters X_0, W_0, and B_0 that represent the initial values at $\xi = 0$ of X_c, W, and B. The solution takes the following simple form when $X_0 = 0$ and $B_0 = 0$ for an input beam:

$$U(X, \xi) = \frac{e^{\xi}}{W_0} \text{sech} \left(\frac{Xe^{\xi}}{W_0} \right) \exp \left(-\frac{i}{2} X^2 + \frac{ie^{2\xi}}{4W_0^2} \right). \tag{6.6.12}$$

The beam's center remains fixed at $X = 0$ for this solution, but its width decreases, and its amplitude increases exponentially because of the beam's amplification inside the GRIN amplifier. It can be realized in practice only when the input beam is launched at $\xi = 0$ with a curved wavefront and the gain of the amplifier is adjusted to ensure $g = b$.

The beam's evolution becomes even more interesting when it is launched with an offset from the waveguide's center. As seen from Eq. (6.6.8), the beam may move toward the waveguide's center located at $X = 0$ or away from it, depending on the magnitude of the beam's initial offset X_0. Figure 6.19 shows this behavior for (a) $X_0 = 6$ and (a) $X_0 = 1$ using $W_0 = 2$ and $B_0 = 0.5$. In case (a), the beam first moves toward the waveguide's center, before moving away from it. In contrast, the beam moves away from the center in case (b) right from the beginning. This behavior can be understood from the expression of X_c in Eq. (6.6.8).

An important question is whether the solutions obtained here are stable against small perturbations. To address the stability issue, Eq. (6.6.5) can be solved numerically with the split-step Fourier method [18], and the results compared with the analytical solution given in Eq. (6.6.6). It was found that the numerical and analytical solutions nearly coincide at short distances but begin to differ for $\xi > 2$ in the region away from the beam's center [68]. The reason for this behavior is related to the radiation emitted by similaritons in the form of dispersive waves, the same mechanism found for the GRIN solitons in Section 5.3.2. In the 2006 study, the radiation level was relatively small at $\xi = 2$ but exceeded 1% of the peak intensity for $\xi > 3$.

6.6.2 Axially Varying Gain

The assumption of a constant gain in Eq. (6.6.2) may be too restrictive for some applications. As we saw earlier, the gain depends on the pump intensity, which varies along the waveguide's length. The transverse variations of the gain may be averaged over the spatial profile of the signal, but its z dependence still remains. We extend the preceding analysis to this situation by modifying Eq. (6.6.5) as

$$i\frac{\partial U}{\partial \xi} + \frac{1}{2}\frac{\partial^2 U}{\partial X^2} + \frac{1}{2}F(\xi)X^2 U + |U|^2 U = \frac{i}{2}G(\xi)U, \tag{6.6.13}$$

where the factor $F(\xi)$ was added to the GRIN term to account for any tapering of the waveguide (see Chapter 7).

It turns out that Eq. (6.6.13) allows self-similar solutions of the form given in Eq. (6.6.6), provided we allow V to depend on a second variable η. The modified solution takes the form [69]

$$U(X,\xi) = \frac{V(\chi,\eta)}{W(\xi)}\exp\left[\frac{i}{2}C(\xi)X^2 + iB(\xi)X + i\theta(\xi)\right], \quad \chi = \frac{X - X_c(\xi)}{W(\xi)}, \tag{6.6.14}$$

where η varies with ξ. The function $V(\chi,\eta)$ satisfies the NLS equation,

$$i\frac{\partial V}{\partial \eta} + \frac{1}{2}\frac{\partial^2 V}{\partial \chi^2} + |V|^2 V = 0, \tag{6.6.15}$$

which has the standard soliton solution $V(\chi,\eta) = \mathrm{sech}(\chi)e^{i\eta/2}$. The parameters of the beam are found to evolve with ξ as

$$\frac{d\eta}{d\xi} = \frac{1}{W^2}, \quad B(\xi) = \frac{B_0}{W}, \quad C(\xi) = \frac{1}{W}\frac{dW}{d\xi}. \tag{6.6.16}$$

The beam's width depends on the tapering function $F(\xi)$ and satisfies the equation,

$$\frac{d^2 W}{d\xi^2} = F(\xi)W. \tag{6.6.17}$$

Once $W(\xi)$ is known, we can find η, B, and C from Eq. (6.6.16). The beam's center evolves with distance as $X_c(\xi) = (X_0 + B_0\xi)/W(\xi)$. The constraint given in Eq. (6.6.7) remains $G = -C$, i.e, G and C vary with ξ in an identical fashion.

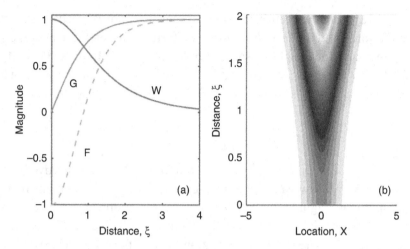

Figure 6.20 (a) Gain, width, and tapering profiles plotted as a function of ξ. (b) Evolution of the similariton for $W_0 = 1$ and $B_0 = 0$.

As a simple example, consider the case of an untapered waveguide with $F(\xi) = 1$. The physically valid solution of Eq. (6.6.13) in this case is $W(\xi) = W_0 e^{-\xi}$. It follows from Eq. (6.6.16) that C must be a constant with the value $C = -1$. In this specific case, the gain must also be a constant with the value $G = 1$, the same value found in Section 6.6.1. It follows that the formation of similaritons in the case of axially varying gain requires the waveguide to be tapered. The properties of such similaritons depend on the tapering profile $F(\xi)$. For a given form of $F(\xi)$, one can find the required $G(\xi)$ by solving for C in Eq. (6.6.16) and using the constraint $G = -C$. For example, for a linear taper with $F(\xi) = 1 + \alpha\xi$, the solution of Eq. (6.6.17) provides $W(\xi)$ is in terms of an Airy function. The width equation can also be solved for an exponential taper with $F(\xi) = e^{-q\xi}$ (see Section 7.2).

An interesting solution of Eq. (6.6.17) is obtained for a specific form of $F(\xi) = 1 - 2\,\text{sech}^2(\xi)$ [69]. The solution is $W(\xi) = W(0)\text{sech}(\xi)$ and its use in Eq. (6.6.16) yields $C(\xi) = -\tanh(\xi)$. The required form of the gain for which a similariton can form is $G(\xi) = \tanh(\xi)$. Figure 6.20 shows how F, G, and W vary ξ and how the similariton evolves inside such a waveguide amplifier using $W_0 = 1$ and $B_0 = 0$. In terms of the original variable, the gain must vary with distance as $g(z) = b\tanh(bz)$. Such variations can be realized in practice by increasing the dopant density along the waveguide's length. However, it may be harder to produce the tapering profile because $F(\xi)$ changes its sign at a certain distance.

6.6.3 Similaritons in GRIN-Fiber Amplifiers

In the case of a GRIN-fiber amplifier, we need to solve Eq. (6.6.1). It turns out that this equation can also be solved with the same technique used in Section 6.6.2

for finding self-similar solutions [71]. Using the normalized variables given in Eq. (6.6.4), we can write it in the following dimensionless form:

$$i\frac{\partial U}{\partial \xi} + \frac{1}{2}\left(\frac{\partial^2 U}{\partial X^2} + \frac{\partial^2 U}{\partial Y^2}\right) + \frac{1}{2}F(\xi)(X^2 + Y^2)U + |U|^2 U - \frac{iG}{2}U = 0, \quad (6.6.18)$$

where $Y = y/w_g$ and a tapering function has been added to the GRIN term for completeness. For a self-similar solution, the amplitude increases with distance (because of amplification), but the width decreases at the same time such that the beam's power remains the same. Thus, a suitable form of the solution is

$$U(X, \xi) = \frac{1}{W}V(\chi, \mu, \eta)\exp\left[\frac{i}{2}C(\xi)(X^2 + Y^2) + i\theta\right], \quad \chi = \frac{X}{W}, \quad \mu = \frac{Y}{W},$$
$$(6.6.19)$$

where χ and η are self-similar variables, and all parameters (W, C, η, and θ) vary with ξ. For simplicity, we have set $B = 0$ and assumed that the beam's center remains fixed at its input position ($X_c = Y_c = 0$).

Substituting Eq. (6.6.19) into Eq. (6.6.18) and following the procedure outlined earlier, we find that V satisfies the following NLS equation [71]:

$$i\frac{\partial V}{\partial \eta} + \frac{1}{2}\frac{\partial^2 V}{\partial \chi^2} + \frac{1}{2}\frac{\partial^2 V}{\partial \mu^2} + |V|^2 V = 0. \quad (6.6.20)$$

The beam's parameters vary with ξ as indicated in Eq. (6.6.16) and (6.6.17). The constraint given in Eq. (6.6.7) remains $G = -C$, i.e, G and C vary with ξ in an identical fashion. As a result, the width W and other parameters of the beam evolve in the same fashion as in the one-dimensional case. However, the beam's intensity profile, governed by $|V(\chi, \mu, \eta)|^2$ varies in two dimensions (X and Y) at any distance ξ.

The soliton-like solutions of Eq. (6.6.20) were found in the 1960s in the context of self-focusing and self-trapping of optical beams inside a nonlinear Kerr medium [73, 74]. The stability of such spatial solitons has also been studied [67]. For a spatial soliton, the solution of Eq. (6.6.20) has the form $V(\chi, \mu, \eta) = u(\chi, \mu)\exp(iK\eta)$, where $u(\chi, \mu)$ governs the soliton's spatial profile that does not change with ξ or η. If we focus on the radially symmetric solitons and write Eq. (6.6.20) in the cylindrical coordinates (ρ, ϕ), we obtain

$$\frac{d^2u}{d\rho^2} + \frac{1}{s}\frac{du}{d\rho} + 2u^3 = 2Ku, \quad (6.6.21)$$

where K plays the role of an eigenvalue. This eigenvalue equation must be solved numerically and is found to have multiple solutions for different values of K that represent the fundamental and higher-order spatial solitons [73, 74]. Even though a closed-form solution does not exist, the shape of the fundamental spatial soliton resembles that of a Gaussian beam.

Consider a GRIN-fiber amplifier with constant gain along the amplifier's length. In this case, self-similar solutions of Eq. (6.6.18) must satisfy the constraint given in Eq. (6.6.7) such that $G = -C = 1$. As $F(\xi) = 1$ in Eq. (6.6.17), its physically valid solution is $W(\xi) = W_0 e^{-\xi}$. The variable η can be found using this solution in Eq. (6.6.18). As a result, the solution given in Eq. (6.6.19) for radially symmetric similaritons can be written as

$$U(\rho, \xi) = \frac{e^{\xi}}{W_0} u\left(\frac{\rho e^{\xi}}{W_0}\right) \exp\left[-\frac{i}{2}\rho^2 + \frac{iK}{2}\frac{e^{2\xi} - 1}{W_0^2}\right]. \qquad (6.6.22)$$

For exciting this similariton, the input beam should be in the form of a fundamental spatial soliton and it should be launched into the GRIN-fiber amplifier with a suitably curved phase front, as indicated in Eq. (6.6.22). After some initial adjustment, this beam will evolve in a self-similar fashion such that its amplitude increases and its width decreases inside the amplifier.

No experiments have yet been done to observe the spatial similaritons discussed in this section. The reason is that it is not easy to satisfy the input conditions and constraints under which they are predicted to exist. For a standard GRIN fiber (no tapering), the gain must be constant with a specific value and satisfy the requirement $g = b$. Practical GRIN amplifiers have a gain that varies both axially and radially. Moreover, its magnitude is typically such that $g < b$. Even if the constraint $g = b$ is satisfied approximately, the excitation of a similariton requires an input beam with a specific shape and a specific phase-front curvature, a situation not easily realized in practice.

References

[1] D. Gloge, *Appl. Opt.* **11**, 2506 (1985); https://doi.org/10.1364/AO.11.002506.
[2] M. S. Sodha and A. K. Ghatak, *Inhomogeneous Optical Waveguides* (Plenum Press, 1977).
[3] R. Olshansky, *Rev. Mod. Phys.* **51**, 341 (1979); https://doi.org/10.1103/RevModPhys .51.341.
[4] Y. Koike, *Fundamentals of Plastic Optical Fibers* (Wiley, 2014).
[5] M. Atef and H. Zimmermann, *Optical Communication over Plastic Optical Fibers* (Springer, 2014).
[6] D. Marcuse, *Light Transmission Optics*, 2nd ed. (Van Nostrand Reinhold, 1982).
[7] M. Premaratne and G. P. Agrawal, *Light Propagation in Gain Media: Optical Amplifiers* (Cambridge University Press, 2011).
[8] G. P. Agrawal, *Applications of Nonlinear Fiber Optics*, 3rd ed. (Academic Press, 2021).
[9] G. P. Agrawal, *Fiber-Optic Communication Systems*, 5th ed. (Wiley, 2021).
[10] E. Desuvire, *Erbium-Doped Fiber Amplifiers: Principles and Applications* (Wiley, 1994).
[11] P. C. Becker, N. A. Olsson, and J. R. Simpson, *Erbium-Doped Fiber Amplifiers: Fundamentals and Technology* (Academic Press, 1999).

[12] N. K. Dutta, *Fiber Amplifiers and Fiber Lasers* (World Scientific, 2014).

[13] V. Ter-Mikirtychev, *Fundamentals of Fiber Lasers and Fiber Amplifiers*, 2nd ed. (Springer, 2019).

[14] R. Ahmad, S. Chatigny, and M. Rochette, *Opt. Express* **18**, 19983 (2010); https://doi.org/10.1364/OE.18.019983.

[15] A. Anuszkiewicz, R. Kasztelanic, A. Filipkowski et al., *Sci. Rep.* **8**, 12329 (2018); https://doi.org/10.1038/s41598-018-30284-1.

[16] N. Choudhury, N. K. Shekhar, A. Dhar, and R. Sen, *Phys. Stat. Solidi A* **216**, 1900365 (2019); https://doi.org/10.1002/pssa.201900365.

[17] R. H. Stolen and E. P. Ippen, *Appl. Phys. Lett.* **22**, 276 (1973); https://doi.org/10.1063/1.1654637.

[18] G. P. Agrawal, *Nonlinear Fiber Optics*, 6th ed. (Academic Press, 2019).

[19] L. F. Mollenauer, R. H. Stolen, and M. N. Islam, *Opt. Lett.* **10**, 229 (1985); https://doi.org/10.1364/OL.10.000229.

[20] L. F. Mollenauer and K. Smith, *Opt. Lett.* **13**, 675 (1988); https://doi.org/10.1364/OL.13.000675.

[21] F. Poletti and P. Horak, *J. Opt. Soc. Am. B* **25**, 1645 (2008); https://doi.org/10.1364/JOSAB.25.001645.

[22] N. B. Terry, T. G. Alley, and T. H. Russell, *Opt. Express* **15**, 17509 (2007); https://doi.org/10.1364/OE.15.017509.

[23] K. S. Chiang, *Opt. Lett.* **17**, 352 (1992); https://doi.org/10.1364/OL.17.000352.

[24] A. E. Siegman, New Developments in Laser Resonators, *Proc. SPIE* **1224**, 2 (1990); https://doi.org/10.1117/12.18425.

[25] A. B. Grudinin, E. M. Dianov, D. V. Korbkin, A. M. Prokhorov, and D. V. Khaidarov, *JETP Lett.* **47**, 356 (1988).

[26] N. R. Islam and K. Sakuda, *J. Opt. Soc. Am. B* **14**, 3238 (1997); https://doi.org/10.1364/JOSAB.14.003238.

[27] T. H. Russell, S. M. Willis, M. B. Crookston, and W. B. Roh, *J. Nonlinear Opt. Phys. Mater.* **11**, 303 (2002); https://doi.org/10.1142/S0218863502001036.

[28] H. Pourbeyram, G. P. Agrawal, and A. Mafi, *Appl. Phys. Lett.* **102**, 201107 (2013); https://doi.org/10.1063/1.4807620.

[29] S. H. Baek and W. B. Roh, *Opt. Lett.* **29**, 153 (2004); https://doi.org/10.1364/OL.29.000153.

[30] S. I. Kablukov, E. I. Dontsova, E. A. Zlobina et al., *Laser Phys. Lett.* **10**, 085103 (2013); https://doi.org/10.1088/1612-2011/10/8/085103.

[31] T. Yao, A. V. Harish, J. K. Sahu, and J. Nilsson, *Appl. Sci.* **5**, 1323 (2015); https://doi.org/10.3390/app5041323.

[32] V. R. Supradeepa, Y. Feng, and J. W. Nicholson, *J. Opt.* **19**, 023001 (2017); https://doi.org/10.1088/2040-8986/19/2/023001.

[33] S. A. Babin, E. A. Zlobina, and S. I. Kablukov, *IEEE J. Sel. Topics Quantum Electron.* **24**, 1400310 (2018); https://doi.org/10.1109/JSTQE.2017.2764072.

[34] Y. Glick, Y. Shamir, A. A. Wolf et al., *Opt. Lett.* **43**, 1027 (2018); https://doi.org/10.1364/OL.43.001027.

[35] C. Fan, Y. Chen, H. Xiao et al., *Opt. Express* **29**, 19441 (2021); https://doi.org/10.1364/OE.427605.

[36] X. Yang, L. Zhang, H. Jiang, T. Fan, and Y. Feng, *Opt. Express* **23**, 19831 (2015); https://doi.org/10.1364/OE.23.019831.

[37] A. G. Kuznetsov, D. S. Kharenko, E. V. Podivilov, and S. A. Babin, *Opt. Express* **23**, 16280 (2021); https://doi.org/10.1364/OE.24.016280.

[38] A. G. Kuznetsov, S. I. Kablukov, Y. A. Timirtdinov, and S. A. Babin, *Photonics* **9**, 539 (2022); https://doi.org/10.3390/photonics9080539.

[39] Y. Chen, J. Leng, H. Xiao, T. Yao, and P. Zhou, *Laser Phys. Lett.* **15**, 085104 (2018); https://doi.org/10.1088/1612-202X/aac428.

[40] Y. Chen, J. Leng, H. Xiao, T. Yao, and P. Zhou, *IEEE Access* **7**, 28334 (2019).

[41] Y. Chen, T. Yao, H. Xiao, J. Leng, and P. Zhou, *High Power Laser Sci. Eng.* **8**, e33 (2020); https://doi.org/10.1017/hpl.2020.33.

[42] Y. Chen, T. Yao, H. Xiao, J. Leng, and P. Zhou, *J. Lightwave Technol.* **39**, 1785 (2021); https://doi.org/10.1109/JLT.2020.3039677.

[43] C. Fan, H. Xiao, T. Yao et al., *Opt. Lett.* **46**, 3432 (2021); https://doi.org/10.1364/OL .431273.

[44] C. Fan, Y. An, and T. Yao, *Opt. Lett.* **46**, 4220 (2021); https://doi.org/10.1364/OL .433750.

[45] A. G. Kuznetsov, S. I. Kablukov, E. V. Podivilov, and S. A. Babin, *OSA Continuum* **4**, 1034 (2021); https://doi.org/10.1364/OSAC.421985.

[46] R. H. Stolen, *IEEE J. Quantum Electron.* **11**, 100 (1975); https://doi.org/10.1109/JQE .1975.1068571.

[47] M. E. Marhic, *Fiber Optical Parametric Amplifiers, Oscillators and Related Devices* (Cambridge University Press, 2007).

[48] K. O. Hill, D. C. Johnson, and B. S. Kawasaki, *Appl. Opt.* **20**, 1075 (1981); https: //doi.org/10.1364/AO.20.001075.

[49] E. Nazemosadat, H. Pourbeyram, and A. Mafi, *J. Opt. Soc. Am. B* **33**, 144 (2016); https://doi.org/10.1364/JOSAB.33.000144.

[50] A. Bendahmane, M. Conforti, O. Vanvincq et al., *Opt. Express* **29**, 30822 (2021); https://doi.org/10.1364/OE.436229.

[51] R. Dupiol, A. Bendahmane, K. Krupa et al., *Opt. Lett.* **42**, 1293 (2017); https:// doi.org/10.1364/OL.43.001293.

[52] A. Bendahmane, K. Krupa, A. Tonello et al., *J. Opt. Soc. Am. B* **35**, 295 (2018); https://doi.org/10.1364/JOSAB.35.000295.

[53] R. Guenard, K. Krupa, R. Dupiol et al., *Opt. Express* **25**, 4783 (2017); https://doi.org/ 10.1364/OE.25.004783.

[54] A. Niang, T. Mansuryan, K. Krupa et al., *Opt. Express* **27**, 24018 (2019); https: //doi.org/10.1364/OE.27.024018.

[55] A. Niang, D. Modello, A. Tonello et al., *IEEE Photon. J.* **12**, 6500308 (2020); https: //doi.org/10.1109/JPHOT.2020.2979938.

[56] M. A. Jima, A. Tonello, A. Niang et al., *J. Opt. Soc. Am. B* **39**, 2172 (2022); https: //doi.org/10.1364/JOSAB.463473.

[57] L. G. Wright, D. N. Christodoulides, and F. W. Wise, *Science* **358**, 94 (2017); https: //doi.org/10.1126/science.aao0831.

[58] L. G. Wright, P. Sidorenko, H. Pourbeyram et al., *Nat. Phys.* **16**, 565 (2020); https: //doi.org/10.1038/s41567-020-0784-1.

[59] Y. Ding, X. Xiao, K. Liu et al., *Phys. Rev. Lett.* **126**, 093901 (2021); https://doi.org/ 10.1103/PhysRevLett.126.093901.

[60] H. Zhang, Y. Zhang, J. Peng et al., *Photon. Res.* **10**, 483 (2022); https://doi.org/ 10.1364/PRJ.444750.

[61] Y. Wu, D. N. Christodoulides, and F. W. Wise, *Opt. Lett.* **47**, 4439 (2022); https: //doi.org/10.1364/OL.471457.

[62] H. Haig, P. Sidorenko, A. Dhar et al., *Opt. Lett.* **47**, 46 (2022); https://doi.org/10.1364/ OL.447208.

[63] B. Cao, C. Gao, Y. Ding et al., *Opt. Lett.* **47**, 4584 (2022); https://doi.org/10.1364/OL .469291.

[64] E. Nazemosadat and A. Mafi, *J. Opt. Soc. Am. B* **30**, 1357 (2013); https://doi.org/ 10.1364/JOSAB.30.001357.

[65] S. Thulasi and S. Sivabalan, *Appl. Opt.* **59**, 7357 (2020); https://doi.org/10.1364/AO .395356.

[66] X. Zhang, Z. Wang, C. Shen, and T. Guo, *Opt. Lett.* **47**, 2081 (2022); https://doi.org/ 10.1364/OL.451832.

[67] Y. S. Kivshar and G. P. Agrawal, *Optical Solitons: From Fibers to Photonic Crystals* (Academic Press, 2003).

[68] S. A. Ponomarenko and G. P. Agrawal, *Phys. Rev. Lett.* **97**, 013901 (2006); https: //doi.org/10.1103/PhysRevLett.97.013901.

[69] S. A. Ponomarenko and G. P. Agrawal, *Opt. Lett.* **32**, 1659 (2007); https://doi.org/ 10.1364/OL.32.001659.

[70] S. A. Ponomarenko and G. P. Agrawal, *J. Opt. Soc. Am. B* **25**, 983 (2008); https: //doi.org/10.1364/JOSAB.35.000983.

[71] L. Wu, J.-F. Zhang, L. Li, Q. Tian, and K. Porsezian, *Opt. Lett.* **16**, 6352 (2008); https://doi.org/10.1364/OE.16.006352.

[72] C. Dai and J. Zhang, *Opt. Lett.* **35**, 2651 (2010); https://doi.org/10.1364/OL.35 .002651.

[73] R. Y. Chiao, E. Garmire, and C. H. Townes, *Phys. Rev. Lett.* **13**, 479 (1964); https: //doi.org/10.1103/PhysRevLett.13.479.

[74] H. A. Haus, *Appl. Phys. Lett.* **8**, 128 (1966); https://doi.org/10.1063/1.1754519.

7

Nonuniform GRIN Media

So far in this book, the refractive index $n(\mathbf{r})$ of a GRIN medium is assumed to depend on the transverse coordinates x and y, but not on the distance z along the medium's propagation direction. This is usually the case in practice, but there are situations in which $n(\mathbf{r})$ may depend on z as well. An example is provided by a GRIN fiber that has been tapered such that its core's diameter varies along its length. In this chapter, we discuss how longitudinal variations of the refractive index affect the propagation of light in such a GRIN medium. Section 7.1 describes the ray-optics and wave-optics techniques that can be used for this purpose. Section 7.2 focuses on tapered GRIN fibers and describes the impact of tapering on the periodic self-imaging for a few different tapering profiles. The analogy between a GRIN medium and a harmonic oscillator is exploited in Section 7.3 by employing several quantum-physics techniques for solving the GRIN problem. Section 7.4 is devoted to the case of periodic variations in the refractive index that are induced by changing the core's radius of a GRIN fiber along its length in a periodic fashion.

7.1 Axially Varying Refractive Index

Tapering of optical waveguides was considered during the 1970s in several different contexts [1–3]. Step-index fibers often need to be tapered for the purpose of splicing, coupling, or making directional couplers. GRIN fibers can also be tapered with the techniques used for them. Tapering introduces axial variations in the refractive index that affect the modes of a GRIN device. As a result, the modal description is not very useful for tapered fibers. In this section, we discuss non-modal techniques for studying propagation of optical beams inside a tapered GRIN medium.

Since 1970, many studies have focused on the propagation of optical beams inside a tapered GRIN fiber [4–9]. To keep the discussion simple, we ignore the polarization effects by neglecting the last term in Eq. (2.1.12) and solving Eq. (2.1.13) given in Section 2.1. We consider a parabolic refractive-index profile but include its axial variations using

$$n^2(\rho, z) = n_0^2[1 - b^2(z)\rho^2], \qquad b(z) = \sqrt{2\Delta}/a(z), \qquad (7.1.1)$$

where b depends on z because the core's radius a varies with z.

The refractive index n_0 at the central axis of the fiber may also vary with z. Such variations can be induced through changes in the density of dopants within a GRIN fiber's core or by exposing the fiber to ultraviolet radiation [12]. They can be included [5] by replacing the phase factor $\exp(in_0 k_0 z)$ in Eq. (2.1.14) with $\exp[ik_0 \int_0^z n_0(z)\, dz]$. For this reason, we treat n_0 as a constant in this chapter.

7.1.1 Geometrical-Optics Approach

When we use the ray equation in Eq. (3.1.3) with the refractive index given in Eq. (7.1.1) and follow the procedure of Section 3.1, the path of an optical ray is governed by

$$\frac{d^2 q}{dz^2} + b^2(z)q = 0, \qquad q = x \text{ or } y. \qquad (7.1.2)$$

This harmonic-oscillator equation differs from those in Eq. (3.1.9) because the spatial frequency b changes with z. To obtain its solution, we multiply it with $2(dq/dz)$ and integrate once to obtain

$$\left(\frac{dq}{dz}\right)^2 = -b^2 q^2 + 2\int_0^z b\frac{db}{dz}q^2(z)\, dz. \qquad (7.1.3)$$

When b does not vary with z, the last term in Eq. (7.1.3) vanishes, and the solution takes the form of Eq. (3.1.10) for a ray entering the GRIN medium at the position (x_0, y_0) with the initial slope (x_0', y_0').

When b varies with z, an analytic solution of Eq. (7.1.2) is not possible for an arbitrary $b(z)$. However, a closed-form solution can be obtained if we make the adiabatic approximation, valid when $b(z)$ varies with z so slowly that the condition $|db/dz| \ll b^2(z)$ holds for all z. As a result, we can neglect the last term in Eq. (7.1.3). The solution of Eq. (7.1.2) can then be written in terms of its two linearly independent solutions as

$$q(z) = c_1 F(z) + c_2 G(z), \qquad (7.1.4)$$

where c_1 and c_2 are constants and the functions $F(z)$ and $G(z)$ are defined as [13]

$$F(z) = \sqrt{\frac{b_0}{b(z)}}\cos\left[\int_0^z b(z')\, dz'\right], \qquad G(z) = \sqrt{\frac{1}{b_0 b(z)}}\sin\left[\int_0^z b(z')\, dz'\right],$$

$$(7.1.5)$$

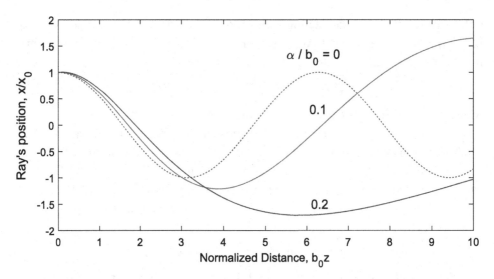

Figure 7.1 Path of a field ray inside an exponentially tapered GRIN medium for several values of α/b_0. Dotted curve shows the untapered case ($\alpha = 0$).

with $b_0 = b(0)$. These functions reduce to $\cos(b_0 z)$ and $\sin(b_0 z)/b_0$ for an untapered GRIN medium with a constant index gradient such that $b(z) = b_0$. One can show that the Wronskian of the functions $F(z)$ and $G(z)$ does not change with z and satisfies the relation

$$W(z) = F(z)G'(z) - G(z)F'(z) = W(0) = 1, \qquad (7.1.6)$$

where a prime denotes the first derivative ($F' = dF/dz$). The function $F(z)$ and $G(z)$ are normalized such that $W(z) = 1$.

The use of Eq. (7.1.5) in Eq. (7.1.4) allows us to calculate the path of any ray inside a tapered GRIN medium. As an example, consider the case where the core's radius $a(z)$ increases exponentially with z. In this situation, $b(z)$ varies with z as $b_0 e^{-\alpha z}$, and the integral in Eq. (7.1.5) can be done analytically to obtain

$$\int_0^z b_0 e^{-\alpha z'}\, dz' = \frac{b_0}{\alpha}(1 - e^{-\alpha z}). \qquad (7.1.7)$$

Figure 7.1 shows how the ray's position, $x(z)/x_0$, varies with distance for several values of α/b_0 when $x_0' = 0$. The ray's path is periodic for $\alpha = 0$ and is confined to the region $|x| \leq x_0$. As α increases, the ray follows a non-periodic trajectory that spreads outside this region. This is expected because the GRIN-induced focusing becomes weaker when the index gradient $b(z)$ decreases with distance.

It is possible to solve Eq. (7.1.2) exactly, without the adiabatic approximation, for a few specific tapering functions. For example, this equation has solutions in terms of the Bessel functions for an exponential taper of the form $b^2(z) = b_1 e^{-2\alpha z} + b_2$, where b_1 and b_2 are constants [6]. The two solutions, $F(z)$ and $G(z)$, of Eq. (7.1.2) involve the Airy functions when $b(z) = b_0(1 + b_1 z)^{1/2}$. Another important case corresponds to periodic axial variations such that $b^2(z) = b_1 - b_2 \cos(2z)$. In this case, Eq. (7.1.2) reduces to the Mathieu equation, and the solutions $F(z)$ and $G(z)$ correspond to two Mathieu functions [7]. In fact, it is possible to find specific forms of the tapering function $b(z)$ that lead to other special functions such as the Legendre and Laguerre functions [8]. However, such solutions are rarely useful in practice.

The solutions in Eq. (7.1.4) can also be used to construct the ABCD matrix, given in Eq. (3.1.16) of Section 3.1 when b is constant. This matrix relates a ray's position and its slope at any distance to their initial values at $z = 0$. Using Eq. (7.1.4) and its derivative, $q'(z) = c_1 F'(z) + c_2 G'(z)$, we can express c_1 and c_2 in terms of the initial values of the elements of the ABCD matrix. These can be used to write the ABCD matrix at any distance in terms of functions $F(z)$ and $G(z)$ [7]:

$$A(z) = F(z)G'(0) - G(z)F'(0), \qquad B(z) = G(z)F(0) - F(z)G(0), \quad (7.1.8)$$
$$C(z) = F'(z)G'(0) - G'(z)F'(0), \qquad D(z) = G'(z)F(0) - F'(z)G(0). \quad (7.1.9)$$

Here we used the Wronskian given in Eq. (7.1.6). Recalling that the same matrix elements appear in the propagation kernel given in Eq. (3.2.16), we can also use them to study the evolution of optical beams inside a tapered GRIN medium.

7.1.2 Wave-Optics Approach

Using the form of $n^2(\rho, z)$ given in Eq. (7.1.1) into Eq. (2.1.13), we need to solve the following equation to study wave propagation inside a tapered GRIN medium:

$$\nabla^2 \tilde{\mathbf{E}} + k^2 [1 - b^2(z)\rho^2] \tilde{\mathbf{E}} = 0, \qquad (7.1.10)$$

where $k = n_0 k_0$. We simplify the problem by making the paraxial approximation and focusing on optical beams that remain cylindrically symmetric inside the GRIN medium. We look for solutions for Eq. (7.1.10) in the form

$$\tilde{\mathbf{E}}(\mathbf{r}) = \hat{\mathbf{p}} A(\rho, z) \exp(ikz), \qquad (7.1.11)$$

where $\hat{\mathbf{p}}$ is the polarization unit vector and the amplitude $A(\rho, z)$ satisfies

$$2ik\frac{\partial A}{\partial z} + \frac{\partial^2 A}{\partial \rho^2} + \frac{1}{\rho}\frac{\partial A}{\partial \rho} - k^2 b^2(z)\rho^2 A = 0. \qquad (7.1.12)$$

The preceding equation is similar to Eq. (2.5.37) with the only difference being that b varies with z. It can also be solved with the eikonal-based technique discussed in Section 2.5.3. Seeking its solution in the form $A = A_0 \exp(ikS)$, we obtain the following two equations for the eikonal S and the intensity $I = A_0^2$:

$$2\frac{\partial S}{\partial z} + \left(\frac{\partial S}{\partial \rho}\right)^2 = -b^2(z)\rho^2 + \frac{1}{k^2 A_0}\left(\frac{\partial^2 A_0}{\partial \rho^2} + \frac{1}{\rho}\frac{\partial A_0}{\partial \rho}\right) \qquad (7.1.13)$$

$$\frac{\partial I}{\partial z} + \frac{\partial I}{\partial \rho}\frac{\partial S}{\partial \rho} + \left(\frac{\partial^2 S}{\partial \rho^2} + \frac{1}{\rho}\frac{\partial S}{\partial \rho}\right)I = 0. \qquad (7.1.14)$$

As discussed in Section 2.5.3, the general solution of Eq. (7.1.13) has the form

$$S(\rho, z) = \frac{\rho^2}{2}h(z) + g(z), \qquad (7.1.15)$$

where $1/h$ is related to the radius of curvature of the phase front. Using it, the solution of Eq. (7.1.14) is found to be [5]

$$I(\rho, z) = \frac{I_0}{f^2(z)}F\left(\frac{\rho}{w_0 f(z)}\right), \qquad (7.1.16)$$

where $F(\cdot)$ is an arbitrary function, w_0 is the initial with of the beam at $z = 0$, and $f(z)$ satisfies the simple differential equation:

$$\frac{df}{dz} = h(z)f(z), \qquad f(0) = 1. \qquad (7.1.17)$$

The scaling factor $f(z)$ depends on the beam's shape and can be found once a shape is specified.

As an example, we assume that a collimated Gaussian beam is launched into a tapered GRIN fiber with the initial intensity $I(\rho, 0) = I_0 \exp(-\rho^2/w_0^2)$. The intensity at a distance z is obtained from Eq. (7.1.16) and has the form

$$I(\rho, z) = \frac{I_0 w_0^2}{w^2(z)} \exp\left(-\frac{\rho^2}{w^2(z)}\right), \qquad (7.1.18)$$

where $w(z) = w_0 f(z)$ is the spot size at the distance z. Using this form in Eq. (7.1.13), we obtain

$$2\frac{\partial S}{\partial z} + \left(\frac{\partial S}{\partial \rho}\right)^2 = -b^2(z)\rho^2 + \frac{1}{k^2 w^2}\left(\frac{\rho^2}{w^2} - 2\right). \qquad (7.1.19)$$

When we use Eq. (7.1.15) for S and equate the coefficient of ρ^2 on both sides of Eq. (7.1.19), we obtain the following equation for $w(z)$:

$$\frac{d^2 w}{dz^2} + b^2(z)w = \frac{1}{k^2 w^3}. \qquad (7.1.20)$$

This equation is similar to Eq. (2.5.48) of Section 2.5.3; the only difference is that b depends on z because of changes in the core's radius.

The last term in Eq. (7.1.20) has its origin in the diffraction phenomenon and remains relatively small until the width w becomes comparable to the wavelength of input beam. If we neglect it in the geometrical-optics approximation, Eq. (7.1.20) becomes identical to the ray equation in Eq. (7.1.2). Following Eq. (7.1.4), its approximate solution can be written as

$$w(z) = w_0 F(z) + w_0' G(z), \qquad (7.1.21)$$

where the functions $F(z)$ and $G(z)$ are given in Eq. (7.1.5) and w_0 is the input width at $z = 0$. The initial slope, $w_0' = dw/dz$, is related to the wavefront curvature as $w_0' = h(0)w_0$; it vanishes for a collimated input beam with a planar phase front.

The solution in Eq. (7.1.21) becomes invalid in the focal region where the beam experiences GRIN-induced focusing in each self-imaging period. The diffraction term in Eq. (7.1.20) must be included to ensure w remains finite. It is useful to normalize this equation using $s = w/w_0$, $b(z) = b_0 h(z)$ and $\xi = b_0 z$. The result is

$$\frac{d^2 s}{d\xi^2} + h^2(\xi)s = \frac{C_f^2}{s^3}, \tag{7.1.22}$$

where $C_f = (w_g/w_0)^2$ and $w_g = (kb_0)^{-1/2}$ are the parameter introduced in Section 2.5.3. As mentioned there, w_g is related to the spot size of the fundamental mode of the GRIN device at $\xi = 0$.

Equation (7.1.22) allows one to study the impact of tapering in terms of a single device parameter C_f for any tapering profile $h(\xi)$. It has been studied in several other contexts and is known as the Ermakov equation [10]. With the initial conditions $s = 1$ and $ds/d\xi = 0$ at $\xi = 0$, its solution can be written in terms of the linearly independent solutions of Eq. (7.1.2) as [11]

$$s^2(\xi) = F^2(\xi) + C_f^2 G^2(\xi), \tag{7.1.23}$$

where the initial conditions are such that $F = 1$ and $dF/d\xi = 0$ at $\xi = 0$, while $G = 0$ and $dG/d\xi = 1$ at $\xi = 0$. Within the adiabatic approximation, we can use the solution in Eq. (7.1.5) for the functions $F(\xi)$ and $G(\xi)$ to obtain

$$s^2(\xi) = h^{-1}(\xi)\left[\cos^2\left(\int_0^\xi h(\xi')d\xi'\right) + C_f^2 \sin^2\left(\int_0^z h(\xi')d\xi'\right)\right]. \tag{7.1.24}$$

This approximate analytic solution of Eq. (7.1.22) allows one to study, for any function $h(\xi)$, how the tapering of a GRIN medium affects the periodic evolution of a Gaussian beam occurring in the absence of tapering.

7.2 Tapered GRIN Fibers

The eikonal approach has shown us that a cylindrically symmetric beam evolves inside a tapered GRIN medium in a self-similar fashion such that its width, amplitude, and phase-front curvature change with z. Changes in the beam's width can be found by solving Eq. (7.1.20) numerically for different tapering profiles. This equation can also be solved analytically in a few specific cases [4–6]. We use the normalized form of Eq. (7.1.20) given in Eq. (7.1.22). If tapering changes the core's radius as $a(z) = a_0 d(z)$, where a_0 is the core's radius at $z = 0$, the function $h(z)$ is related to $d(z)$ inversely ($h = 1/d$).

7.2.1 Linear Tapering

As a simple example of tapering, we choose $d(\xi) = 1 + \xi/\xi_0$. In this case, the core's radius increases linearly for positive values of ξ_0 and doubles when $\xi = \xi_0$. The radius decreases linearly for negative values of ξ_0. In the latter case, the condition $\xi < \xi_0$ must hold because the radius vanishes at $\xi = \xi_0$.

Equation (7.1.22) can be solved analytically for a linearly tapered GRIN medium [5]. With the transformation $\eta = 1 + \xi/\xi_0$, it becomes

$$\frac{d^2s}{d\eta^2} + \frac{\xi_0^2}{\eta^2}s = \frac{C_f^2\xi_0^2}{s^3}. \tag{7.2.1}$$

A further change of variables, $W = s/\eta^{1/2}$ and $u = \ln \eta$ converts this equation into the following one:

$$\frac{d^2W}{du^2} + \gamma^2 W = \frac{C_f^2\xi_0^2}{W^3}, \tag{7.2.2}$$

where $\gamma = (\xi_0^2 - 1/4)^{1/2}$ is a positive quantity for $\xi_0 > \frac{1}{2}$. In practice, ξ_0 is expected to be larger than 1.

The preceding equation has the form of Eq. (7.1.20), but $b(z)$ has been replaced with a constant. Its solution has the form given in Eq. (2.5.33) of Section 2.5.2 and can be written as

$$W^2(\xi) = 1 + (C_f^2\xi_0^2 - 1)\sin^2(\gamma u). \tag{7.2.3}$$

In terms of the original variables, we can write this solution as

$$s(\xi) = (1 + \xi/\xi_0)^{1/2}\{1 + (C_f^2\xi_0^2 - 1)\sin^2[\gamma \ln(1 + \xi/\xi_0)]\}^{1/2}. \tag{7.2.4}$$

Figure 7.2 shows how $s = w/w_0$ varies with $\xi = b_0z$ for several values of ξ_0 using $C_f = 0.2$. In general, the beam's width oscillates but does not vary in a periodic fashion because of tapering. For $\xi_0 = 100$, the evolution becomes almost periodic because the core's radius is nearly constant. For $\xi_0 = 20$, beam's width exceeds w_0 at self-imaging distances because the core's radius is increasing with distance. The width contracts for $\xi_0 = -20$ because the core becomes smaller at longer distances.

One may ask how the exact solution in Eq. (7.2.4) differs from the solution obtained in the adiabatic approximation and given in Eq. (7.1.24). Using $h(\xi) = (1 + \xi/\xi_0)^{-1}$, it is easy to show that the two solutions match when we make the approximation $\gamma = (\xi_0^2 - 1/4)^{1/2} \approx \xi_0$, that is, ξ_0 must be much larger than one half. This is generally the case in practice.

7.2.2 Other Tapering Profiles

It is relatively straightforward to solve Eq. (7.1.22) numerically for any taper profile $h(z)$. One can also use the approximate analytic solution given in Eq. (7.1.24) for any

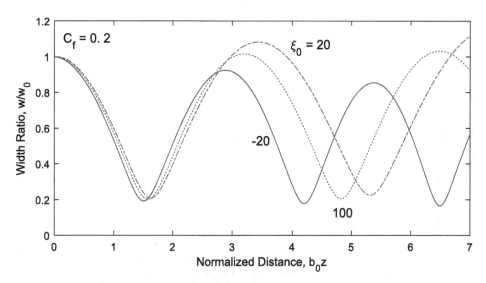

Figure 7.2 Variations in the width of a Gaussian beam inside a linearly tapered GRIN medium for several values of the parameter ξ_0.

taper profile. As an example, we consider an exponential form of the taper, governed by the function $h(\xi) = e^{\xi/\xi_0}$, which corresponds to an exponential reduction in the core's diameter with distance for $\xi_0 > 0$. Its use in Eq. (7.1.22) leads to

$$\frac{d^2 s}{du^2} + \xi_0^2 e^{2u} s = \frac{C_f^2 \xi_0^2}{s^3}, \tag{7.2.5}$$

where we made the transformation $u = \xi/\xi_0$. From Eq. (7.1.23), we can write its exact solution as

$$s(u) = [F^2(u) + C_f^2 \xi_0^2 G^2(u)]^{1/2}, \tag{7.2.6}$$

where $F(u)$ and $G(u)$ are the solutions of

$$\frac{d^2 Q}{du^2} + \xi_0^2 e^{2u} Q = 0, \qquad Q = F \text{ or } G, \tag{7.2.7}$$

obtained with the initial conditions $F = 1$ and $dF/d\xi = 0$ at $\xi = 0$, while $G = 0$ and $dG/d\xi = 1$ at $\xi = 0$.

Equation (7.2.7) has analytic solutions in terms of the Bessel functions, which can be used to obtain an exact solution [7]. However, it is much simpler to use the approximate form given in Eq. (7.1.24). The integral in this equation can easily be done to obtain the result

$$s^2(\xi) = e^{-\xi/\xi_0} \{\cos^2[\xi_0(e^{\xi/\xi_0} - 1)] + C_f^2 \sin^2[\xi_0(e^{\xi/\xi_0} - 1)]\}. \tag{7.2.8}$$

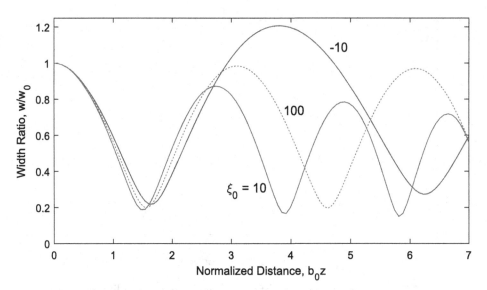

Figure 7.3 Variations in the width of a Gaussian beam inside an exponentially tapered GRIN medium for several values of the parameter ξ_0. The effects of diffraction are included using $C_f = 0.2$.

Figure 7.3 shows the results obtained for an exponential taper by plotting w/w_0 as a function of $\xi = b_0 z$ for three values of ξ_0 with $C_f = 0.2$. Similar to the case shown in Figure 7.2, the beam's width oscillates but does not vary in a periodic fashion because of tapering. An almost periodic variation of the width occurs for $\xi_0 = 100$ because the core's radius does not change much over the distance range covered in Figure 7.3. As ξ_0 is reduced, the core's size decreases exponentially with ξ. For $\xi_0 = 10$, the core becomes considerably narrower as the distance increases. As a result, the width becomes smaller than its input value each time self-imaging occurs. The opposite happens for $\xi_0 = -10$ because the core's size in this case increases with distance. Note that the self-imaging distance is also affected because it depends on the local value of $h(z)$.

7.3 Quantum-Mechanical Techniques

As we discussed in Section 2.5, the GRIN problem is identical to a two-dimensional harmonic oscillator. In the case of a tapered GRIN medium. the situation corresponds to an oscillator whose frequency changes with time. As such a quantum harmonic oscillator has been studied extensively [14–18], we make use of the relevant tools developed for it to understand the propagation of optical beams inside a tapered GRIN medium.

7.3.1 Solution in the Adiabatic Approximation

Equation (7.1.10) describing wave propagation inside a tapered GRIN medium can be written in the form of a Schrödinger equation, as was done in Section 2.5.1 in the untapered case. Writing $\tilde{E}$ in the form of Eq. (7.1.11), we obtain the paraxial equation:

$$2ik\frac{\partial A}{\partial z} + \frac{\partial^2 A}{\partial x^2} + \frac{\partial^2 A}{\partial y^2} - k^2 b^2(z)\rho^2 A = 0. \tag{7.3.1}$$

We write $b(z)$ as $b(z) = b_0 h(z)$ and introduce the normalized variables as

$$q_1 = \sqrt{kb_0}\,x, \qquad q_2 = \sqrt{kb_0}\,y, \qquad \xi = b_0 z. \tag{7.3.2}$$

In terms of the new variables, we can write Eq. (7.3.1) as the following Schrödinger equation (with $\hbar = 1$):

$$i\frac{dA}{d\xi} = \hat{H}A, \tag{7.3.3}$$

where the Hamiltonian is given by

$$\hat{H} = \frac{1}{2}\left(\hat{p}_1^2 + \hat{p}_2^2\right) + \frac{1}{2}h^2(\xi)\left(\hat{q}_1^2 + \hat{q}_2^2\right), \tag{7.3.4}$$

and the momentum operators are defined as $\hat{p}_j = -i(d/dq_j)$ for $j = 1, 2$. Recalling that ξ plays the role of time, the Hamiltonian in Eq. (7.3.4) is identical to that of a two-dimensional harmonic oscillator of unit mass, whose frequency $h(\xi)$ varies with time.

As the Hamiltonian is separable in its two dimensions, the wave function can be factored as $A(q_1, q_2, \xi) = \Psi(q_1, \xi)\Psi(q_2, \xi)$. This feature allows us to consider the simpler one-dimensional problem [14–16]:

$$i\frac{d}{d\xi}|\Psi\rangle = \hat{H}|\Psi\rangle, \qquad \hat{H} = \frac{1}{2}\hat{p}^2 + \frac{1}{2}h^2(\xi)\hat{q}^2, \tag{7.3.5}$$

where we adopted the Dirac notation for a quantum state and dropped the subscript 1 or 2 to simplify the notation. Before solving the problem exactly, we consider its solution in the adiabatic approximation [19]. This approximation can be used when $|dh/d\xi| \ll h^2(\xi)$.

In the adiabatic approximation, we can use the concept of local eigenstates. At any distance ξ, we write the solution of Eq. (7.3.5) in the form $|\Psi(\xi)\rangle = e^{-iE\xi}|\psi(\xi)\rangle$, where $|\psi(\xi)\rangle$ is obtained by solving the eigenvalue equation,

$$\hat{H}(\xi)|\psi(\xi)\rangle = E(\xi)|\psi(\xi)\rangle. \tag{7.3.6}$$

For a harmonic oscillator, all eigenstates involve a Hermite–Gauss function:

$$\psi_m(q, \xi) = \langle q|\psi_m\rangle = N_m H_m(hq)\exp(-h^2 q^2/2), \tag{7.3.7}$$

with the eigenvalue $E_m(\xi) = (m + \frac{1}{2})h(\xi)$ and the normalization factor N_m.

The wave function $|\Psi(\xi)\rangle$ at any distance ξ can be expanded in terms of the local eigenstates as

$$|\Psi(\xi)\rangle = \sum_k c_k(\xi)|\psi_m(\xi)\rangle. \tag{7.3.8}$$

Using it in Eq. (7.3.5), the expansion coefficients in the adiabatic approximation are found to evolve as [19]

$$i\frac{dc_k}{d\xi} \approx \left(E_k - i\langle\psi_k|\psi_k'\rangle\right) c_k, \tag{7.3.9}$$

where $\psi_k' = d\psi_k/d\xi$. The preceding equation can be integrated and has the solution [19]

$$c_k(\xi) = c_k(0) \exp\left(-i\int_0^\xi E_k(\xi')d\xi\right) \exp\left(-i\int_0^\xi v_k(\xi')d\xi\right), \tag{7.3.10}$$

where $v_k = i\langle\psi_k|\psi_k'\rangle$ is a real quantity. This result shows that the initial value of c_k is modified by two phase factors. The first one is the dynamical phase, and the second one is the geometrical phase, also known as the Berry phase [20].

The preceding adiabatic solution has an important property. When the system is initially in a specific eigenstate [$c_m(0) = 1$ and $c_k(0) = 0$ for all $k \neq m$], it remains in the same eigenstate, even though both the wave function and the eigenvalue change with ξ in a continuous manner. For example, if the LP_{01} mode of an adiabatically tapered GRIN fiber is excited at $z = 0$, the same mode persists as the beam propagates inside the fiber, but its spot size, $w_g(z) = [kb(z)]^{-1/2}$, evolves continuously with z as dictated by the local value of $b(z)$.

7.3.2 Invariant-Based Solution

The Schrödinger equation in Eq. (7.3.5) can be solved, without the adiabatic approximation, using the concept of an invariant [14]. In this approach, one looks for an invariant operator $\hat{I}$ that does not change with ξ. Following Ref. [15], this invariant has the form

$$\hat{I} = \frac{1}{2}\left[\frac{1}{s^2}\hat{q}^2 + \left(s\hat{p} - \frac{ds}{d\xi}\hat{q}\right)^2\right], \tag{7.3.11}$$

where the function $s(\xi)$ is the solution of

$$\frac{d^2s}{d\xi^2} + h^2(\xi)s = \frac{1}{s^3}, \tag{7.3.12}$$

with the initial conditions $s = 1$ and $ds/d\xi = 0$ at $\xi = 0$. Equation (7.3.12) is similar to that in Eq. (7.1.22), and its general solution is given by

$$s(\xi) = |h(\xi)|^{-1/2} \left[\sqrt{1 + C^2} + C \sin(2h\xi + \phi_0) \right], \qquad (7.3.13)$$

where C and ϕ_0 are two constants of integration.

The eigenstates and eigenvalues of the invariant operator $\hat{I}$ are found by introducing the annihilation and creation operators as

$$\hat{a} = [\hat{q}/s + i(s\hat{p} - s'\hat{q})]/\sqrt{2}, \qquad \hat{a}^\dagger = [\hat{q}/s - i(s\hat{p} - s'\hat{q})]/\sqrt{2}, \qquad (7.3.14)$$

where $s' = ds/d\xi$. These operators satisfy the standard commutation relation, $[\hat{a}, \hat{a}^\dagger] = 1$, and allow us to write the operator I in the form $I = \hat{a}^\dagger \hat{a} + \frac{1}{2}$. Thus, the problem has been reduced to a standard harmonic oscillator, and the eigenstates of the operator $\hat{I}$ are the number states $|n\rangle$ with the eigenvalue $n + \frac{1}{2}$, where $n \geq 0$ is an integer. We can use these states as the basis for calculating various averages. The Hamiltonian is not diagonal in this basis, and its diagonal elements are given by [15]

$$\langle n|\hat{H}|n\rangle = \frac{1}{2}[s'^2 + h^2(\xi)s^2 + s^{-2}](n + \frac{1}{2}). \qquad (7.3.15)$$

We can also use the number states as the basis for expanding the wave function Ψ as

$$|\Psi(\xi)\rangle = \sum_k c_n e^{i\phi_n(\xi)}|n\rangle, \qquad (7.3.16)$$

where the phase ϕ_n is given by [17]

$$\phi_n(\xi) = -(n + \frac{1}{2}) \int_0^\xi \frac{d\xi'}{s^2(\xi')}. \qquad (7.3.17)$$

The variable $\phi_n(\xi)$ is related to the local eigenvalue of the quantum state. In the limit $h = 1$ for all ξ (no tapering), $s(\xi')$ becomes 1 in Eq. (7.3.17), and $\phi_n = -(n + \frac{1}{2})\xi$. This feature can be used to calculate the propagation constant β_n as a function of z for a tapered GRIN fiber.

7.3.3 Solution in the Heisenberg Picture

In another approach to solving Eq. (7.3.5), the Heisenberg picture is employed. In this case, any operator evolves with ξ as $\hat{X}(\xi) = \hat{U}^\dagger \hat{X}(0)\hat{U}$, where the evolution operator $\hat{U}$ satisfies [16]

$$i\frac{d\hat{U}}{d\xi} = \hat{H}(\xi)\hat{U}. \qquad (7.3.18)$$

The operators $\hat{q}$ and $\hat{p}$ at a distance ξ can be related to their initial values at $\xi = 0$ through a transfer matrix T as

$$\begin{pmatrix} \hat{q}(\xi) \\ \hat{p}(\xi) \end{pmatrix} = \begin{pmatrix} T_{11} & T_{12} \\ T_{21} & T_{22} \end{pmatrix} \begin{pmatrix} \hat{q}(0) \\ \hat{p}(0) \end{pmatrix}, \tag{7.3.19}$$

where T_{ij} varies with ξ. Clearly, $T_{11}(0) = T_{22}(0) = 1$ and $T_{11}(0) = T_{21}(0) = 0$. To preserve the commutation relations, the matrix T must be unitary, and its determinant should satisfy $T_{11}T_{22} - T_{12}T_{21} = 1$ for all ξ. For the Hamiltonian given in Eq. (7.3.5), $T_{21} = dT_{11}/d\xi$ and $T_{22} = dT_{12}/d\xi$. Thus, we only need to know two matrix elements, $T_{11}(\xi)$ and $T_{12}(\xi)$, to find the transfer matrix T at any distance ξ.

With the Heisenberg formalism, we can calculate the expectation value of any operator at a distance ξ by using the quantum state at $\xi = 0$. As $h(\xi) = 1$ at $\xi = 0$, the initial state can be taken to be a number state $|n\rangle$ (see Appendix A). For example, the second moment of $\hat{q}$ in the number state $|n\rangle$ is found to be

$$\langle \hat{q}^2(\xi) \rangle_n = \langle n|\hat{q}^2(\xi)|n\rangle = (T_{11}^2 + T_{12}^2)(n + \tfrac{1}{2}). \tag{7.3.20}$$

As another example, we consider an exponential form of the taper governed by the function $h(\xi) = e^{\alpha\xi}$. In this case, the Hamiltonian in Eq. (7.3.5) leads to the dynamical equations:

$$\frac{d\hat{q}}{d\xi} = \frac{\partial H}{\partial \hat{p}} = \hat{p}, \qquad \frac{d\hat{p}}{d\xi} = -\frac{\partial H}{\partial \hat{q}} = -e^{2\alpha\xi}\hat{q}. \tag{7.3.21}$$

These two equations can be combined to obtain

$$\frac{d^2\hat{q}}{d\xi^2} + e^{2\alpha\xi}\hat{q} = 0. \tag{7.3.22}$$

This equation is identical to the ray equation in Eq. (7.1.2), and we can use the solution given in Section 7.1. In the present notation, the solution can be written as $\hat{q}(\xi) = T_{11}\hat{q}(0) + T_{12}\hat{p}(0)$ with

$$T_{11}(\xi) = e^{-\alpha\xi/2}\cos[(e^{\alpha\xi} - 1)/\alpha], \qquad T_{12}(\xi) = e^{-\alpha\xi/2}\sin[(e^{\alpha\xi} - 1)/\alpha]. \tag{7.3.23}$$

The remaining two elements of the transfer matrix are found using $T_{21} = dT_{11}/d\xi$ and $T_{22} = dT_{12}/d\xi$. The resulting matrix T can be used for the evolution of any optical beam inside an exponentially tapered GRIN medium.

The transfer matrix T is related to the ABCD matrix found in Section 3.1 in the context of the rays. This follows from the classical analog of Eq. (7.3.19) after noting that $\hat{p}$ is related to the slope of a ray. We can identify the four elements of the ABCD matrix as $T_{11} = A$, $T_{12} = B$, $T_{21} = C$, and $T_{22} = D$. In particular, Eq. (7.3.23) provides us with the ABCD matrix elements for an exponentially tapered GRIN medium. It can be verified that they are identical to those obtained using

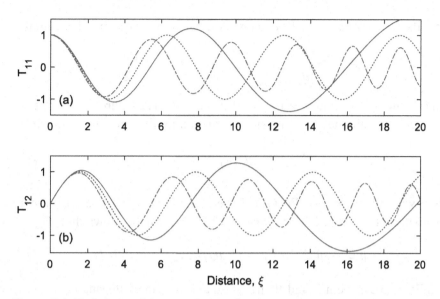

Figure 7.4 Variations of the transfer-matrix elements (a) T_{11} and (b) T_{12} with distance ξ for an exponentially tapered GRIN medium in three cases: $\alpha = -0.05$ (solid), $\alpha = 0.05$ (dashed), and $\alpha = 0$ (dotted).

Eqs. (7.1.8) and (7.1.9). Figure 7.4 shows how T_{11} and T_{22} vary with distance for $\alpha = -0.05$, 0, and 0.05. As expected, periodic evolution occurs only for $\alpha = 0$. When $\alpha = 0.05$, both the amplitude and the period of oscillations are reduced because the core becomes narrower with increasing distance for positive values of α. The opposite occurs for negative values of α because the core becomes wider with increasing distance.

7.3.4 Solution Based on Group Theory

Another approach to solving Eq. (7.3.5) makes use of group theory and a time-ordering technique. We first write its solution as $|\Psi(\xi)\rangle = \hat{U}(\xi)|\Psi(0)\rangle$, where the evolution operator $\hat{U}(\xi)$ is formally given by

$$\hat{U}(\xi) = \hat{T}\exp\left[-i\int_0^\xi \hat{H}(\xi')d\xi'\right], \tag{7.3.24}$$

with $\hat{T}$ denoting the time-ordering operator. Using properties of the SU(1, 1) group, it is found that $\hat{U}(\xi)$ can be factored as [18]

$$\hat{U}(\xi) = \exp\left[\frac{i}{2}u(\hat{q}\hat{p} + \hat{p}\hat{q})\right]\exp\left(\frac{i}{2}v\hat{q}^2\right)\exp\left(-\frac{i}{2}w\hat{p}^2\right), \tag{7.3.25}$$

where the functions u, v, w depend on ξ and satisfy the differential equations

$$\frac{du}{d\xi} = -ve^{2u}, \qquad \frac{dv}{d\xi} = v^2e^{2u} - h^2(\xi)e^{-2u}, \qquad \frac{dw}{d\xi} = e^{2u}, \qquad (7.3.26)$$

with the initial condition $u = v = w = 0$ at $\xi = 0$. These three equations can be combined to obtain the following single equation for the function $s = du/d\xi$:

$$\frac{ds}{d\xi} = s^2 + h^2(\xi), \qquad s(0) = 0. \qquad (7.3.27)$$

It may be difficult to solve Eq. (7.3.27) because of its nonlinear nature. Introducing two auxiliary functions, $U = e^{-u}$ and $W = we^{-u}$, it can be shown that [18]

$$u = -\ln U, \qquad v = U(dU/d\xi), \qquad w = W/U. \qquad (7.3.28)$$

From Eq. (7.3.27), both U and W are found to satisfy the following linear equation:

$$\frac{d^2Q}{d\xi^2} + h^2(\xi)Q = 0, \qquad Q = U \text{ or } W. \qquad (7.3.29)$$

The initial conditions at $\xi = 0$ are $U = 1$ with $dU/d\xi = 0$ and $W = 0$ with $dW/d\xi = 1$. Equation (7.3.29) is identical to the ray equation and is easier to solve. For a given form of the taper function $h(\xi)$, we only need to solve Eq. (7.3.29), and use the solution in Eq. (7.3.28) to obtain u, v, w as a function of ξ.

Let us use the evolution operator in Eq. (7.3.25) to find the wave function at a distance ξ in the q representation, assuming that it is known at $\xi = 0$:

$$\Psi(q, \xi) = \exp\left[\frac{i}{2}u(\hat{q}\hat{p} + \hat{p}\hat{q})\right]\exp\left(\frac{i}{2}v\hat{q}^2\right)\exp\left(-\frac{i}{2}w\hat{p}^2\right)\Psi(q, 0), \qquad (7.3.30)$$

The three exponential operators in Eq. (7.3.30) can be applied to $\psi(q, 0)$ by recalling $\hat{p} = -i(d/dq)$ and employing the known relations [21, 22]:

$$\exp\left(a\frac{d}{dx}\right)f(x) = f(x + a), \qquad \exp\left(ax\frac{d}{dx}\right)f(x) = f(e^a x), \qquad (7.3.31)$$

$$\exp\left(a\frac{d^2}{dx^2}\right)f(x) = \frac{1}{\sqrt{4\pi a}}\int_{-\infty}^{\infty}\exp\left[-\frac{(x - x')^2}{4a}\right]f(x')\,dx'. \qquad (7.3.32)$$

Using Eq. (7.3.32) in Eq. (7.3.30), $\Psi(q, \xi)$ becomes

$$\Psi(q, \xi) = \frac{e^{u/2}}{\sqrt{2\pi iw}}\exp\left(uq\frac{d}{dq}\right)\exp\left(\frac{i}{2}vq^2\right)\int_{-\infty}^{\infty}\exp\left[\frac{i}{2w}(q - q')^2\right]\Psi(q', 0)\,dq'.$$

$$(7.3.33)$$

Using the relations in Eq. (7.3.31), we obtain the final result

$$\Psi(q,\xi) = \frac{e^{u/2}}{\sqrt{2\pi i w}} \exp\left(\frac{i\chi}{2w}q^2\right) \int_{-\infty}^{\infty} \exp\left[\frac{i}{2w}(q'^2 - 2qq'e^u)\right] \Psi(q',0)dq',$$

(7.3.34)

where $\chi = (1 + vw)e^{2u}$ is a function of ξ because u, v, and w vary with ξ.

Let us use Eq. (7.3.34) to study the evolution of optical beams inside a standard GRIN fiber (no tapering). For $h(\xi) = 1$, Eq. (7.3.29) is easily solved to obtain $U(\xi) = \cos\xi$ and $W(\xi) = \sin\xi$. Using them in Eq. (7.3.28), we find

$$u(\xi) = -\ln(\cos\xi), \qquad v(\xi) = -\sin\xi\cos\xi, \qquad w(\xi) = \tan\xi. \qquad (7.3.35)$$

Employing these functions in Eq. (7.3.34), we obtain the following well-known result for a harmonic oscillator [15, 18]:

$$\Psi(q,\xi) = \int_{-\infty}^{\infty} K(q,q',\xi)\Psi(q',0)\,dq', \qquad (7.3.36)$$

where the "propagator" is given by

$$K(q,q',\xi) = \frac{1}{\sqrt{2\pi i \sin\xi}} \exp\left(\frac{i}{2\sin\xi}[(q^2 + q'^2)\cos\xi - 2qq']\right). \qquad (7.3.37)$$

If we convert to the original variables and extend the analysis to both transverse dimensions, the form of the propagator agrees with the propagation kernel found in Eq. (3.2.10) of Section 3.2 for a GRIN medium.

The important question is how the propagator in Eq. (7.3.37) changes for a tapered GRIN medium. The answer is provided by Eq. (7.3.34), which can be written in the form of Eq. (7.3.36) with the propagator given as

$$K(q,q',\xi) = \frac{e^{u/2}}{\sqrt{2\pi i w}} \exp\left(\frac{ie^u}{2w}[(\chi q^2 + q'^2)e^{-u} - 2qq')]\right), \qquad (7.3.38)$$

This form of K implies that the kernel of a tapered GRIN medium retains the same functional form given in Eq. (7.3.37) with different coefficients. We can also write the kernel in the form of Eq. (3.2.16) using the elements of an ABCD matrix; tapering only modifies the ξ dependence of these elements [8]. In the case of Eq. (7.3.38), we can identify the required elements as

$$A = e^{-u}, \qquad B = we^{-u}, \qquad D = \chi e^{-u} = (1 + vw)e^u. \qquad (7.3.39)$$

The element C is found from the condition $AD - BC = 1$, set by the requirement that the ABCD matrix is unitary.

In terms of the ABCD-matrix elements, the propagator in Eq. (7.3.38) takes the form

$$K(q,q',\xi) = \left(\frac{1}{2\pi i B}\right)^{1/2} \exp\left(\frac{ik}{2B}[Dq^2 + Aq'^2 - 2qq']\right). \qquad (7.3.40)$$

We can write the ABCD matrix elements in terms of the functions $U(\xi)$ and $W(\xi)$ that are solutions of the ray equation in Eq. (7.3.29). From Eq. (7.3.28), we find

$$A = U(\xi), \qquad B = W(\xi), \qquad D = \left(1 + W\frac{dU}{d\xi}\right)\frac{1}{U}. \qquad (7.3.41)$$

As a quick check, for a standard GRIN fiber (no tapering), the functions U and W are found using $h(\xi) = 1$ in Eq. (7.3.29) and are $U = \cos(\xi)$ and $W = \sin(\xi)$. It is easy to show that $D = \cos(\xi)$ so that Eq. (7.3.40) reduces to Eq. (7.3.37) in this specific case.

It is important to stress that the kernel in Eq. (7.3.40) reduces the problem of wave propagation inside a tapered GRIN medium to solving the ray equation given in Eq. (7.3.29). As discussed in Section 7.1.1, this equation can be solved in the adiabatic approximation, providing the following expression for $U(\xi)$ and $W(\xi)$:

$$U(\xi) = h^{-1/2}(\xi)\cos\left(\int_0^\xi h(\xi')d\xi'\right), \qquad W(\xi) = h^{-1/2}(\xi)\sin\left(\int_0^z h(\xi')d\xi'\right). \qquad (7.3.42)$$

As an example, for an exponential taper for which $h(\xi') = e^{\alpha\xi}$, these functions take the form

$$U(\xi) = e^{-\alpha\xi/2}\cos[(e^{\alpha\xi} - 1)/\alpha], \qquad W(\xi) = e^{-\alpha\xi/2}\sin[(e^{\alpha\xi} - 1)/\alpha]. \quad (7.3.43)$$

We can use them in Eq. (7.3.41) and employ the kernel in Eq. (7.3.40) to study how an optical beam evolves inside such a GRIN medium. As an example, Figure 7.5 shows the evolution of a Gaussian beam inside an exponentially tapered GRIN medium for $\alpha = \pm0.05$. The initial width of the beam corresponds to $C_f = 0.3$. The beam oscillates in both cases but without self-imaging. It becomes narrower with increasing distance and its oscillation period becomes smaller for $\alpha = 0.05$; the opposite occurs for $\alpha = -0.05$. We can understand this behavior by recalling that the core size increases with distance for negative values of α and decreases for its positive values. Recall that Figure 7.4 shows how the matrix elements A and B evolve inside such a GRIN medium ($A = T_{11}$ and $B = T_{12}$).

7.3.5 Solution Based on Squeezing

In an interesting approach, the evolution operator $\hat{U}(\xi)$ is related to an operator responsible for squeezing of noise in some quantum systems [23]. The Hamiltonian in Eq. (7.3.5) is first written in terms of the annihilation and creation operators of a standard harmonic oscillator [$h(\xi) = 1$] using

$$\hat{a} = (\hat{q} + i\hat{p})/\sqrt{2}, \qquad \hat{a}^\dagger = (\hat{q} - i\hat{p})/\sqrt{2}. \qquad (7.3.44)$$

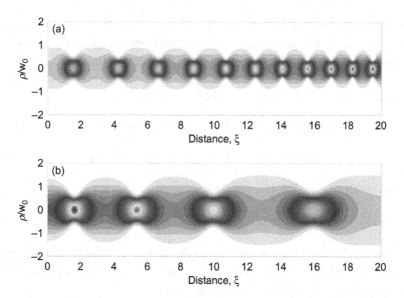

Figure 7.5 Evolution of a Gaussian beam inside an exponentially tapered GRIN medium for (a) $\alpha = 0.05$ and (b) $\alpha = -0.05$. The initial width of the beam corresponds to $C_f = 0.3$.

The Hamiltonian in Eq. (7.3.5) can then be expressed as

$$\hat{H} = f_0(\xi)\hat{K}_0 + f(\xi)\hat{K}_+ + f^*(\xi)\hat{K}_-, \tag{7.3.45}$$

where $f_0(\xi) = (h^2 + 1)$, $f(\xi) = \frac{1}{2}(h^2 - 1)$, and the three operators are defined as

$$\hat{K}_0 = \frac{1}{2}(\hat{a}^\dagger\hat{a} + \tfrac{1}{2}), \qquad \hat{K}_+ = \frac{1}{2}\hat{a}^2, \qquad \hat{K}_- = \frac{1}{2}\hat{a}^{\dagger 2}. \tag{7.3.46}$$

The evolution operator associated with the Hamiltonian in Eq. (7.3.45) can be found using theory behind the SO(2,1) group and is given by [23]

$$\hat{U}(\xi) = \exp[ig(\xi)\hat{K}_0]\exp[\chi(\xi)\hat{K}_+ - \chi^*(\xi)\hat{K}_-], \tag{7.3.47}$$

where the functions $g(\xi)$ and $\chi(\xi)$ are defined as

$$g(\xi) = -\int_0^\xi [f_0(\xi) + f^*(\xi)Q + f(\xi)Q^*]d\xi, \qquad \chi(\xi) = Qe^{-ig}, \tag{7.3.48}$$

and $Q(\xi)$ is the solution of the differential equation

$$\frac{dQ}{d\xi} + i(f + f_0Q + f^*Q^2) = 0, \qquad Q(0) = 0. \tag{7.3.49}$$

The second exponential term in Eq. (7.3.47) generates the so-called two-photon coherent states [24]. It is also responsible for squeezing of quantum noise [21]. For this reason, it is known as the squeezing operator and is defined as

$$\hat{S}(\chi) = \exp[\chi \hat{K}_+ - \chi^* \hat{K}_-]. \tag{7.3.50}$$

For complex values of $\chi = |\chi| e^{i\theta}$, the squeezing operator can be factored as

$$\hat{S}(\chi) = \exp(\tfrac{1}{2} i\theta \hat{a}^\dagger \hat{a}) \hat{S}(|\chi|) \exp(-\tfrac{1}{2} i\theta \hat{a}^\dagger \hat{a}). \tag{7.3.51}$$

Using this result in Eq. (7.3.47), we obtain

$$\hat{U}(\xi) = \exp[i(g + \theta) \hat{K}_0] S(|\chi|) \exp[-i\theta \hat{K}_0]. \tag{7.3.52}$$

Let us apply the preceding formalism to the untapered case by setting $h(\xi) = 1$, resulting in $f_0 = 2$ and $f = 0$. The solution of Eq. (7.3.49) is found to be $Q = \exp(-2i\xi) - 1$. Its use in Eq. (7.3.48) then yields

$$g(\xi) = -2\xi, \qquad \chi(\xi) = 1 - e^{2i\xi}. \tag{7.3.53}$$

It is easy to show that $\chi(\xi) = -2i \sin \xi \, e^{i\xi}$, resulting in $|\chi| = 2 \sin \xi$ and $\theta = \xi$ (ignoring a constant phase). Using these values, the evolution operator becomes

$$\hat{U}(\xi) = \exp(-i\xi \hat{K}_0) S(2 \sin \xi) \exp(-i\xi \hat{K}_0). \tag{7.3.54}$$

This form can be used to study how an initial quantum state evolves with ξ. In the position or q basis, the three terms affect the state differently. The exponential operator corresponds to a fractional Fourier transform, expressed as a convolution with a kernel of the $\exp[id(q - q')^2]$, where d depends on ξ. The squeezing operator scales the field such that

$$\langle q | S(|\chi|) \psi \rangle = \exp(-\tfrac{1}{2} |\chi|) \psi(q e^{-|\chi|}). \tag{7.3.55}$$

Using these results, one can show that the initial state evolves as indicated in Eq. (7.3.36) with the kernel given in Eq. (7.3.37).

7.4 Periodic Axial Variations

Periodic axial variations of the refractive index constitute an important special case. For example, the refractive index n_0 of a fiber's core is modulated periodically to make fiber Bragg gratings that have a multitude of applications [12]. In the case of a GRIN fiber, the core's radius can be modulated along its length in a periodic fashion [7]. In this section we focus on this latter situation.

7.4.1 Sinusoidal Modulation of Core's Radius

When the core's radius is modulated sinusoidally along a GRIN fiber's length, we can write the radius as $a(z) = a_0[1 + \eta \sin(Kz + \theta)]$, where a_0 is the average value, η is the amplitude and K is the spatial frequency of modulation. The paraxial equation given in Eq. (7.1.12) in this case takes the form

$$2ik\frac{\partial A}{\partial z} + \nabla_T^2 A - (kb_0)^2 h^2(z)\rho^2 A = 0, \tag{7.4.1}$$

where $b_0 = \sqrt{2\Delta}/a_0$ and $h(z) = [1 + \eta \sin(Kz + \theta)]^{-1}$. Similar to Section 7.1.2, we can use the eikonal-based technique to find the self-similar solutions of Eq. (7.4.1). This approach leads to the following equation that describes how the width $w(z)$ of the beam evolves along a sinusoidally modulated GRIN fiber (or rod):

$$\frac{d^2w}{dz^2} + b_0^2 h^2(z)w = \frac{1}{k^2 w^3}. \tag{7.4.2}$$

As we did in Section 7.2, it is useful to write Eq. (7.4.2) in a normalized form using $\xi = b_0 z$ and $s = w/w_0$, where $w_0 = w(0)$ is the initial width of an input beam. The resulting equation becomes

$$\frac{d^2 s}{d\xi^2} + h^2(\xi)s = \frac{C_f^2}{s^3}, \qquad h(\xi) = [1 + \eta \sin(\mu\xi + \theta)]^{-1}, \tag{7.4.3}$$

where $C_f = (kb_0 w_0^2)^{-1}$ and $\mu = K/b_0 = L_p/L_m$ with $L_m = 2\pi/K$ and $L_p = 2\pi/b$. The parameter μ represents the ratio of length scales associated with self-imaging and axial modulation. The parameter C_f first appeared in Eq. (2.5.33). As discussed in Section 7.1.2, the general solution of this equation has the form given in Eq. (7.1.23). Moreover, when the adiabatic approximation holds, the solution can be written in an analytic form as

$$s^2(\xi) = h^{-1}(\xi)\left[\cos^2\left(\int_0^\xi h(\xi')d\xi'\right) + C_f^2 \sin^2\left(\int_0^z h(\xi')d\xi'\right)\right]. \tag{7.4.4}$$

The integral appearing in this equation can be done in a closed form using $h(\xi)$ from Eq. (7.4.3), and the result for $\theta = 0$ is [25]

$$\int_0^\xi h(\xi')d\xi' = \frac{2}{\mu\sqrt{1-\eta^2}}\left[\tan^{-1}\left(\frac{\tan(\mu\xi/2 + \eta)}{\sqrt{1-\eta^2}}\right) - \tan^{-1}\left(\frac{\tan(\eta)}{\sqrt{1-\eta^2}}\right)\right]. \tag{7.4.5}$$

To go beyond the adiabatic approximation, one can solve Eq. (7.4.3) numerically to study how periodic variations in the core's size of a GRIN medium affect the width of a Gaussian beam. Figure 7.6 shows the numerical results by choosing $\mu = 1$, $\theta = 0$, and $C_f = 0.2$. The width ratio w/w_0 is plotted as a function of ξ for three values of η. A nearly periodic variation of the width occurs for $\eta = 0.1$,

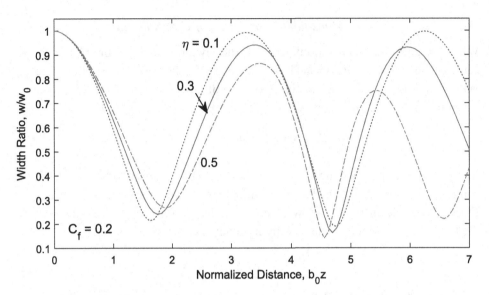

Figure 7.6 Variations in the width of a Gaussian beam inside a sinusoidally tapered GRIN medium for several values of the parameter η. The effects of diffraction are included using $C_f = 0.2$.

indicating that sinusoidal modulations of the core's size have a minor effect for small modulation amplitudes. For larger values of η, the width's evolution becomes non-periodic and deviates considerably from the periodic solution expected for uniform GRIN fibers with a constant core radius.

In practice, the modulation period L_m is expected to be longer than the self-imaging period L_p, whose typical values are ~ 1 mm for most GRIN devices. In this case, one should expect an oscillatory behavior with the period L_p such that the amplitude of oscillations changes on longer length scales. One way to study such long-scale variations is to employ the approximation $\eta \ll 1$ and use $h^2(\xi) \approx 1 - 2\eta \sin(\mu\xi + \theta)$. In this case, Eq. (7.4.3) takes the form

$$\frac{d^2 s}{d\xi^2} + s = \frac{C_f^2}{s^3} + 2\eta \sin(\mu\xi + \theta)s. \qquad (7.4.6)$$

For $\eta = 0$, we recover the standard equation for the beam's width leading to periodic self-imaging. For a small value of η, the last term in Eq. (7.4.6) acts as a periodic perturbation.

When η is relatively large and the condition $\eta \ll 1$ is not valid, one must solve Eq. (7.4.3) numerically. As an example, Figure 7.7 shows the width variations with distance for $\eta = 0.3$ and $\eta = 0.5$ after choosing $\mu = 0.1$, $\theta = 0$, and $C_f = 0.2$. In both cases, the width does not follow a periodic evolution pattern expected for a uniform GRIN medium. It is evident from Figure 7.7 that the period of the

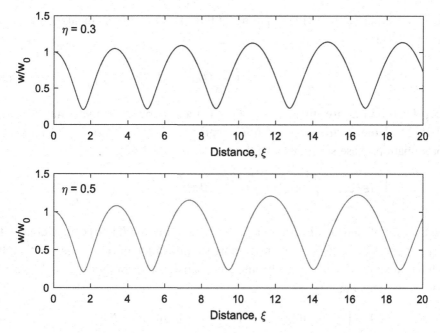

Figure 7.7 Variations in the width of a Gaussian beam inside a sinusoidally tapered GRIN medium for two values of the parameter η when $C_f = 0.2$ and $\mu = 0.1$. For this value of μ, one modulation cycle is 10 times longer than the initial self-imaging period.

oscillations slowly increases with distance. More importantly, the amplitude of oscillations also increases, indicating that the beam becomes wider after each cycle.

7.4.2 Quantum Approach

We can apply the quantum techniques of Section 7.3 to the case of sinusoidal modulation of the core's radius of a GRIN fiber. As we saw there, this situation corresponds to a harmonic oscillator whose frequency is modulated in a sinusoidal fashion [26–28]. To study this problem, we write Eq. (7.4.1) in the form of a Schrödinger equation (7.3.3), after normalizing the variables as in Eq. (7.3.2). We then separate the Hamiltonian into two parts and solve the one-dimensional problem. From Eq. (7.3.5), we need to solve

$$i\frac{d}{d\xi}|\Psi\rangle = \hat{H}|\Psi\rangle, \qquad \hat{H} = \frac{1}{2}[\hat{p}^2 + h^2(\xi)\hat{q}^2], \tag{7.4.7}$$

As we saw in Section 7.3.4, any quantum state for this Hamiltonian evolves as

$$\Psi(q,\xi) = \int_{-\infty}^{\infty} K(q,q',\xi)\Psi(q',0)\,dq', \tag{7.4.8}$$

where the Kernel can be written in terms of the ABCD-matrix elements as indicated in Eq. (7.3.38):

$$K(q, q', \xi) = \left(\frac{1}{2\pi iB}\right)^{1/2} \exp\left(\frac{ik}{2B}[Dq'^2 + Aq'^2 - 2qq']\right). \tag{7.4.9}$$

Here, A, B, and D are given in Eq. (7.3.41) and depend on the solutions $U(\xi)$ and $W(\xi)$ of the harmonic-oscillator equation in Eq. (7.3.29). In the adiabatic approximation, these solutions take the form of Eq. (7.3.42):

$$U(\xi) = h^{-1/2}(\xi)\cos\left(\int_0^\xi h(\xi')d\xi'\right), \qquad W(\xi) = h^{-1/2}(\xi)\sin\left(\int_0^z h(\xi')d\xi'\right). \tag{7.4.10}$$

In the case of sinusoidal modulation, the integral in Eq. (7.4.10) can be done with the result given in Eq. (7.4.5). Considerable simplification occurs if assume that the modulation amplitude is relatively small and satisfies the condition $\eta \ll 1$. We can then approximate $h(\xi)$ as

$$h(\xi) = [1 + \eta\sin(\mu\xi + \theta)]^{-1} \approx 1 - \eta\sin(\mu\xi + \theta). \tag{7.4.11}$$

Using this form, the integral in Eq. (7.4.10) becomes

$$\int_0^\xi h(\xi')d\xi' = \xi + \frac{\eta}{\mu}[\cos(\mu\xi + \theta) - \cos(\theta)]. \tag{7.4.12}$$

This result can be used to find the matrix elements A, B, and D appearing in Eq. (7.4.9) and calculate $\Psi(q, \xi)$ at any distance from Eq. (7.4.9), once the input state $\Psi(q, 0)$ has been specified.

As an example, Figure 7.8 shows the evolution of a Gaussian beam inside a sinusoidally modulated GRIN medium for $\mu = 0.1$ and $\mu = 0.2$ using $\eta = 0.2$ and $\theta = 0$. The initial width of the beam is matched to the width of the fundamental mode of the fiber. In a uniform GRIN fiber ($\eta = 0$), such a beam would propagate without change in its width or shape, as it excites only the LP_{01} mode of the GRIN fiber. However, as seen in Figure 7.8, the beam width changes in a periodic fashion when the core's radius is modulated sinusoidally. As one may expect, the period is related to the modulation frequency as $\xi_m = 2\pi/\mu$.

The beam's evolution does not remain adiabatic when μ is close to 1. We can use the approach of Section 7.3.5 based on the squeezing operator in the non-adiabatic regime. In the case of weak modulation ($\eta \ll 1$), we can write the Hamiltonian as

$$\hat{H} = \frac{1}{2}(\hat{p}^2 + \hat{q}^2) + \eta\sin(\mu\xi + \theta)\hat{q}^2, \tag{7.4.13}$$

where the first term corresponds to an unperturbed harmonic oscillator and the second term acts as a perturbation resulting from sinusoidal modulations of the

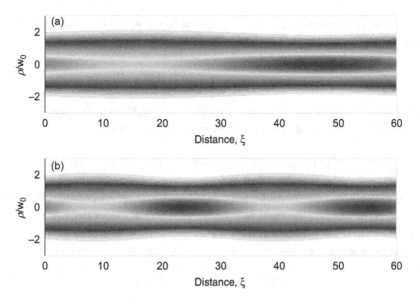

Figure 7.8 Evolution of a Gaussian beam inside a sinusoidally modulated GRIN medium for (a) $\mu = 0.1$ and (b) $\mu = 0.2$ with $\eta = 0.2$ and $\theta = 0$. The initial width of the beam is matched to the width of the fundamental mode of the fiber at $\xi = 0$.

frequency of this oscillator. When we introduce the standard annihilation and creation operators using

$$\hat{a} = (\hat{q} + i\hat{p})/\sqrt{2}, \qquad \hat{a}^\dagger = (\hat{q} - i\hat{p})/\sqrt{2}, \tag{7.4.14}$$

the Hamiltonian in Eq. (7.4.13) can be expressed as

$$\hat{H} = f_0(\xi)\hat{K}_0 + f(\xi)(\hat{K}_+ + \hat{K}_-), \tag{7.4.15}$$

where the operators K_0 and $K_\pm$ are defined in Eq. (7.3.46) and the function s $f_0(\xi)$ and $f(\xi)$ are related to the modulation parameters as

$$f_0(\xi) = 2[1 + \eta \sin(\mu\xi + \theta)] \approx 2, \qquad f(\xi) = \eta \sin(\mu\xi + \theta). \tag{7.4.16}$$

The evolution operator associated with the preceding Hamiltonian is given in Eq. (7.3.47). The functions $g(\xi)$ and $\chi(\xi)$ depend on $Q(\xi)$ as indicated in Eq. (7.3.48), where $Q(\xi)$ is the solution of Eq. (7.3.49). In the case of sinusoidal modulation, this equation becomes

$$\frac{dQ}{d\xi} + 2iQ = -\eta \sin(\mu\xi + \theta)(1 + Q^2). \tag{7.4.17}$$

When $\eta = 0$, Q varies with ξ as $e^{-2i\xi}$. For $\eta \neq 0$, we write the solution in the form $Q(\xi) = y(\xi)e^{-2i\xi}$, where $y(\xi)$ satisfies

$$\frac{dy}{d\xi} = -\eta \sin(\mu\xi + \theta)(e^{2i\xi} + y^2 e^{-2i\xi}). \tag{7.4.18}$$

At this point, we make the so-called rotating-wave approximation and keep on the right side only the terms that vary slowly with ξ. The result is

$$\frac{dy}{d\xi} = \frac{i\eta}{2}\left(y^2 \exp[i(\mu - 2)\xi + i\theta] - \exp[-i(\mu - 2)\xi - i\theta]\right). \tag{7.4.19}$$

This equation shows that the value $\mu = 2$ represents a kind of parametric resonance. For this specific choice, the right side of Eq. (7.4.19) does not depend on ξ.

It turns out that the problem can be solved analytically for $\mu = 2$ in the rotating-wave approximation [26], and the evolution operator takes the form [28]

$$\hat{U}(\xi) = \exp[-2i\xi\hat{K}_0]\hat{S}(\chi), \tag{7.4.20}$$

where the squeezing operator S, defined in Eq. (7.3.50), is given as

$$\hat{S}(\chi) = \exp[\chi\hat{K}_+ - \chi^*\hat{K}_-], \qquad \chi(\xi) = (\eta\xi/4)e^{-i\theta}. \tag{7.4.21}$$

We can simplify Eq. (7.4.20) further using Eq. (7.3.51) in Eq. (7.4.21). After some algebra, the evolution operator can be written in the form

$$\hat{U}(\xi) = \exp[-i(\xi - \theta/2)\hat{H}_0]S(|\chi|)\exp[-i(\theta/2)\hat{H}_0]. \tag{7.4.22}$$

Each factor in this equation has a distinct physical interpretation. The exponential containing $\hat{H}_0$ corresponds to a convolution as indicated in Eq. (7.4.8). The squeezing operator $\hat{S}(|\chi|)$ scales the coordinate q such that

$$\langle q|\hat{S}(|\chi|)|\psi\rangle = \exp(-|\chi|/2)\psi(e^{-|\chi|}q). \tag{7.4.23}$$

For an unmodulated GRIN fiber ($\eta = 0$), χ becomes zero. As $\hat{S}(|\chi|) = 1$ in this situation, we recover the result expected for a uniform GRIN fiber.

The important question is how the sinusoidal modulation of the core's radius affects the propagation of an optical beam incident on such a GRIN fiber. As we can expand any optical field in terms of the Hermite–Gauss eigenmodes of the unmodulated GRIN fiber (obtained in Section 2.2), we consider how a specific mode, $\psi_n(q, 0) = \langle q|n\rangle$, evolves inside the fiber. Using the evolution operator in Eq. (7.4.22) and choosing $\theta = 0$ to simplify the following discussion, we obtain

$$\psi_n(q, \xi) = \langle q|\hat{U}(\xi)|n\rangle, \qquad \hat{U}(\xi) = \exp[-i\xi\hat{H}_0]S(\eta\xi/4) \tag{7.4.24}$$

Using Eq. (7.4.23) and the result, $\hat{H}_0|n\rangle = (n + \frac{1}{2})|n\rangle$, we obtain

$$\psi_n(q, \xi) = \sqrt{S}\psi_n(Sq, 0)\exp[iC_2 q^2 - i(n + \tfrac{1}{2})C_1], \tag{7.4.25}$$

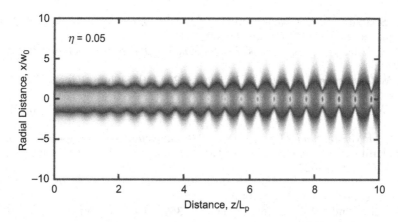

Figure 7.9 Evolution of a Gaussian beam inside a sinusoidally modulated GRIN medium under the parametric resonance conditions ($\mu = 2$) using $\eta = 0.05$ and $\theta = 0$. The initial width of the beam is matched to the width of the fundamental mode of the fiber at $\xi = 0$. (After Ref. [28]; Courtesy S. A. Wadood.)

where S, C_1, C_2 vary with distance ξ and are defined as

$$S(\xi) = e^{-\eta\xi/4}/\sqrt{D}, \quad C_1(\xi) = \tan^{-1}(e^{-\eta\xi/2}\tan\xi), \quad C_2 = \sin\xi(e^{-\eta\xi} - 1)/4D,$$

$$(7.4.26)$$

with $D = 1 + \sin^2\xi(e^{-\eta\xi} - 1)$.

Figure 7.9 shows the evolution of a Gaussian beam inside a sinusoidally modulated GRIN medium over a distance of $10L_p$ for $\mu = 2$ using $\eta = 0.05$ and $\theta = 0$ [28]. The initial width of the beam is matched to the width of the fundamental mode of the fiber ($C_f = 1$). In a uniform GRIN fiber ($\eta = 0$), such a beam would excite the LP_{01} mode of the GRIN fiber and propagate without change in its width or the shape. However, as seen in Figure 7.9, the beam's width changes in a periodic fashion when the core's radius is modulated sinusoidally at twice the self-imaging frequency. Even though the radius is modulated by only 5% ($\eta = 0.05$), the amplitude of width's oscillations keeps increasing with distance because of the parametric resonance occurring for $\mu = 2$. This situation is an analog of the phenomenon of parametric amplification in which a signal becomes amplified when pumped at twice its own frequency.

References

[1] W. K. Burns, A. F. Milton, and A. B. Lee, *Appl. Phys. Lett.* **30**, 28 (1977); https://doi.org/10.1063/1.89199.
[2] B. S. Kawasaki and K. O. Hill, *Appl. Opt.* **16**, 1794 (1977); https://doi.org/10.1364/AO.16.001794.
[3] J. C. Campbell, *Appl. Opt.* **18**, 900 (1979); https://doi.org/10.1364/AO.18.000900.

[4]   M. S. Sodha, A. K. Ghatak, and D. P. S. Malik, *J. Opt. Soc. Am.* **61**, 1492 (1971); https://doi.org/10.1364/OSA.61.001492.

[5]   M. S. Sodha and A. K. Ghatak, *Inhomogeneous Optical Waveguides* (Plenum Press, 1977).

[6]   L. W. Casperson, *J. Lightwave Technol.* **3**, 264 (1985); https://doi.org/10.1109/JLT .1985.1074203.

[7]   L. W. Casperson, *Appl. Opt.* **24**, 4395 (1985); https://doi.org/10.1109/JLT.1985 .1074176.

[8]   J. N. McMullin, *Appl. Opt.* **25**, 2184 (1986); https://doi.org/10.1364/AO.25.002184.

[9]   D. Bertilone, A. Ankiewicz, and C. Pask, *Appl. Opt.* **26**, 2213 (1987); https://doi.org/ 10.1364/AO.26.002213.

[10] P. G. L. Leach and K. Andriopoulos, *Appl. Anal. Discrete Math.* **2**, 146 (2008); https://doi.org/8/AADM0802146L.

[11] D. Barral and J. Liñares, *Opt. Commun.* **359**, 61 (2016); https://doi.org/10.1016/ j.optcom.2015.09.052.

[12] R. Kashyap, *Fiber Bragg Gratings*, 2nd ed. (Academic Press, 2009).

[13] C. Gomez-Reino, M. V. Perez, and C. Bao, *Gradient Index Optics: Fundamentals and Applications* (Springer, 2003).

[14] H. R. Lewis, *Phys. Rev. Lett.* **18**, 510 (1967); https://doi.org/10.1103/PhysRevLett.18 .510.

[15] H. R. Lewis and W. B. Riesenfeld, *J. Math. Phys.* **10**, 1458 (1969); https://doi.org/ 10.1063/1.1664991.

[16] L. F. Landovitz, A. M. Levine, and M. Schreiber, *Phys. Rev. A* **20**, 1162 (1978); https://doi.org/10.1103/PhysRevA.20.1162.

[17] J. G. Hartley and J. R. Ray, *Phys. Rev. D* **25**, 382 (1982); https://doi.org/10.1103/ PhysRevD.25.382.

[18] G. Dattoli, S. Solimeno, and A. Torre, *Phys. Rev. A* **34**, 2646 (1986); https://doi.org/ 10.1103/PhysRevA.34.2646.

[19] J. J. Sakurai and J. Napolitano, *Modern Quantum Mechanics*, 3rd ed. (Cambridge University Press, 2020).

[20] M. V. Berry, *Proc. R. Soc. Lond. A*, **392**, 45 (1984); https://doi.org/10.1098/rspa.1984 .0023.

[21] M. M. Nieto, *Quantum Semiclass. Opt.* **8**, 1061 (1996); https://doi.org/10.1088/ 1355-5111/8/5/011.

[22] K. Zhu, X. Sun, K. Wang, H Tang, and Y. Peng, *Opt. Commun.* **221**, 1 (2003); https://doi.org/10.1016/S0030-4018(03)01402-0.

[23] J. M. Cervero and J. D. Lejarreta, *Quantum Opt.* **2**, 333 (1990); https://doi.org/ 10.1088/0954-8998/2/5/001.

[24] H. P. Yuen, *Phys. Rev. A* **13**, 2226 (1976); https://doi.org/10.1103/PhysRevA.13.2226.

[25] I. S. Gradshteyn and I. M. Ryzhik, *Table of Integrals, Series, and Products*, 8th ed. (Academic Press, 2015).

[26] C. C. Gerry, *Phys. Rev. A* **35**, 2146 (1987); https://doi.org/10.1103/PhysRevA.35.2146.

[27] J. R. Choi, *Opt. Commun.* **283**, 3720 (2009); https://doi.org/10.1016/j.optcom.2009 .06.023.

[28] S. A. Wadood, K. Liang, G. P. Agrawal et al., *Opt. Express* **29**, 21240 (2021); https: //doi.org/10.1364/OE.425921.

8

Vortex Beams

Vortices in optics result from phase singularities, and their study falls in the domain of structured light or *singular optics* [1]. The vortex-like features of a phase singularity were first noticed in 1967 [2], and vortex solitons were found later in the context of nonlinear optics [3]. Vortex beams have attracted considerable attention since the 1990s, especially in the context of the orbital angular momentum associated with such beams [4, 5]. This chapter focuses on the propagation of vortex beams inside a GRIN medium. After an overview in Section 8.1 of the state of polarization (SOP) associated with an optical beam, the concept of a phase singularity is discussed in Section 8.2. This concept is used to form specific combinations of the modes that act as vortices with different SOPs. In Section 8.3, we discuss the techniques used for generating different types of vortex beams. Section 8.4 shows that a vortex beam also exhibits the self-imaging property during its propagation inside a GRIN medium. The impact of random mode-coupling is also discussed in this section. Vortex-based applications of GRIN fibers are covered in Section 8.5.

8.1 Basic Polarization Concepts

The modes of a GRIN medium discussed in Section 2.2 are linearly polarized because they are based on an approximate form of the wave equation given in Eq. (2.1.13). The incident beam, however, can have an SOP that deviates from being linear. In this section, we review the basic concepts such as what is meant by the SOP of an optical beam and how it can be represented on the Poincaré sphere through a Stokes vector [6–8].

8.1.1 Stokes Vector and Poincaré Sphere

The electric field $\mathbf{E}(\mathbf{r}, t)$ of an optical beam propagating along the z axis is a vector, whose direction can be different at different spatial points $\mathbf{r}$ and can also change with time t at a specific spatial point. In some cases, the SOP may be spatially uniform at all points, but the direction of **E** vector can still change with time.

It is common to ignore the longitudinal component of the electric field when defining the SOP (because it vanishes or is relatively small in most cases) and write the electric field oscillating at a specific frequency ω_0 as

$$\mathbf{E}(\mathbf{r}, t) = \left(\hat{\mathbf{x}} E_x + \hat{\mathbf{y}} E_y\right) e^{-i\omega_0 t}, \tag{8.1.1}$$

where $E_x = a_1 e^{i\psi_1}$ and $E_y = a_2 e^{i\psi_2}$ represent the two transverse components of the electric field. Their ratio, $E_y/E_x = re^{i\psi}$ is a complex number that determines the SOP of the optical beam at a spatial point. Here, $r = a_2/a_1$ is the amplitude ratio and $\psi = \psi_2 - \psi_1$ is the relative phase of the two components. An optical field becomes linearly polarized for specific values of r and ψ. It is linearly polarized in the x direction when $E_y = 0$ ($r = 0$) and in the y direction when $E_x = 0$ ($r = \infty$). It becomes linearly polarized in a different direction for other values of r when $\psi = 0$ or π.

For arbitrary values of r and ψ, the vector **E** at any point **r** traces an ellipse in the x–y plane as time evolves [6]. The ellipse is characterized by the orientation of its major axis from the x axis, its ellipticity, and the sense of rotation of the electric field along it. When $r = 1$ and $\psi = \pm\pi/2$, the ellipse is reduced to a circle, and the SOP of the beam is called circular. The SOP is said to right-circularly polarized (RCP) for $\psi = \pi/2$ and left-circularly polarized (LCP) for $\psi = -\pi/2$. Two different conventions are used for specifying the direction of rotation, leading often to confusion. We adopt the convention that **E** vector rotates clockwise for the RCP state when looking at an incoming beam [7]. The opposite convention in which clockwise rotation is assigned to the LCP state is also used [8].

A common way of identifying the SOP of an optical beam makes use of the Stokes vector **S**, whose three components are defined as

$$S_1 = (|E_x|^2 - |E_y|^2), \qquad S_2 + iS_3 = 2E_x^* E_y. \tag{8.1.2}$$

The magnitude of this vector is related to the total intensity as $S_0 = |E_x|^2 + |E_y|^2$, i.e., $S_1^2 + S_2^2 + S_3^2 = S_0^2$. The SOP at any point corresponds to the direction of the Stokes vector on a sphere whose three axes show the magnitudes of S_1, S_2, and S_3 components. This sphere is called the Poincaré sphere. It is common to normalize the Stokes vector such that it becomes a unit vector. Figure 8.1 shows the surface of the Poincaré sphere, with the SOP indicated by the tip of the Stokes vector on this sphere. The north and south poles of the sphere correspond to the RCP and LCP states, respectively, and all linear SOPs lie on its equator. The SOP is elliptical at all other points of this sphere's surface.

The two-dimensional picture of the polarization rotation in one transverse plane is oversimplified because it ignores the z direction in which the optical wave is

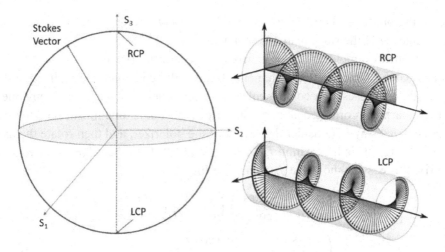

Figure 8.1 Surface of a Poincaré sphere at which the SOP of an optical beam can be indicated by the tip of the Stokes vector. Helical rotation of the field vector is shown on the right for the RCP and LCP states.

advancing. The three-dimensional view of the RCP and LCP states, shown in Figure 8.1, makes it clear that the SOP is actually linear at any moment in time, and it is the direction of the electric field that rotates as the phase front advances in the z direction, returning to its original direction after one cycle (at a distance of one wavelength). One can also understand this feature through the spin angular momentum (SAM) associated with circularly polarized beams. Photons in such beams carry a SAM of $\sigma \hbar$ such that $\sigma = +1$ for the LCP state and -1 for the RCP state.

The SOP of an optical beam changes as it passes through optical elements such as a birefringent wave plate. A formalism based on the Jones matrices is employed to study such SOP changes. In this approach, transmission of an optical beam through any optical device is written in a matrix form as

$$\begin{pmatrix} E'_x \\ E'_y \end{pmatrix} = \begin{pmatrix} J_{11} & J_{12} \\ J_{21} & J_{22} \end{pmatrix} \begin{pmatrix} E_x \\ E_y \end{pmatrix}, \tag{8.1.3}$$

where J_{ij} with $i,j = 1,2$ are the elements of a Jones matrix. In terms of a column vector $\mathbf{V}$ with E_x and E_y its two elements, this equation can be written in a compact matrix form as $\mathbf{V}' = J\mathbf{V}$. As a simple example, the Jones matrix for a wave plate of thickness d is a diagonal matrix such that

$$J = \begin{pmatrix} \exp(i\delta_x) & 0 \\ 0 & \exp(i\delta_y) \end{pmatrix}, \tag{8.1.4}$$

where the phase shifts, $\delta_x = n_x k_0 d$ and $\delta_y = n_y k_0 d$, are different because of the birefringence of the wave plate ($n_x \neq n_y$).

As another example, the Jones matrix for a polarizer, oriented along the x axis, is a diagonal matrix J_p, with the elements 1 and 0 along its diagonal, indicating that such a device passes the x component of the electric field but blocks its y component. If the polarizer is oriented at an angle θ from the x axis, one must rotate the two Cartesian axes by θ, consider the action of the polarizer, and then rotate the axes back. This results in the relation $J_p(\theta) = R(-\theta)J_p(0)R(\theta)$, where $R(\theta)$ is the rotation matrix. It is easy to show that $J_p(\theta)$ has the form

$$
\begin{aligned}
J_p(\theta) &= \begin{pmatrix} \cos\theta & -\sin\theta \\ \sin\theta & \cos\theta \end{pmatrix} \begin{pmatrix} 1 & 0 \\ 0 & 0 \end{pmatrix} \begin{pmatrix} \cos\theta & \sin\theta \\ -\sin\theta & \cos\theta \end{pmatrix} \\
&= \begin{pmatrix} \cos^2\theta & \sin\theta\cos\theta \\ \sin\theta\cos\theta & \sin^2\theta \end{pmatrix}.
\end{aligned}
\tag{8.1.5}
$$

8.1.2 Linear and Circular Polarization Basis

When we write the electric field vector in the form, $\mathbf{E} = E_x\hat{\mathbf{x}} + E_y\hat{\mathbf{y}}$, we are making use of a linear basis in which an arbitrary SOP is decomposed into two orthogonal linear SOPs pointing along the x and y directions. This can be seen more clearly in terms of the Jones vector associated with $\mathbf{E}$, which can be written as

$$
\begin{pmatrix} E_x \\ E_y \end{pmatrix} = E_x \begin{pmatrix} 1 \\ 0 \end{pmatrix} + E_y \begin{pmatrix} 0 \\ 1 \end{pmatrix}.
\tag{8.1.6}
$$

The Jones vectors on the right side correspond to the linear SOPs in the x and y directions.

In the case of a circularly polarized electric field, E_x and E_y only differ only in phase and can be written as $E_x = A$ and $E_y = A\exp(\pm i\pi/2) = \pm iA$, where $\pm$ signs stand for the RCP and LCP states. As a result, $\mathbf{E}_\pm = A(\hat{\mathbf{x}} \pm i\hat{\mathbf{y}})$. This result suggests that we should define two complex unit vectors as

$$
\hat{\mathbf{e}}_\pm = \frac{1}{\sqrt{2}}(\hat{\mathbf{x}} \pm i\hat{\mathbf{y}}),
\tag{8.1.7}
$$

where $\hat{\mathbf{e}}^+$ and $\hat{\mathbf{e}}^-$ represent the RCP and LCP states, respectively. The factor of $\sqrt{2}$ ensures that both vectors are of unit length. The Jones vectors for the two circularly polarized states can thus be written as

$$
\hat{\mathbf{e}}_+ = \frac{1}{\sqrt{2}} \begin{pmatrix} 1 \\ i \end{pmatrix}, \qquad \hat{\mathbf{e}}_- = \frac{1}{\sqrt{2}} \begin{pmatrix} 1 \\ -i \end{pmatrix}.
\tag{8.1.8}
$$

Similar to the linear polarization basis, we can decompose the SOP of any optical field into its two circularly polarized components as

$$\mathbf{E} = E_+\hat{\mathbf{e}}_+ + E_-\hat{\mathbf{e}}_-, \qquad E_\pm = (E_x \mp iE_y)/\sqrt{2}. \tag{8.1.9}$$

Such a decomposition is useful when we consider vortex beams that carry orbital angular momentum (OAM), in addition to the SAM. We turn to such beams in the next section.

8.2 Vortex Beams with OAM

It has been known for some time that any spatial point within an optical beam where intensity vanishes represents a phase singularity [9–12]. It is easy to understand the origin of this singularity. The complex amplitude of an optical beam at a specific frequency ω and at a point $\mathbf{r}$ can be written as

$$A(\mathbf{r}) = \sqrt{I(\mathbf{r})}\exp[i\psi(\mathbf{r})], \tag{8.2.1}$$

where $I(\mathbf{r})$ is the local intensity at the point $\mathbf{r}$ and ψ is the phase at that point. If we write A in the form $A_r + iA_i$, we have the usual relations

$$I = A_r^2 + A_i^2, \qquad \psi = \tan^{-1}(A_i/A_r). \tag{8.2.2}$$

At any point where $I = 0$, the phase is undefined because both the real and imaginary parts of A vanish. However, the phase is not zero in the vicinity of this singular point. If we write A as $\eta e^{i\psi}$ close to this point, where η is a small quantity, and we go around the singular point in the transverse plane, ψ must change by a multiple of 2π to recover its original value after one full circle. As the phase is finite at all nearby points except where $I = 0$, the location where the intensity of an optical beam vanishes represents a phase singularity, which is also called a vortex in analogy with the superfluid vortices [13].

8.2.1 Linearly Polarized Vortex Modes

Many of the Laguerre–Gauss modes, found in Section 2.2.2 for a GRIN fiber, exhibit a phase singularity at $\rho = 0$. This can be seen in Eq. (2.2.20), where the spatial distribution $U_{\ell m}(\rho, \phi)$ is given for an arbitrary LP$_{\ell m}$ mode in terms of the cylindrical coordinates ρ and ϕ. It is evident from the factor ρ^ℓ in this expression that $U_{\ell m}$ vanishes at the core's center (where $\rho = 0$) for all modes with $\ell \neq 0$. However, the even and the odd versions of these modes depend on the azimuthal angle ϕ through the factors $\cos(\ell\phi)$ and $\sin(\ell\phi)$, and neither of them contains a vortex at $\rho = 0$.

However, recall from Section 2.2.2 that each LP$_{\ell m}$ mode with $\ell \neq 0$ is four-fold degenerate, i.e., all four modes have the same propagation constant and travel at the same speed inside a GRIN medium. For this reason, any linear combination of these four modes is also a mode. The simplest situation occurs when we combine the even

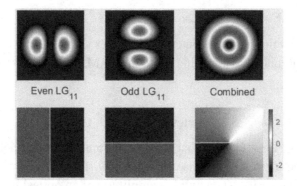

Figure 8.2 Intensity (top) and phase (bottom) profiles for the even and odd versions of the LP_{11} mode, and the vortex mode formed by combining them linearly with a $\pi/2$ phase shift.

and odd versions of a linearly polarized $LP_{\ell m}$ mode with a 90° relative phase shift between them to form a new mode such that

$$V_{\ell m} = \frac{1}{\sqrt{2}} \left[U_{\ell m}^{(e)} \pm i U_{\ell m}^{(o)} \right] = \frac{1}{\sqrt{2}} R_{\ell m}[\cos(\ell\phi) \pm i \sin(\ell\phi)] = \frac{1}{\sqrt{2}} R_{\ell m} e^{\pm i\ell\phi}, \quad (8.2.3)$$

where $R_{\ell m}$ is the radial part of the $LP_{\ell m}$ mode and the factor $1/\sqrt{2}$ ensures correct normalization. Noting that the phase of $V_{\ell m}$ is related to ℓ as $\psi = \pm\ell\phi$, it is easy to see that it changes by $2\pi\ell$ after one rotation around the point $\rho = 0$ (where $R_{\ell m}$ vanishes for all $\ell > 0$). All such $V_{\ell m}$ modes contain a vortex at their center. The integer ℓ is called the topological charge of the vortex. The same integer appears in the OAM (as $\hbar\ell$) carried by the photons of such beams [14]. It can take a positive or negative value, depending on the choice of $\pm$ sign in Eq. (8.2.3).

Figure 8.2 corresponds to the simplest case of Eq. (8.2.3) for $\ell = 1$ and $m = 1$. It shows the spatial intensity and phase profiles for the $U_{11}^{(e)}$ and $U_{11}^{(o)}$ modes, and for the vortex mode V_{11} formed by combining them linearly with a plus sign in Eq. (8.2.3). As seen there, phase takes only two values, 0 and π, for the $U_{11}^{(e)}$ and $U_{11}^{(o)}$ modes. However, it varies in a continuous fashion by 2π for the vortex mode V_{11} around the zero-intensity point at $\rho = 0$. If the minus sign is used in Eq. (8.2.3) to form the vortex mode, its phase varies with ϕ as $e^{-i\phi}$, and changes in the phase occur in the reverse fashion compared to those seen in Figure 8.2. It is common to use a negative value $\ell = -1$ for this mode, where ℓ indicates the topological charge (or the OAM) of the vortex.

A similar behavior occurs for other vortex modes with $|\ell| > 1$, whose phase changes by $2\pi\ell$ after one rotation around the point $\rho = 0$. Figure 8.3 shows the intensity and phase patterns for seven $V_{\ell 1}$ modes for which ℓ ranges from -3 to 3 [15]. The $\ell = 0$ case is special because the odd version of the LP_{01} mode does not

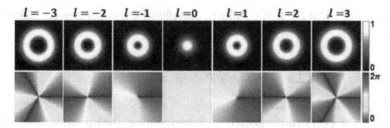

Figure 8.3 Intensity (top) and phase (bottom) profiles of the vortex modes formed by combining the even and odd versions of LP$_{lm}$ modes with $m = 1$. The $\ell = 0$ case is special because its odd version does not exist, and no vortex is formed. (After Ref. [15]; ©2021 CC BY.)

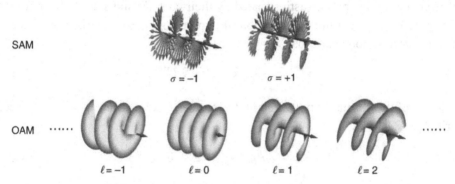

Figure 8.4 Rotation of the electric-field vector for circularly polarized light with SAM (top). Phase fronts of optical beams carrying OAM such that $\ell = -1, 0, 1$ and $\ell = 2$. Phase front becomes helical for all vortex modes with $\ell \neq 0$. (After Ref. [5]; ©2019 CC BY.)

exist in Eq. (8.2.3). As seen in Figure 8.3, this mode does not contain a vortex, and its intensity does not vanish at $\rho = 0$. Its phase is also constant across the entire transverse plane. Other modes in Figure 8.3 have vortices of different topological charges and OAMs such that the phase varies by $2\pi\ell$ after one rotation around the point $\rho = 0$. Even though the $V_{\ell 1}$ vortex modes exhibit a single bright ring (for the choice $m = 1$), the size of the dark spot becomes larger as $|\ell|$ increases because higher-order LP$_{\ell 1}$ modes are used in Eq. (8.2.3) to form the vortex mode.

An important property of a vortex mode is that its phase front evolves in a helical fashion around the direction of propagation. This can be understood by noting that the total phase of a vortex mode becomes $\psi_{lm} = \beta_{lm} z + \ell\phi$ when we include the propagation factor $\exp(i\beta_{lm} z)$ in Eq. (8.2.3). As z increases, the angle ϕ must change to maintain the same value of the phase. The resulting phase fronts are shown in Figure 8.4 for four modes with $\ell = -1, 0, 1$ and $\ell = 2$. For $\ell = 0$, the phase front is planar because the ϕ term disappears, resulting in $\psi_{01} = \beta_{01} z$. In contrast, the phase front is helical for $\ell \neq 0$ because the phase $\psi_{\ell 1} = \beta_{\ell 1}z + \ell\phi$ changes with both z and ϕ.

8.2.2 Circularly Polarized Vortex Modes

As we saw in Section 2.2.2, all LP$_{\ell m}$ modes of a GRIN medium exhibit polarization degeneracy. For every mode polarized linearly along one direction, another mode exists that is polarized orthogonally but has the same spatial distribution and the same propagation constant. It is common to choose the x and y axes along these two polarization directions.

For $\ell \neq 0$, the LP$_{\ell m}$ modes are four-fold degenerate because each mode also has its even and odd versions distinguished by their $\cos(\ell\phi)$ and $\sin(\ell\phi)$ factors. It is thus possible to form the combination of two orthogonally polarized modes and obtain a vector mode such that

$$\mathbf{V}_{\ell m}^{(+)} = \frac{1}{\sqrt{2}}\left[\hat{\mathbf{x}}\, U_{\ell m}^{(e)} + \hat{\mathbf{y}}\, U_{\ell m}^{(o)}\right]. \tag{8.2.4}$$

In terms of the circularly polarized unit vectors $\hat{\mathbf{e}}_{\pm}$, defined in Eq. (8.1.7), we can write Eq. (8.2.4) as

$$\mathbf{V}_{\ell m}^{(+)} = \frac{1}{\sqrt{2}}R_{\ell m}\left[\hat{\mathbf{e}}_{+}\, e^{-i\ell\phi} + \hat{\mathbf{e}}_{-}\, e^{i\ell\phi}\right]. \tag{8.2.5}$$

The first term in the preceding expression represents a RCP vortex mode with an OAM of $-l$ (in units of $\hbar$). The second term corresponds to a LCP mode with an OAM of $+l$. Recall that circularly polarized light contains photons that also have a SAM of $\sigma = \pm 1$ (in units of $\hbar$) for the LCP and RCP states, respectively. Clearly, the two vortex modes in Eq. (8.2.5) contain photons exhibiting both the OAM and the SAM. Following the atomic physics convention, we can define their total angular momentum as the sum $J = \ell + \sigma$, because the OAM and SAM vectors are oriented along the same z axis. With this notation, $J = \pm(\ell + 1)$ for the two terms in Eq. (8.2.5). Notice that if we choose $\ell = -1$ in Eq. (8.2.5), the total angular momentum J vanishes for the resulting vector modes.

If we combine the two modes in Eq. (8.2.4) with a $-$ sign in place of the $+$ sign, we obtain the vector mode

$$\mathbf{V}_{lm}^{(-)} = \frac{1}{\sqrt{2}}\left[\hat{\mathbf{x}}\, U_{lm}^{(e)} - \hat{\mathbf{y}}\, U_{lm}^{(o)}\right]. \tag{8.2.6}$$

Writing this equation in terms of the right- and left-circular polarization states, we obtain

$$\mathbf{V}_{lm}^{(-)} = \frac{1}{\sqrt{2}}R_{lm}\left[\hat{\mathbf{e}}_{+}\, e^{i\ell\phi} + \hat{\mathbf{e}}_{-}\, e^{-i\ell\phi}\right]. \tag{8.2.7}$$

This combination corresponds to two circularly polarized vortex modes with the total angular momentum $J = \pm(l - 1)$. Again, the total angular momentum J vanishes if the vortex carries an OAM of $\ell = 1$.

8.2.3 Radial and Azimuthal Polarizations

Vortex modes discussed so far have a uniform SOP over their entire spatial profile. It is possible to produce modes whose SOP varies in the transverse plane. Such optical beams have attracted considerable attention in recent years and are known as vector beams or Poincaré beams [16, 17].

Consider the SOP of the mode $\mathbf{V}_{11}^{(+)}$ ($\ell = m = 1$), formed by a linear combination of the even and odd versions of the LP_{11} mode as indicated in Eq. (8.2.4):

$$\mathbf{V}_{11}^{(\rho)} = \frac{1}{\sqrt{2}} \left[\hat{\mathbf{x}} \, U_{11}^{(e)} + \hat{\mathbf{y}} \, U_{11}^{(o)} \right]. \tag{8.2.8}$$

Using U_{11} from Eq. (2.2.20), we can write this combination as

$$\mathbf{V}_{11}^{(\rho)} = \frac{1}{\sqrt{2}} R_{11} \left[\hat{\mathbf{x}} \cos \phi + \hat{\mathbf{y}} \sin \phi \right] = \frac{1}{\sqrt{2}} R_{11} \hat{\rho}, \tag{8.2.9}$$

where $\hat{\rho} = \hat{\mathbf{x}} \cos \phi + \hat{\mathbf{y}} \sin \phi$ is a unit vector pointing in the radial direction in the x–y plane. Such an optical beam is said to be radially polarized. Beams with a nonuniform but radially symmetric SOP across their spatial profile are called cylindrical vector beams [16]. As indicated in Eq. (8.2.5), this radially polarized mode is composed of two orthogonally polarized vortex modes, carrying topological charges of ± 1. However, the total angular momentum J vanishes for this combination.

It is possible to combine the even and odd versions of the LP_{11} mode such that the combination exhibits azimuthal polarization:

$$\mathbf{V}_{11}^{(\phi)} = \frac{1}{\sqrt{2}} \left[\hat{\mathbf{y}} \, U_{11}^{(e)} - \hat{\mathbf{x}} \, U_{11}^{(o)} \right]. \tag{8.2.10}$$

Using the expression for U_{11} in Eq. (2.2.20), we can write this combination as

$$\mathbf{V}_{11}^{(\phi)} = \frac{1}{\sqrt{2}} R_{11} \left[\hat{\mathbf{y}} \cos \phi - \hat{\mathbf{x}} \sin \phi \right] = \frac{1}{\sqrt{2}} R_{11} \hat{\phi}, \tag{8.2.11}$$

where $\hat{\phi} = \hat{\mathbf{y}} \cos \phi - \hat{\mathbf{x}} \sin \phi$ is a two-dimensional vector pointing in the azimuthal direction in the x–y plane. As $\hat{\rho} \cdot \hat{\phi} = 0$, $\hat{\phi}$ points in a direction orthogonal to that of the $\hat{\rho}$ vector. Cylindrical vector beams, with a radial or azimuthal SOP, have attracted considerable attention and have found interesting applications in recent years [18–20]. Such beams can also be produced in a pulsed form. In a 2021 experiment, a cylindrical vector beam contained pulses of 2 ps duration that were referred to as spatiotemporal optical vortices [21].

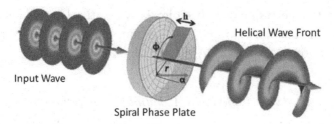

Figure 8.5 A spiral phase plate generates a vortex beam by converting the planar wave front of an incoming beam into a helical one.

8.3 Generation of Vortex Beams

A vortex mode cannot be excited by simply launching a Gaussian beam into a GRIN medium. The reason can be understood from the discussion in Section 2.4, where it was found that the on-axis launch of a Gaussian beam excites only the radially symmetric modes with $\ell = 0$. Modes with $\ell \neq 0$ can be excited if the beam is offset from the core's center. Although a radially polarized mode could be excited under specific conditions [22], it is difficult to ensure that the input beam couples efficiently into a single vortex mode. In practice, the formation of a vortex beam inside a GRIN medium requires an incident beam that either contains a vortex or is prepared such that its phase is not spatially uniform. In this section, we discuss several techniques that have been developed for generating vortex beams [23–26].

8.3.1 *Structured Optical Elements*

As we saw in Section 8.2, all vortex beams have a helical phase front. Thus, any optical element that can transform a planar wave front into a helical one should be able to produce a vortex beam. A spiral phase plate was first used in 1994 for this purpose [27]. Figure 8.3.1 shows such a phase plate [14]. Its thickness increases linearly with the azimuthal angle ϕ such that it is thicker after one rotation by $h = \ell\lambda/(n_p - 1)$, where n_p is the refractive index of the plate's material at the wavelength λ. It is not easy to fabricate such plates at wavelengths in the visible spectral region. In the 1994 demonstration, the plate was immersed in an index-matched fluid, whose temperature was controlled to tune h to its precise value. Two years later, a spiral phase plate was made to work at wavelengths ~ 1 mm [28]. By 2004, it became possible to fabricate spiral phase plates capable of working at visible wavelengths [29].

The phase-front conversion seen in Figure 8.5 is easy to understand. As the incoming light passes through the spiral phase plate, it acquires a phase shift, whose magnitude at any point depends on the local thickness of the plate at that point. The difference in the phase shifts after one rotation at the locations of maximum and minimum thicknesses is $2\pi\ell$, just the value required for a vortex with the topological charge ℓ. Although several other techniques have been developed in recent years, the use of a spiral phase plate remains a viable option [30].

More compact devices are fabricated by etching a nanoscale pattern that modifies incoming light to produce a desired phase distribution at a specific wavelength. The technique makes use of the fact that any phase value can be reduced to lie in the range 0–2π. As a result, thickness of the material has to change by at most one wavelength to produce any phase value. Examples of such devices are diffractive optical elements acting as lenses or gratings [31]. With the digitization of the phase, a computer can also be used to produce the required phase distribution, resulting in a computer-generated hologram.

The use of a computer-generated hologram was proposed in 1992 for producing vortex beams [11]. The underlying idea is simple. When two plane waves with the electric fields $E_j = A_j \exp(i\psi_j)$ interfere, the total intensity depends on their phase difference as

$$I = |E_1 + E_2|^2 = A_1^2 + A_2^2 + 2A_1 A_2 \cos(\psi_1 - \psi_2), \tag{8.3.1}$$

resulting in a pattern of bright and dark fringes. In the case of a vortex beam with a curved wavefront, the phase can be written in the cylindrical coordinates as $\psi_1 = k\rho^2/2R_c + \ell\phi$, where R_c is its radius of curvature of the wavefront and ℓ is its topological charge. Assuming that the other beam is a plane wave with $\psi_2 = 0$ and is aligned with the incoming wave, the boundary between a bright fringe and a dark fringe corresponds to

$$\psi_1 - \psi_2 = k\rho^2/(2R_c) + \ell\phi = (m + \tfrac{1}{2})\pi, \tag{8.3.2}$$

where m is an integer. This equation provides a spatial pattern for the hologram in the form of a spiral zone plate.

When the plane wave is not aligned with the vortex beam and propagates at an angle α, its transverse phase must be included in Eq. (8.3.2). Using its wave vector in the form $\mathbf{k} = k_x x + k_z z$, where $k_x = k\sin\alpha$, we have $\psi_2 = k_x x$. Using $x = \rho\cos\phi$, Eq. (8.3.2) becomes

$$k\rho^2/(2R_c) + \ell\phi - k\rho\sin\alpha\cos\phi = (m + \tfrac{1}{2})\pi. \tag{8.3.3}$$

The term linear in ρ produces a fringe pattern that has a fork at the point of phase singularity. Figure 8.6 shows the spiral and forked patterns for three vortex beams with $\ell = 1$, 2, and 3. Such interference patterns can be transferred to a thin film with a suitable technique. The resulting hologram is used to produce a vortex beam.

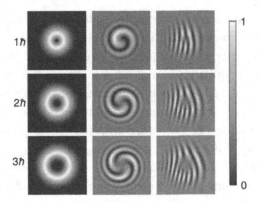

Figure 8.6 Spiral (middle column) and forked (right column) interference patterns created by interfering with a vortex beam (left column) with a collinear or a tilted plane wave. Three rows correspond to the choice $\ell = 1$, 2, and 3. (After Ref. [5]; ©2019 CC BY.)

8.3.2 Liquid Crystals and Metasurfaces

Liquid crystals are anisotropic materials containing molecules that can be oriented by applying an electric field. The alignment of molecules creates birefringence that can be used to change the SOP of incoming light. This tunable birefringence is exploited in a commercial device known as the *spatial light modulator*, which can be programmed to modulate, pixel by pixel, either the amplitude or the phase of incoming light. Amplitude modulation is used for making computer and television screens known as liquid-crystal displays. Phase modulation is employed to generate spatial phase patterns that can be used for creating vortex beams. When controlled by a computer, a spatial light modulator can impose any desired phase distribution on an incoming optical beam [32]. Although versatile, such devices are bulky and require an energy source.

It is possible to use liquid crystals for making a passive device that is capable of producing a vortex beam with a specific topological charge. Such a device is known as the q-plate, where q stands for its topological charge. A q-plate is a wave plate with an inhomogeneous distribution of birefringence in its transverse plane, which alters the SOP of light incident on the q-plate in a specific way. Figure 8.7 shows schematically its operation. When an LCP beam with a plane wave front ($\sigma = 1$ and $\ell = 0$) traverses a q-plate with $q = 1$, it is converted into a vortex beam with $\ell = 2$ and becomes RCP at the same time ($\sigma = -1$). It is easy to verify that the total angular momentum, $J = \ell + \sigma$, is conserved during this conversion. The whole process can be viewed as the SAM-to-OAM conversion performed by the q-plate [33].

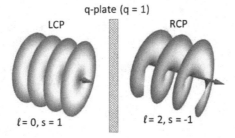

Figure 8.7 Schematic showing how a *q*-plate converts the planar phase front of an RCP beam into a helical one.

The *q*-plate technology has advanced considerably since its advent in 2006 [33]. A photo-alignment technique was developed around 2011 that allowed fabrication of tunable *q*-plates with arbitrary topological charges [34]. More details about this technology can be found in a 2019 review [35].

In recent years, metasurfaces have been employed for producing vortex beams [36–38]. A metasurface makes use of nanoscale elements, made of metallic or dielectric materials and deposited on the surface of a suitable substrate (see Section 10.1). A plasmonic metasurface was first used in 2011 to generate a vortex beam by reflecting a Gaussian beam from it [36]. It contained a large number of V-shape optical antennas (made of gold) on a silicon substrate, which were patterned such that the phase of reflected light varied from 0 to 2π around a central point. A birefringent metasurface was used in 2014 for creating a vortex beam in the same way as a *q*-plate does [37]. It contained identical nanosize antennas (made of gold) whose dimensions and separation were chosen to introduce a π phase shift between the orthogonally polarized components of an optical beam. The operation of this device is similar to that of a *q*-plate and is shown in Figure 8.7. When an LCP Gaussian beam reflects from such a metasurface, it becomes RCP and is converted into a vortex beam with an OAM of +2.

By 2019, dielectric metasurfaces have been made that worked in transmission over a wide wavelength range (1300 to 1700 nm) and were able to create vortex beams with OAM values as large as $\ell = 20$ [38]. Figure 8.8 shows schematically such a metasurface, containing many silicon resonators fabricated on top of a silica substrate. Each resonator is a nanoscale cylinder (with an elliptical base) that acts as a birefringent pixel whose transmission is polarization dependent. The relative phase shift introduced between the orthogonally polarized components of an optical beam depends on the dimensions and orientation of each pixel and allows one to generate vortex beams with different values of topological charge. Several other designs have been proposed for such metasurfaces [39–41]. Cylindrical vector beams with

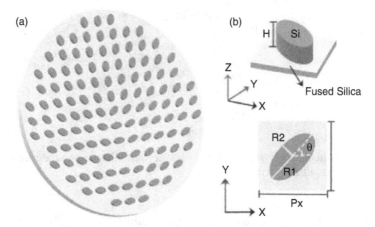

Figure 8.8 (a) Schematic of a dielectric metasurface with silicon resonators fabricated on top of a silica substrate. (b) Shape of each resonator with an elliptical base and the four design parameters involved. (After Ref. [38]; ©2019 Optica.)

a nonuniform SOP can also be generated using a spatial light modulator or other devices discussed in this section [42–44].

8.3.3 Fiber-Based Devices

The devices discussed so far produce vortex beams in air (or free space). If we are interested in exciting the vortex modes of a GRIN fiber, a device based on optical fibers is the most practical. Optical fibers designed with a square or ring-shaped core can be employed [45, 46], but their use is not always practical. Adjustment of the input SOP and the launch conditions for a two-mode standard fiber can excite a radially polarized vortex mode [47], but such a technique is not reliable. For this reason, several fiber-based devices have been developed in recent years [48–52].

In a 2015 experiment, a two-mode fiber was placed on a mechanical long-period grating that was designed to couple its LP_{01} and LP_{11} modes [48]. It was found that the LP_{01} mode entering from a single-mode fiber (spliced to the two-mode fiber) could be converted to the $\ell = \pm 1$ vortex mode, when suitable adjustments were applied mechanically. The period of the long-period grating (about 0.83 mm at 1550 nm) was chosen to maximize the coupling between the LP_{01} and LP_{11} modes. The fiber was also stressed to remove the degeneracy of the LP_{11} mode. In another experiment, the single-mode and two-mode fibers were fused together to form a coupler that was placed inside the cavity of a mode-locked fiber laser [49]. With proper adjustments, the laser produced a vortex beam containing pulses as short as 140 fs.

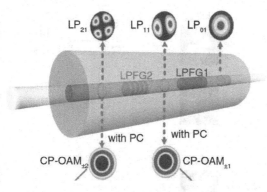

Figure 8.9 Two long-period fiber gratings (LPFG1 and LPFG1) were used to produce the circularly polarized (CP) vortex modes with $\ell = \pm 1$ and $\ell = \pm 2$. (After Ref. [50]; ©2018 CC BY.)

The potential of long-period fiber gratings was realized in a 2018 experiment in which vortex beams could be produced in a controllable fashion [50]. The fiber supported four mode groups: LP_{01}, LP_{11}, LP_{21}, and LP_{02}. Figure 8.9 shows how two long-period fiber gratings were employed to produce circularly polarized vortex modes with $\ell = \pm 1$ and $\ell = \pm 2$. The first grating converts the LP_{01} mode into a linear combination of two LP_{11} modes. The second grating then converts this mode into a vortex mode with $\ell = \pm 2$.

Birefringence can also break the degeneracy of the four members of the LP_{11} mode group. This was demonstrated in a 2020 experiment in which a high-birefringence Yb-doped fiber was employed to make a laser [51]. Two long-period fiber gratings were employed inside the laser's cavity to convert the LP_{01} mode into the x and y-polarized components of the even LP_{11} mode. A fiber coupler at the output end combined these two components to form a vortex beam with an OAM of $\ell = \pm 1$.

GRIN fibers have also been used for coupling incoming light into a vortex mode. In a 2018 experiment, a two-mode GRIN fiber was fabricated with a relatively large index difference ($\Delta n = 0.0363$) between its core (radius 4.8 μm) and cladding [52]. As a result, the degeneracy of the LP_{11} mode was lifted. Figure 8.10 shows schematically the scalar LP modes and their underlying vector modes [46]. The LP_{11} mode is composed of four vector modes (TE_{01}, TM_{01}, HE_{21}^e, and HE_{21}^o) that are polarized in different ways [46]. The TE_{01} and TM_{01} modes are polarized in an azimuthal fashion, while the other two are radially polarized. The linear combination of the even and odd HE_{21} modes with a $\pi/2$ phase shift produces vortex modes with $\ell = \pm 1$. To excite these modes, the GRIN fiber was fused with a single-mode fiber to form a mode-selective coupler. When circularly polarized light was launched

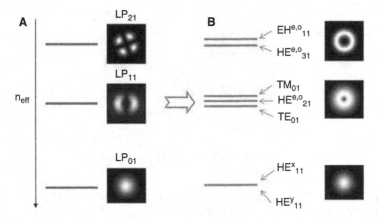

Figure 8.10 Effective indices and spatial shapes of (A) the scalar LP modes and of (B) the vector modes of a GRIN fiber. (After Ref. [46]; ©2013 CC BY.)

into the single-mode fiber, the coupler transferred its power to a circularly polarized vortex mode of the GRIN fiber with relatively high purity.

It was discovered recently that GRIN lenses can be used to produce vortex beams [53]. This is possible because the ion-exchange process that is used to make a GRIN lens also induces birefringence that increases with ρ such that the slow axis always points in the radial direction. Such spatially varying birefringence produces a beam at the output of a GRIN lens whose polarization is nonuniform and depends on the SOP of the input beam. When the input SOP is circular, a vortex beam can be produced by passing the output of a GRIN lens through a polarization-state analyzer, consisting of a quarter-wave plate and a polarizer. Such a GRIN-lens device works in the same way as a q-plate in Figure 8.7. When an RCP beam with no OAM is incident, it is converted into a vortex beam with an OAM of $+2$, while becoming lCP at the same time. In contrast with a q-plate, the GRIN-lens technique is much more versatile as it can be used to produce a variety of vector beams by cascading multiple GRIN lenses with other optical components.

8.3.4   Vortex Mode Sorters

A new type of device, known as a mode sorter or transformer, has been developed in recent years [54–57]. It makes use of two phase plates that are designed to split multiple overlapping vortex beams into spatially separated beams, sorted on the basis of the OAM associated with them. Figure 8.11 shows the operation of such a device schematically. The first phase plate performs a log-polar coordinate transformation that maps azimuthal phase variations of a vortex beam into a tilted plane wave, and the second corrects for the phase distortions induced by the first

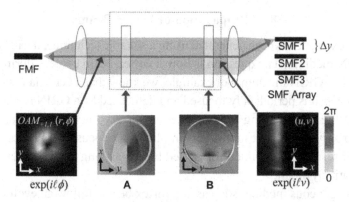

Figure 8.11 Schematic showing how a vortex-mode sorter works. Two phase plates (A and B) perform a transformation such that vortex beams with different ℓ values are separated spatially at the focal plane of the output lens. (After Ref. [57]; ©2015 CC BY.)

plate. As a result, at the focal plane of the output lens, vortex beams with different OAM values are separated physically. Conceptually, such a device functions similar to a prism or a grating, which sorts out different wavelengths contained within the incident beam.

In the original 2010 experiment [54], two spatial light modulators were used to create the two phase plates that performed the required log-polar transformation, $u = a\phi$, $v = -a\ln(\rho/b)$, where a and b are two scale factors, on the incoming signal. By 2012, both phase plates could be fabricated in the form of two refractive optical elements using the free-form design approach [55]. Such a mode sorter was used in a 2015 experiment [57], both as a multiplexer and as a demultiplexer, to demonstrate data transmission over a 5-km-long GRIN fiber using four vortex modes (see Section 6.4).

The coordinate-transformation idea has been extended in several directions. In one approach, a tunable mode filter was realized by placing a programmable mirror array at the output lens's focal plane so that only one selected beam goes back through the same device [56]. The use of a beam splitter at the input end then allows one to filter out a specific vortex mode from multiple vortices. In another advance, both phase plates of the mode sorter were fabricated with three-dimensional (3D) laser printing [58]. Intense femtosecond pulses from a laser were used to build the entire structure layer by layer, resulting in a device whose physical dimensions were relatively small (< 1 mm). In a more advanced design, the log-polar transformation was replaced with a new design that mapped spirals into parallel lines [59]. It was found that such a transformation sorted vortex modes of different OAMs with better resolution. In a 2020 experiment [60], such a mode sorter was used to demonstrate high-capacity data transmission over a 50-km-long GRIN fiber.

8.4 Propagation of Vortex Beams

After a vortex beam has been generated using a suitable technique, it can be launched into a GRIN medium. The question is how a vortex beam evolves during its propagation inside a GRIN medium. For example, we know that a Gaussian beam exhibits self-imaging as it is periodically focused and defocused by a GRIN medium. Should one expect the self-imaging to happen even for a vortex beam? This section is intended to answer such questions. For simplicity, we focus on CW beams, but note that similar results are obtained for pulsed beams as long as the dispersive effects remain relatively small.

In a homogeneous medium such as air, vortex beams follow an evolution pattern similar to that of a Gaussian beam because of the Laguerre-Gauss modes associated with the paraxial wave equation [1]. Propagation of an array of optical vortices nested within a Gaussian beam was studied in 1993 in the paraxial regime [61]. When all vortices had the same OAM, the entire array expanded in size because of diffraction, without affecting the relative positions of vortices. When vortices had different OAMs, those with opposite topological charges attracted each other and were annihilated. These results suggest that a single optical vortex should survive inside a GRIN fiber.

8.4.1 Laguerre–Gauss Beams

Several types of input fields have been used for studying the evolution of vortex beams inside a GRIN fiber. They all have a phase singularity with OAM in the form $e^{i\ell\phi}$ but their spatial distribution at the input end may correspond to a Laguerre–Gauss, Airy–Gauss, Bessel, or Gaussian beam [62–66].

Consider first the case of a Laguerre–Gauss beam launched on-axis into a GRIN fiber. Its initial field distribution has the same functional form as the corresponding mode of a GRIN fiber but with a different spot size w_0:

$$\mathbf{E}(\rho, \phi, 0) = \hat{\mathbf{p}} A_0 R_{lm}(\rho) e^{i\ell\phi}, \tag{8.4.1}$$

where $\hat{\mathbf{p}}$ is the polarization unit vector, A_0 is the initial amplitude, and the radial part $R_{lm}(\rho)$ is given by [see Eq. (2.2.20)]

$$R_{lm}(\rho) = N_{lm}\rho^l L_{m-1}^l(\rho^2/w_0^2) \exp(-\rho^2/2w_0^2). \tag{8.4.2}$$

A comparison with Eq. (2.2.20) shows that the corresponding spot size for the GRIN fiber is $w_g = a/\sqrt{V}$, where a is the core's radius. For standard GRIN fibers with $a = 25\ \mu m$ and V close to 25, a typical value of w_g is 5 μm.

If the input beam is mode-matched to the GRIN fiber by choosing $w_0 = w_f$, its power will couple into one specific mode of this fiber and will remain in that mode over lengths for which mode coupling is negligible. For longer fibers, mode coupling will transfer some power to other degenerate modes in that specific mode group, reducing the purity of the vortex beam. This coupling can be reduced by removing the modal degeneracy through a suitable design [46]. Birefringence of the fiber can also be used for this purpose [51].

The situation is more complex when the initial spot size is not mode-matched to the GRIN fiber ($w_0 \neq w_f$). In this case, the input beam will excite multiple vortex modes of the fiber, and we should employ an expansion technique similar to that used in Section 2.4.1. It is possible to calculate the coupling coefficients in an analytic form [62], but such an approach becomes cumbersome and does not provide much physical insight.

8.4.2 Gaussian Vortex Beams

An alternative approach makes use of the propagation kernel derived in Section 3.2 by using the mode expansion and summing over all modes. We illustrate this method for a Gaussian vortex beam [66], whose spatial distribution at the input plane of the GRIN fiber has the form:

$$E(\rho, \phi, 0) = A_0 \exp(-\rho^2/2w_0^2)\rho^\ell e^{i\ell\phi}, \tag{8.4.3}$$

where the integer ℓ represents the OAM or topological charge. The electric field at a distance z within the GRIN fiber is obtained from Eq. (3.2.5):

$$E(x, y, z) = \iint_{-\infty}^{\infty} K(x, x'; y, y')E(x', y', 0)\, dx'\, dy', \tag{8.4.4}$$

using the propagation kernel given in Eq. (3.2.10).

The calculation is simpler in the Cartesian coordinates because the kernel can be factored as $K(x, x'; y, y') = F(x, x')F(y, y')$. Moreover, $F(x, x')$ can be written in terms of the elements of an $ABCD$ matrix as

$$F(x, x') = \left(\frac{ke^{ikz}}{2\pi iB}\right)^{1/2} \exp\left(\frac{ik}{2B}[Dx^2 - 2xx' + Ax'^2]\right), \tag{8.4.5}$$

where $A = D = \cos(bz)$ and $B = \sin(bz)/b$. In the Cartesian coordinates, the input field in Eq. (8.4.3) takes the form

$$E(x, y, 0) = A_0 \exp[-(x^2 + y^2)/2w_0^2](x \pm iy)^{|\ell|}, \tag{8.4.6}$$

where the minus sign is chosen for negative values of ℓ. The last term is not separable in x and y. However, if we make use of the binomial expansion,

$$(x \pm iy)^{|\ell|} = \sum_{n=0}^{|\ell|} \frac{|\ell|!(\pm i)^n}{n!(|\ell| - n)!} x^{|\ell|-n} y^n, \tag{8.4.7}$$

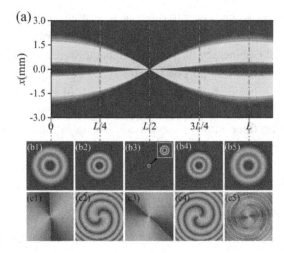

Figure 8.12 (a) Evolution of a vortex beam ($\ell = 2$) over one self-imaging period. Panels (b) and (c) show the intensity and phase profiles at five locations as indicated. (After Ref. [66]; ©2020 Elsevier.)

it is possible to express the final result in a series form involving Hermite polynomials. We refer to Ref. [66] for details. The $\ell = 0$ case corresponds to the Gaussian-beam case discussed in Section 3.2.3.

As an example, Figure 8.12 shows the evolution of a Gaussian vortex beam (wavelength 633 nm and $\ell = 2$) inside a GRIN rod over a distance $L = L_p/2$ using $b = 0.3$ mm^{-1} and $w_0 = 0.7$ mm. Here $L_p = 2\pi/b$ is the self-imaging period encountered in Section 3.3 with its value close to 2 cm for the GRIN rod. Part (a) in Figure 8.12 shows that a vortex beam exhibits self-imaging and evolves in a periodic fashion (just as a Gaussian beam does for $\ell = 0$). During each period, the vortex beam undergoes focusing and acquires a minimum width at a distance of $L_f = L_p/4$. The parameter C_f defined in Eq. (2.5.34) is about 0.06 for the parameters used, and compression by a factor of 16 is expected. As seen in panel (b), the vortex beam evolves in a self-similar fashion and maintains its spatial shape at all locations, except for a scaling factor. In contrast, panel (c) shows that its phase does not evolve in a self-similar fashion. The direction of phase rotation flips at the focal point. It is remarkable that the phase distribution at the distance $L = L_p/2$ does not recover fully to its original shape.

8.4.3 *Airy and Bessel Vortex Beams*

The propagation of two other types of vortex beams, known as the Airy and Bessel beams, has been investigated [63–65]. Such beams are of interest because of their non-diffracting nature in free space. Airy beams were discovered in 1979 [67], and the Bessel beams a few years later [68]. In the latter case, it was found that the free-space Helmholtz equation has exact solutions in cylindrical coordinates of the form

$$E(\rho, \phi, z) = E_0 J_\ell(\sqrt{k^2 - \beta^2}\rho)e^{i\ell\phi}e^{i\beta z}, \tag{8.4.8}$$

where $\beta < k$, ℓ is an integer, and $J_\ell(x)$ is the Bessel function of order ℓ. When $\ell \neq 0$, a Bessel beam contains a vortex at $\rho = 0$ with the OAM $\hbar\ell$. Using the integral representation of the Bessel function, Eq. (8.4.8) can be written in the Cartesian coordinates as

$$E(x, y, z) = \frac{E_0 e^{i\beta z}}{2\pi i^\ell} \int_0^{2\pi} \exp[ik_r(x\cos\phi + y\sin\phi) + i\ell\phi]\, d\phi, \tag{8.4.9}$$

where $k_r = \sqrt{k^2 - \beta^2}$. The spatial intensity pattern of a Bessel beam contains multiple circular rings and appears similar to that of LP_{lm} modes in Figure 2.5. An important difference is that the ring pattern of a Bessel beam extends over an infinite range. Bessel beams realized in practice are of finite size. Despite this limitation, such beams remain nearly diffraction-free as long as their size is much larger than their wavelength.

When a Bessel beam is launched into a GRIN fiber, its spatial pattern evolves as indicated in Eq. (8.4.4). All integrals can be carried out analytically, and the electric field at a distance z is found to be [65]:

$$E(x, y, z) = \frac{E_0 e^{i\beta z}}{2\pi i^\ell A} J_\ell\left(\frac{k_r\rho}{A}\right)\exp\left(\frac{ikD}{2B}\rho^2\right)\exp\left[\frac{ik}{2AB}(\rho^2 + B^2 k_r^2/k^2)\right], \tag{8.4.10}$$

where $A = D = \cos(bz)$ and $B = \sin(bz)/b$. Figure 8.13 shows the evolution of a second-order Bessel vortex beam ($\ell = 2$) inside a GRIN rod over a distance $L = L_p$ using $b = 0.233$ mm^{-1} and $w_0 = 0.1$ mm. Part (a) shows that the Bessel beam exhibits self-imaging and evolves in a periodic fashion, just as a Gaussian beam does. Parts (b) and (c) show the spatial intensity patterns at distances close to the input and the focal point at $L_p/4$. The beam is compressed at the focal point by a factor of 15, but its structure follows the pattern dictated by the second-order Bessel function in Eq. (8.4.10). When a Bessel beam is launched off-center, its behavior remains identical to that of an off-center Gaussian beam (see Section 3.2.3). This is not surprising because the discussion of the self-imaging phenomenon in Section 3.2 applies to any input beam. The evolution of a Bessel–Gaussian beam has also been analyzed and shows similar features [69].

8.4.4 Impact of Random Mode Coupling

The preceding analysis is based on a perfect GRIN fiber with no coupling among its modes. As a result, it may apply for relatively short fibers (length < 10 m) but

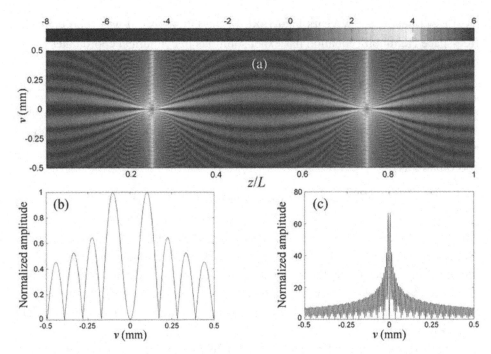

Figure 8.13 (a) Evolution of a Bessel beam ($\ell = 2$) over one self-imaging period. Panels (b) and (c) show the intensity profiles close to the input and near the focal point at $L_p/4$. (After Ref. [65]; ©2018 Optica.)

cannot be used for fibers longer than 100 m. As discussed in Section 4.3, every fiber suffers from random variations along its length that lead to random coupling among the modes, and the degenerate modes are most strongly coupled. As we saw in Section 8.1, vortex modes represent specific linear combinations of four degenerate modes belonging to the LP_{11} mode group. Any coupling among these modes will lead to intermodal power transfer, thereby reducing the purity of any vortex mode excited inside a GRIN fiber. The only solution is to design GRIN fibers such that the modal degeneracy is removed, or reduced as much as possible.

As was discussed in Section 8.3.3, birefringence can be used to break the degeneracy of the four members of the LP_{11} group [51]. These are denoted as $LP_{11}^{e,x}$, $LP_{11}^{e,y}$, $LP_{11}^{o,x}$, and $LP_{11}^{o,y}$ in the scalar theory, where the superscripts (e, o) stand for the even and mode versions and (x, y) indicate the directions of orthogonally polarized modes, In the vector theory, these modes are denoted as TE_{01}, TM_{01}, HE_{21}^e, and HE_{21}^o, where the first and last two exhibit azimuthal and radial SOPs, respectively [46]. Birefringence can be induced by making the shape of the core elliptical, or by stressing the core in one direction so that the refractive index becomes larger in that direction (stress-induced birefringence). It changes the effective index of the four

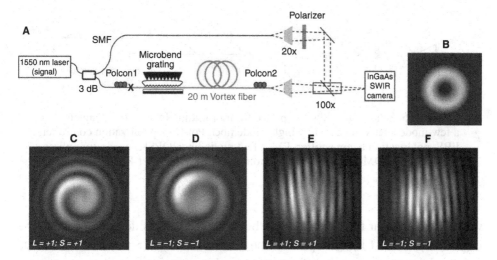

Figure 8.14 (A) Experimental setup; (B) Intensity pattern; Spiral (C and D) and fork (E and F) interference patterns for the LCP and RCP vortices. (After Ref. [46]; ©2013 CC BY.)

modes such that it is the largest for the TE_{01} mode and the smallest for the TM_{01} mode, with a difference of about $\Delta n = 10^{-4}$. The effective index of the remaining two modes remains the same and lies in the middle of this range.

Modal degeneracy can also be broken by modifying the index profile of a GRIN fiber. In one type of fiber, sometimes called the vortex fiber [46], the refractive index is made largest, not at the center where $\rho = 0$, but along a ring at a distance $\rho = a_1$, where a_1 is less than the core's radius a. The index profile of such a fiber is far from being parabolic, which helps in removing the mode degeneracy and forming a vortex mode. Figure 8.14 shows the experimental setup (part A) used to propagate the $\ell = 1$ vortex mode over one such 20-m-long fiber. The 1550-nm input was in the form of a circularly polarized LP_{01} mode with $\ell = 0$. A long-period grating was used as a mode-selective coupler to excite and propagate a vortex mode inside the GRIN fiber with $\ell = \pm 1$. The fiber's output was characterized through a standard interference technique. Part B shows a donut-shape spatial intensity pattern, as expected. Parts C and D show the spiral interference patterns for the LCP and RCP vortices. Parts E and F show the forked patterns in these two cases when an off-axis Gaussian beam is employed for the interference. Clearly, all features of the vortex modes are preserved with high purity even after 20 m of propagation. Experiments done with a 1-km-long fiber revealed that purity of the vortex modes remained high (<10% degradation) over such distances [46].

In a 2014 study, a new type of GRIN fiber was fabricated with an "inverse" parabolic profile such that the refractive index had a dip at the core's center and

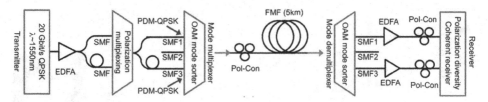

Figure 8.15 Schematic of the setup used for transmitting four vortex channels over a few-mode GRIN fiber. SMF: single mode fiber, Pol-Con: polarization controller, PBS, polarization beam splitter, FMF: few-mode fiber, QPSK: quadrature phase-shift keying, PDM: polarization division multiplexing. (After Ref. [57]; ©2015 CC BY.)

was the largest near the core-cladding boundary [70]. This design also lifted the degeneracy of the LP_{11} group, and resulted in Δn values $\sim 10^{-4}$. Experiments indicated that vortices with different OAM values remained intact even when the fiber was more than 1 km long.

8.5 Applications

Vortex beams carrying OAM are used for a variety of applications including optical communication, sensing, and imaging. In the case of optical communication systems, they can enhance a system's capacity through OAM-domain multiplexing by using different OAM values for different data channels. In the context of imaging, a GRIN fiber can be used for focusing a vortex beam. It can also be used as a fiber-optic tweezer for trapping tiny particles. We discuss a few such applications in this section.

Channel multiplexing is invariably employed for enhancing the capacity of fiber-optic communication systems [71]. As we saw in Section 4.5.1, multiplexing in the wavelength domain was developed during the 1990s, and it was followed by space-division multiplexing (SDM) after the year 2010. In one type of SDM, the LP_{lm} modes of a GRIN fiber are used as different spatial channels. The use of OAM provides another dimension for multiplexing because different OAM values can be employed for different data channels. Indeed, several such experiments have been carried in recent years out to demonstrate the basic concept [72–74].

In a 2013 experiment, four OAM modes were used to transmit data at 400-Gb/s over 1.1-km-long vortex fiber by employing a single wavelength [72]. Data rates could be increased four-fold by multiplexing ten channels at different wavelengths over the same fiber. Figure 8.15 shows the experimental setup in which four OAM channels (with $\ell = \pm 1$ and $\sigma = \pm 1$) were transmitted over a 5-km-long GRIN fiber, designed to support only a few low-order mode groups [57]. Each channel carried quaternary phase-shift keying (QPSK) data at a bit rate of 20 Gb/s. A vortex mode sorter (see Section 8.3.4) was used to convert four linearly polarized data channels into four OAM beams. The same device was used at the receiver side in the reverse direction to convert OAM modes into linearly polarized data channels. This scheme allowed the use of a standard polarization-diversity coherent receiver to recover the transmitted data [71].

Considerable progress has been made since 2015 in the area of OAM-domain multiplexing. In a 2018 experiment six OAM modes with $\ell = -1$, 0, 1 and $\sigma = \pm 1$ were employed to transmit a 120-Gb/s signal over a standard, 8.8-km-long, GRIN fiber [74]. Only four modes had vortices because the modes with $\ell = 0$ belong to the LP_{01} mode group. Random mode coupling occurring along the fiber introduced linear crosstalk among the four vortex channels. The effects of this coupling were compensated at the receiver through digital signal processing. In a 2020 experiment, data transmission with the total capacity of 2.56 Tb/s was realized over 100 km of a ring-core vortex fiber by sending ten WDM channels, each carrying eight OAM modes [75].

Another potential application of vortex modes in GRIN fibers is as a fiber-optic tweezer for trapping and manipulating tiny particles (see Section 3.4.4). Although optical beams without any OAM are generally employed, the use of a vortex beam can improve such fiber-optic probes by providing the ability to rotate the trapped particles. Their design is similar to conventional GRIN-based tweezers, but the step-index fiber before the GRIN fiber has a wider core and supports several low-order spatial modes (see Figure 3.13). A mode-selective coupler is used to excite a suitable vortex mode inside the step-index fiber, which guides it and launches it into the GRIN fiber [76]. The GRIN fiber reduces the spatial size of the vortex mode through the self-imaging phenomenon. It is necessary to tailor the length of the GRIN fiber to ensure that it acts as a focusing lens. Numerical simulations were used to show that the working distance depends on this length and can be maximized by optimizing it. The predicted working distance was close to 200 μm for vortices with $\ell = 1$ and 2, when the GRIN fiber's length was 350 μm.

References

[1] G. J. Gbur, *Singular Optics* (CRC Press, 2016).

[2] A. Boivin, J. Dow, and E. Wolf, *J. Opt. Soc. Am.* **57**, 1171 (1967); https://doi.org/10.1364/JOSA.57.001171.

[3] G. A. Swartzlander and C. T. Law, *Phys. Rev. Lett.* **69**, 2503 (1992); https://doi.org/10.1103/PhysRevLett.69.2503.

[4] L. Allen, M. W. Beijersbergen, R. J. C. Spreeuw, and J. P. Woerdman, *Phys. Rev. A* **45**, 8185 (1992); https://doi.org/10.1103/PhysRevA.45.8185.

[5] Y. Shen, X. Wang, Z. Xie et al., *Light Sci. Appl.* **8**, 90 (2019); https://doi.org/10.1038/s41377-019-0194-2.

[6] M. Born and E. Wolf, *Principles of Optics*, 7th ed. (Cambridge University Press, 1999).

[7] A. Kumar and A. Ghatak, *Polarization of Light with Applications in Optical Fibers* (SPIE Press, 2011).

[8] M. Chekhova and P. Banzer, *Polarization of Light: In Classical, Quantum, and Nonlinear Optics* (De Gruyter, 2021).

[9] M. Berry, J. Nye, and F. Wright, *Phil. Trans. R. Soc. Lond.* **291**, 453 (1979); https://doi.org/10.1098/rsta.1979.0039.

[10] J. M. Vaughan and D. V. Willetts, *Opt. Commun.* **30**, 263 (1979); https://doi.org/10.1016/0030-4018(79)90350-X.

[11] N. R. Heckenberg, R. McDuff, C. P. Smith, and A. G. White, *Opt. Lett.* **17**, 221 (1992); https://doi.org/10.1364/OL.17.000221.

[12] G. P. Karman, M. W. Beijersbergen, A. van Duijl, and J. P. Woerdman, *Opt. Lett.* **22**, 1503 (1997); https://doi.org/10.1364/OL.22.001503.

[13] P. Coullet, G. Gil, and F. Rocca, *Opt. Commun.* **73**, 403 (1989); https://doi.org/10.1016/0030-4018(89)90180-6.

[14] A. M. Yao and M. J. Padgett, *Adv. Opt. Photon.* **3**, 161 (2011); https://doi.org/10.1364/AOP.3.000161.

[15] J. Wang, S. Chen, and J. Liu, *APL Photon.* **6**, 060804 (2021); https://doi.org/10.1063/5.0049022.

[16] Q. Zhan, *Adv. Opt. Photon.* **1**, 1 (2009); https://doi.org/10.1364/AOP.1.000001.

[17] A. M. Beckley, T. G. Brown, and M. A. Alonso, *Opt. Express* **18**, 10777 (2010); https://doi.org/10.1364/OE.18.010777.

[18] S. N. Khonina, A. V. Ustinov, S. A. Fomchenkov, and A. P. Porfirev, *Sci. Rep.* **8**, 14320 (2018); https://doi.org/10.1038/s41598-018-32469-0.

[19] H. Moradi, V. Shahabadi, E. Madadi, E. Karimi, and F. Hajizadeh, *Opt. Express* **27**, 7266 (2019); https://doi.org/10.1364/OE.27.007266.

[20] Y. Zhang, T. Hou, H. Chang et al., *Opt. Express* **29**, 5259 (2021); https://doi.org/10.1364/OE.417038.

[21] J. Chen, C. Wan, A. Chong, and Q. Zhan, *Nanophotonics* **18**, 4489 (2021); https://doi.org/10.1515/nanoph-2021-0427.

[22] T. Grosjean, D. Courjon, and M. Spajer, *Opt. Commun.* **203**, 1 (2002); https://doi.org/10.1016/S0030-4018(02)01122-7.

[23] X. W. Wang, Z. Nie, Y. Liang et al., *Nanophotonics* **7**, 1533 (2018); https://doi.org/10.1515/nanoph-2018-0072.

[24] L. Zhu and J. Wang, *Front. Optoelectron.* **12**, 52 (2019); https://doi.org/10.1007/s12200-019-0910-9.

[25] K. Zhang, Y. Wang, Y. Yuan, and S. N. Burokur, *Appl. Sci.* **10**, 1015 (2020); https://doi.org/10.3390/app10031015.

[26] J. Wang and Y. Liang, *Front. Phys.* **9**, 688284 (2021); https://doi.org/10.3389/fphy.2021.688284.

[27] M. W. Beijersbergen, R. Coerwinkel, M. Kristensen, and J. P. Woerdman, *Opt. Commun.* **112**, 321 (1994); https://doi.org/10.1016/0030-4018(94)90638-6.

[28] G. A. Turnbull, D. A. Roberson, G. M. Smith, L. Allen, and M. J. Padgett, *Opt. Commun.* **127**, 183 (1996); https://doi.org/10.1016/0030-4018(96)00070-3.

[29] S. Oemrawsingh, J. van Houwelingen, E. Eliel et al., *Appl. Opt.* **43**, 688 (2004); https://doi.org/10.1364/AO.43.000688.

[30] S. G. Ghebjagh and S. Sinzinger, *Appl. Opt.* **59**, 4618 (2020); https://doi.org/10.1364/AO.392746.

[31] V. A. Soifer, *Diffractive Optics and Nanophotonics* (CRC Press, 2017).

[32] A. Forbes, A. Dudley, and M. McLaren, *Adv. Opt. Photon.* **8**, 200 (2016); https://doi.org/10.1364/AOP.8.000200.

[33] L. Marrucci, C. Manzo, and D. Paparo, *Appl. Phys. Lett.* **88**, 221102 (2006); https://doi.org/10.1063/1.2207993.

[34] S. Slussarenko, A. Murauski, T. Du et al., *Opt. Express* **19**, 4085 (2011); https://doi.org/10.1364/OE.19.004085.

[35] A. Rubano, F. Cardano, B. Piccirillo, and L. Marrucci, *J. Opt. Soc. Am. B* **36**, D70 (2019); https://doi.org/10.1364/OSAB.36.000D70.

[36] N. Yu, P. Genevet, M. A. Kats et al., *Science* **339**, 333 (2011); https://doi.org/10.1126/science.1210713.

[37] E. Karimi, S. A Schulz, I. De Leon et al., *Light Sci. Appl.* **3**, e167 (2019); https://doi.org/10.1038/lsa.2014.48.

[38] H. Zhou, J. Yang, C. Gao, and S. Fu, *Opt. Mater. Express* **9**, 2699 (2019); https://doi.org/10.1364/OME.9.002699.

[39] J. Yang, H. Zhou, and T. Lan, *Opt. Mater. Express* **9**, 3594 (2019); https://doi.org/10.1364/OME.9.003594.

[40] Q. Dai, Z. Li, L. Deng et al., *Opt. Lett.* **45**, 3773 (2020); https://doi.org/10.1364/OL.398286.

[41] X. Wang, C. Wang, M. Cheng et al., *J. Lightwave Technol.* **39**, 2830 (2021); https://doi.org/10.1109/JLT.2021.3064560.

[42] J. Liu, X. Chen, Y. He et al., *Res. Phys.* **19**, 103455 (2020); https://doi.org/10.1016/j.rinp.2020.103455.

[43] M. A. Olvera-Santamariaa, J. Garcia, A. Tlapale-Aguilar et al., *Opt. Commun.* **467**, 125693 (2020); https://doi.org/10.1016/j.optcom.2020.125693.

[44] J. Qi, W. Yi, and M. Fu, *Opt. Express* **29**, 25365 (2021); https://doi.org/10.1364/OE.433897.

[45] Y. Yan, L. Zhang, J. Wang et al., *Opt. Lett.* **37**, 3294 (2012); https://doi.org/10.1364/OL.37.003294.

[46] S. Ramachandran and P. Kristensen, *Nanophotonics* **2**, 455 (2013); https://doi.org/10.1515/nanoph-2013-0047.

[47] N. K. Viswanathan and V. V. G. Krishna Inavalli, *Opt. Lett.* **34**, 1189 (2009); https://doi.org/10.1364/OL.34.001189.

[48] S. Li, Q. Mo, X. Hu, C. Du, and J. Wang, *Opt. Lett.* **40**, 4376 (2015); https://doi.org/10.1364/OL.40.004376.

[49] T. Wang, F. Wang, F. Shi et al., *J. Lightwave Technol.* **35**, 2161 (2017); https://doi.org/10.1109/JLT.2017.2676241.

[50] Y. Han, Y. Liu, Z. Wang et al., *Nanophotonics* **7**, 287 (2018); https://doi.org/10.1515/nanoph-2017-0047.

[51] Z. Dong, Y. Zhang, H. Li et al., *Opt. Express* **28**, 9988 (2020); https://doi.org/10.1364/OE.389466.

[52] X. Heng, J. Gan, Z. Zhang et al., *Opt. Express* **26**, 17429 (2018); https://doi.org/10.1364/OE.26.017429.

[53] C. He, J. Chang, Q. Hu et al., *Nat. Commun.* **10**, 4264 (2019); https://doi.org/10.1038/s41467-019-12286-3.

[54] G. C. G. Berkhout, M. P. J. Lavery, J. Courtial, M. W. Beijersbergen, and M. J. Padgett, *Phys. Rev. Lett.* **105**, 153601 (2010); https://doi.org/10.1103/PhysRevLett.105.153601.

[55] M. P. J. Lavery, D. J. Robertson, G. C. G. Berkhout et al., *Opt. Express* **20**, 2110 (2012); https://doi.org/10.1364/OE.20.002110.

[56] H. Huang, Y. Ren, G. Xie et al., *Opt. Lett.* **39**, 1689 (2014); https://doi.org/10.1364/OL.39.001689.

[57] H. Huang, G. Milione, M. P. J. Lavery et al., *Sci. Rep.* **5**, 14931 (2015); https://doi.org/10.1038/srep14931.

[58] S. Lightman, G. Hurvitz, R. Gvishi, and A. Arie, *Optica* **4**, 605 (2017); https://doi.org/10.1364/OPTICA.4.000605.

[59] Y. Wen, I. Chremmos, Y. Chen et al., *Phys. Rev. Lett.* **120**, 193904 (2018); https://doi.org/10.1103/PhysRevLett.120.193904.

[60] Y. Wen, I. Chremmos, Y. Chen et al., *Optica* **7**, 254 (2020); https://doi.org/10.1364/OPTICA.385590.

[61] G. Indebetouw, *J. Mod. Opt.* **40**, 73 (1993); https://doi.org/10.1038/s41467-019-12286-3.

[62] N. I. Petrov, *J. Opt. Soc. Am. A* **33**, 1363 (2016); https://doi.org/10.1364/JOSAA.33.001363.

[63] R. Zhao, F. Deng, W. Yu, J. Huang, and D. Deng, *J. Opt. Soc. Am. A* **33**, 1025 (2016); https://doi.org/10.1364/JOSAA.33.001025.

[64] X. Peng, Y. Peng, D. Li et al., *Opt. Express* **25**, 13527 (2017); https://doi.org/10.1364/OE.25.013527.

[65] Z. Cao, C. Zhai, S. Xu, and Y. Chen, *J. Opt. Soc. Am. A* **35**, 230 (2018); https://doi.org/10.1364/JOSAA.35.000230.

[66] S. Yang, J. Wang, M. Guo, Z. Qin, and J. Li, *Opt. Commun.* **465**, 125559 (2020); https://doi.org/10.1016/j.optcom.2020.125559.

[67] M. V. Berry and N. L. Balazs, *Am. J. Phys.* **47**, 264 (1979); https://doi.org/10.1119/1.11855.

[68] J. Durnin, *J. Opt. Soc. Am. A* **4**, 651 (1987); https://doi.org/10.1364/JOSAA.4.000651.

[69] S. Pei, S. Xu, F. Cui, Q. Pan, and Z. Cao, *Appl. Opt.* **58**, 920 (2019); https://doi.org/10.1364/AO.58.000920.

[70] B. Ung, P. Vaity, L. Wang et al., *Opt. Express* **22**, 18044 (2014); https://doi.org/10.1364/OE.22.018044.

[71] G. P. Agrawal, *Fiber-Optic Communication Systems*, 5th ed. (Wiley, 2021).

[72] N. Bozinovic, Y. Yue, Y. Ren et al., *Science* **340**, 1545 (2013); https://doi.org/10.1126/science.1237861.

[73] A. E. Willner, H. Huang, Y. Yan et al., *Adv. Opt. Photon.* **7**, 66 (2015); https://doi.org/10.1364/AOP.7.000066.

[74] A. Wang, L. Zhu, L. Wang et al., *Opt. Express* **26**, 10038 (2018); https://doi.org/10.1364/OE.26.010038.

[75] J. Zhang, J. Liu, L. Shen et al., *Photon. Res.* **8**, 1236 (2020); https://doi.org/10.1364/PRJ.394864.

[76] W. Ren, Y. Gong, Z. Zhang, and K. Li, *Appl. Opt.* **60**, 7634 (2021); https://doi.org/10.1364/AO.431057.

9

Photonic Spin-Orbit Coupling

It was seen in Chapter 8 that the state of polarization (SOP) plays an important role for vortex beams. However, the theory of GRIN media developed so far is based on Eq. (2.1.13), which ignores the polarization-coupling term in the actual wave equation of Section 2.1. In this chapter, we include this term and show that it corresponds to a photonic analog of the spin-orbit coupling of electrons in an atomic medium. Section 9.1 describes two physical mechanisms that can produce changes in the SOP of an optical beam. The vectorial form of the wave equation is solved in Section 9.2 to introduce a path-dependent geometrical phase. The photonic analog of the spin-orbit coupling and its implications are also discussed in this section. The modes of a GRIN Fiber are also affected by the last term in Eq. (2.1.12). Section 9.3 considers how the scalar LP_{lm} modes, obtained in Section 2.2, change when this term is taken into account. We treat this term first as a perturbation and then obtain the exact vector modes of a GRIN medium. A quantum approach is used in Section 9.4 to discuss various polarization-dependent effects.

9.1 Mechanisms for Polarization Changes

The modes of a GRIN medium obtained in Section 2.2 are linearly polarized and preserve their SOP because they are based on an approximate form of Eq. (2.1.12) given in Eq. (2.1.13). As a result of this approximation, the SOP of any incident beam remains unchanged during its propagation inside a GRIN medium. In practice, several physical mechanisms can induce changes in the SOP of an optical field. This section focuses on two such mechanisms.

9.1.1 Anisotropy and Birefringence

It is well known that anisotropy of a medium produces birefringence that changes the SOP of an optical beam inside that medium [1]. It was noticed as early as 1975 that some GRIN fibers exhibited linear birefringence, which led to changes in the SOP of light transmitted through them [2]. The measured value of this birefringence was larger ($\sim 10^{-5}$) for GRIN fibers compared to a typical single-mode fiber. The physical origin of linear birefringence lies in the anisotropy introduced through stress in a specific direction or by a non-circular core of a GRIN fiber. GRIN rods, used as lenses, also exhibit birefringence that is often spatially nonuniform [3, 4]. The origin of this birefringence lies in the ion-exchange technique used to fabricate such rods.

It is easy to understand why the SOP changes in the case of a constant linear birefringence. The degeneracy found in Section 2.2 for the two orthogonally polarized modes in a mode group is lifted such that the propagation constants, β_x and β_y, become slightly different for the modes polarized in the x and y directions. This difference is related to the magnitude of the birefringence, defined as $n_x - n_y = |\beta_x - \beta_y|/k_0$, where n_x and n_y are the modal refractive indices for the two orthogonally polarized modes. As the phase velocities become different for the two modes, the x and y directions are referred to as the slow and fast axes, respectively, assuming $n_x > n_y$. These two axes are also called the principal axes.

To see why the SOP of incident light changes inside such a GRIN medium, consider a Gaussian beam whose width is matched to that of the LP_{01} mode. We can decompose its initial SOP in a linear polarization basis along the slow and fast axes of the GRIN medium and write its initial electric field at $z = 0$ in the form

$$\mathbf{E}(x,y,0) = \left(\hat{x}A_x + \hat{y}A_y\right) U_{01}(x,y). \tag{9.1.1}$$

After propagating inside the GRIN medium, the electric field of the Gaussian beam at a distance z is given by

$$\mathbf{E}(x,y,0) = \left[\hat{x}A_x \exp(i\beta_x z) + \hat{y}A_y \exp(i\beta_y z)\right] U_{01}(x,y). \tag{9.1.2}$$

The two components of the electric field acquire different phase shifts because of the linear birefringence exhibited by the GRIN medium. Their relative phase difference $\psi = (\beta_x - \beta_y)z$ dictates the SOP of the beam at any distance z. Figure 9.1 shows schematically how the SOP of a linearly polarized beam changes inside such a GRIN medium. The beam becomes elliptically polarized within a short distance, and its SOP changes to circular at a distance at which $\psi = \pi/2$. After that, the SOP reverts back to being elliptical until ψ becomes π. At that distance, the SOP becomes linear in a direction that is orthogonal to the input SOP of the Gaussian beam. The process repeats until ψ becomes 2π, and the beam recovers its initial SOP. This specific distance is known as the beat length L_b, which depends on the birefringence as $L_b = \lambda/|n_x - n_y|$, where we used $k_0 = 2\pi/\lambda$.

Figure 9.1 also shows the evolution of SOP on the Poincaré sphere. As seen there, the SOP evolves along a great circle. It starts at the equator, passes through the

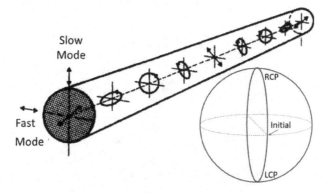

Figure 9.1 Schematic evolution of the SOP of an optical beam inside a birefringent GRIN fiber when the input beam is linearly polarized at 45° from the slow axis. Evolution on the Poincaré sphere is also shown (vertical circle).

north and south poles, and ends at the equator. Over a distance of one beat length, the Stokes vector completes one rotation on the Poincaré sphere. After that distance, the entire sequence of the SOP evolution is repeated in a periodic fashion with the period L_b. At a wavelength of 1 μm, the beat length is only 1 cm for a birefringence such that $|n_x - n_y| = 10^{-4}$.

The situation becomes more complicated when the incident optical beam excites several modes of a GRIN medium. For an arbitrarily polarized optical beam, the electric field at the input of a GRIN medium can be written as

$$E(x, y, 0) = \sum_j (\hat{x} C_j + \hat{y} D_j) U_j(x, y), \qquad (9.1.3)$$

where C_j and D_j are the expansion coefficients and a single subscript s sued to sum over all excited modes. At a distance z inside the medium, the electric field becomes

$$E(x, y, z) = \sum_j (\hat{x} C_j \exp(i\beta_{jx} z) + \hat{y} D_j \exp(i\beta_{jy} z)) U_j(x, y), \qquad (9.1.4)$$

where the phase shifts are different for two orthogonally polarized components of each mode because of linear birefringence. As before, the SOP of each mode evolves periodically such that it traces a great circle on the Poincaré sphere. However, the beat lengths can be different for different modes. It follows from Eq. (9.1.4) that the SOP will evolve in a complex fashion when the incident beam excites multiple modes of a GRIN medium. If the beat length is the same for all modes, the evolution of SOP becomes much simpler. However, modal interference still leads to a complex spatial pattern for the total intensity.

One more complication should be considered for GRIN fibers. In most optical fibers, birefringence does not remain constant along the fiber's length and changes

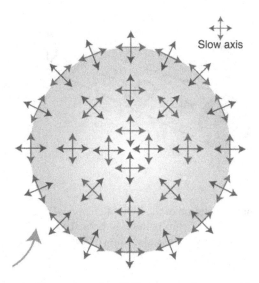

Figure 9.2 Schematic illustration of radially symmetric, nonuniform birefringence in a GRIN rod. Its magnitude is negligible near $\rho = 0$ but becomes $\sim 10^{-5}$ near $\rho = a$. Arrows indicate the directions of slow and fast axes. (After Ref. [10]; ©2019 CC BY.)

randomly because of fluctuations in the core's shape and size introduced during its fabrication. Even the principal axes can rotate in a random fashion along the fiber. Such birefringence changes become an issue for optical communication systems when short pulses are transmitted over long lengths [5]. The reason is that the two polarization components of the optical filed travel with different group velocities. As these velocities change randomly in response to random changes in the fiber's birefringence (analogous to a random-walk problem), the extent of pulse distortion also becomes random. This phenomenon, referred to as *polarization-mode dispersion*, has been studied extensively because of its importance for long-haul telecommunication systems [6–8].

In the case of GRIN rods, measurements of birefringence show that it is spatially nonuniform in the plane transverse to the rod's central axis [4]. This nonuniformity has its origin in the ion-exchange technique used to make the GRIN rod. As shown schematically in Figure 9.2, the ion-exchange process induces birefringence such that the slow axis of the GRIN rod always points in the radial direction, and the magnitude of birefringence increases with ρ. Clearly, such a spatially varying birefringence would produce a beam at the output of a GRIN rod whose SOP is spatially nonuniform and depends on the SOP of the input beam.

Considerable work has been done in recent years to understand the evolution of the SOP of an optical beam inside a birefringent GRIN rod. In a 2001 study, the Jones matrix of the rod was found by considering the tangential and radial components of the electric field along the path of a meridional ray [9]. More recently, a detailed theory has been developed for such rods acting as GRIN lenses that can be used to produce different types of vortex beams [10].

We can use the ray picture to understand the origin of complex polarization changes in a birefringent GRIN rod. It is evident that the paths will become different for the ordinary and extraordinary rays, and their separation will be different for different sets of such rays. It is useful to divide a ray's trajectory into multiple segments of length d, each segment acting as a wave plate. If we use the cylindrical coordinates (ρ, ϕ) for the trajectory, we can write the Jones matrix of a segment by rotating the Cartesian axes by ϕ, using the Jones matrix in Eq. (8.1.4) for the wave plate, and then rotating the axes back. This results in the relation $M = R(-\phi)JR(\phi)$, where $R(\phi)$ is the rotation matrix. We use the form of the J and R matrices given in Section 8.1.1,

$$J = \begin{pmatrix} \exp(i\delta_x) & 0 \\ 0 & \exp(i\delta_y) \end{pmatrix}, \qquad R(\phi) = \begin{pmatrix} \cos\phi & \sin\phi \\ -\sin\phi & \cos\phi \end{pmatrix}, \qquad (9.1.5)$$

where the phase shifts δ_x and δ_y are different because of the birefringence of the wave plate. Using them, the Jones matrix M is given by [9]

$$M(\rho, \phi) = e^{i\psi} \begin{pmatrix} \cos\delta + i\cos 2\phi \sin\delta & i\sin 2\phi \sin\delta \\ i\sin 2\phi \sin\delta & \cos\delta - i\cos 2\phi \sin\delta \end{pmatrix}, \qquad (9.1.6)$$

where $\psi = (\delta_x + \delta_y)/2$, $\delta = (\delta_x - \delta_y)/2$, and $\delta_j = n_j k_0 d$ for $j = x, y$. As this matrix changes along the path of any ray, the SOP at the output of any GRIN lens will vary in a complex fashion in the transverse plane of the optical beam. It can be controlled by cascading several GRIN lenses with other phase plates in between [10].

As an example, Figure 9.3 shows the SOP distribution in two cases for a cascade of two GRIN lenses. In the first case, SOP is initially circular everywhere. At the output, it remains nearly circular at the beam's center, but changes as one moves away from the center in any direction while maintaining radial symmetry. On the Poincaré sphere, the Stokes vector traverses the upper hemisphere as ρ increases, becomes horizontal (linear SOP) at the equator, and crosses the lower hemisphere after that. The situation changes dramatically when a birefringent element is inserted between the two GRIN rods. In the second case shown on the right in Figure 9.3, the initial Gaussian beam is linearly polarized, and the final SOP is also relatively uniform. It was found in Ref. [10] that cascaded GRIN lenses can be used to produce vortex beams with different OAM values. Such a device works in the same way as a q-plate in Figure 8.7 of Section 8.3.2. When an RCP beam with no OAM is incident, it is converted into a vortex beam with an OAM of $+2$, while becoming LCP at the same time. In contrast with a q-plate, the GRIN-lens technique is much more versatile and can be used to produce a variety of vector beams.

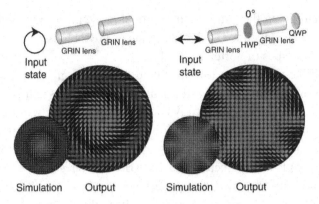

Figure 9.3 Spatial SOP variations across the intensity profile of an initially Gaussian beam in two configurations shown on top. Different shapes correspond to different SOPs and color coding indicates the sense of rotation. Input SOP of the Gaussian beam is spatially uniform but circular in the first case and linear in the second case. (After Ref. [10]; ©2019 CC BY.)

9.1.2 Origin of Photonic Spin-Orbit Coupling

The concept of a geometrical phase is well known in optics since the works of Pancharatnam and Berry [11, 12]. This phase arises when the direction in which light is propagating changes such that the wave vector **k** moves on the surface of a sphere in the **k** space. The geometrical phase leads to a rotation of the plane of polarization of light propagating inside a single-mode fiber along a helical trajectory [13, 14]. Such polarization rotation also occurs in GRIN fibers, where the direction of the propagation vector changes in a periodic fashion because of radial variations in the refractive index [15–17]. It can be attributed to the circular birefringence induced by the gradient of the refractive index in a GRIN fiber.

A reverse effect can also occur, where the SOP of an optical beam inside a GRIN fiber affects the trajectory and width of that beam. This phenomenon is known as the optical spin-Hall effect [18–20], and its origin also lies in the non-planar trajectory of an optical beam. The physical process behind such polarization-dependent effects is the photonic analog of the well-known atomic spin-orbit interaction. As we saw in Section 8.2, photons in a vortex beam carry both the spin angular momentum (SAM), related to the beam's SOP, and the orbital angular momentum (OAM), related to the helical shape of the beam's phase front. These two types of angular momenta become coupled inside a GRIN medium because of radial variations in its refractive index.

To understand the origin of spin-orbit coupling in GRIN media, we need to consider the exact wave equation, Eq. (2.1.12), derived in Section 2.1.1 from Maxwell's equations:

$$\nabla^2 \tilde{\mathbf{E}} + n^2(\mathbf{r})k_0^2\tilde{\mathbf{E}} + \nabla\left[\tilde{\mathbf{E}} \cdot \nabla(\ln n^2)\right] = 0, \tag{9.1.7}$$

where $\tilde{\mathbf{E}}(\mathbf{r}, \omega)$ is the Fourier transform of the electric field, $k_0 = \omega/c = 2\pi/\lambda_0$, and $n(\mathbf{r})$ is the refractive index of the GRIN medium at the location $\mathbf{r}$. When finding the modes of a GRIN medium in Section 2.2, we neglected the last term in this equation, claiming its contribution was negligible compared to the second term containing k_0^2. It turns out that all trajectory-dependent polarization effects are produced by this term. It is thus necessary to solve Eq. (9.1.7) in its full vector form.

Two approaches have been adopted for solving Eq. (9.1.7). In one approach, vector modes of a GRIN medium are found by retaining the last term in this equation (see Section 9.3). In the second approach, Eq. (9.1.7) is simplified first by making the geometrical-optics approximation, and the resulting equation is solved for the path taken by an optical ray inside a GRIN medium. We follow such an approach in this section.

The geometrical-optics approximation is based on the smallness of the optical wavelength λ_0 relative to other dimensions of interest. It amounts to assuming that the second term in Eq. (9.1.7) dominates because k_0 is relatively large at optical wavelengths. The solution of Eq. (9.1.7) is sought in the form $\tilde{\mathbf{E}} = \mathbf{E_0} \exp(ik_0 S)$. As we found in Section 3.1, the phase function $S(\mathbf{r})$ satisfies the eikonal equation [1]:

$$(\nabla S)^2 = n^2(\mathbf{r}). \tag{9.1.8}$$

The surface $S(\mathbf{r}) = C$, where C is a constant, represents the phase front of the wave. The unit vector, $\mathbf{u} = (\nabla S)/n$, is normal to this phase front and points in the direction of the path taken by a geometrical ray. When the last term in Eq. (9.1.7) is ignored, the ray's path is governed by the following two equations [1]:

$$\frac{d\mathbf{r}}{ds} = \mathbf{u}, \qquad \frac{d\mathbf{u}}{ds} = \nabla \ln n - \mathbf{u}(\mathbf{u} \cdot \nabla \ln n), \tag{9.1.9}$$

where ds is the length of a small segment along the ray's path shown schematically in Figure 9.4.

The important question is how the preceding two equations are modified by the last term in Eq. (9.1.7). It was answered in a 1992 study, where it was found that a polarization-dependent term appears in the first equation such that [16]

$$\frac{d\mathbf{r}}{ds} = \mathbf{u} - \frac{\sigma}{nk_0}(\mathbf{u} \times \nabla \ln n), \tag{9.1.10}$$

where $\sigma = \pm 1$ represents the SAM of the optical wave. The presence of σ in Eq. (9.1.10) is a consequence of the photonic spin-orbit coupling introduced by the last term in Eq. (9.1.7). It shows that the left- and right-circularly polarized waves behave differently inside a GRIN medium because of the spatially varying nature

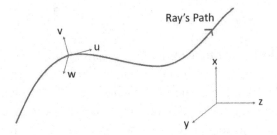

Figure 9.4 Schematic showing a ray's path in three dimensions inside a GRIN medium. The three unit vectors ($\mathbf{u}$, $\mathbf{v}$, $\mathbf{w}$) i the local frame change directions along this path.

of its refractive index. It should be stressed that the change introduced is relatively small as it scales inversely with k_0 (and linearly with the wavelength λ_0).

Let us apply Eq. (9.1.10) to a GRIN fiber with a parabolic profile for the refractive index. Using $p = 2$ in Eq. (1.2.4), the refractive index varies as

$$n(\mathbf{r}) = n_0 \left[1 - \frac{b^2}{2}(x^2 + y^2) \right], \qquad b^2 = \frac{2\Delta}{a^2}. \qquad (9.1.11)$$

In the paraxial approximation, the ray's path remains close to the fiber's axis, taken to be the z direction. We denote the ray's position at a distance z by a two-dimensional vector $\mathbf{r}_t = (x, y)$ in the transverse plane. Following Eq. (3.1.5), we replace the s derivative with a z derivative in Eq. (9.1.10) and obtain the following two equations:

$$\frac{dx}{dz} = u_x - \frac{b^2\sigma}{n_0 k_0}y, \qquad \frac{dy}{dz} = u_y + \frac{b^2\sigma}{n_0 k_0}x. \qquad (9.1.12)$$

We use Eq. (9.1.9) to obtain the equations for u_x and u_y:

$$\frac{du_x}{dz} = -b^2 x, \qquad \frac{du_y}{dz} = -b^2 y. \qquad (9.1.13)$$

The preceding four equations can be combined to obtain a single second-order differential equation for $\mathbf{r}_t$ [16]:

$$\frac{d^2\mathbf{r}_t}{dz^2} + b^2\mathbf{r}_t = \frac{d\mathbf{r}_t}{dz} \times \vec{\Omega}, \qquad (9.1.14)$$

where the vector $\vec{\Omega}$ points in the z direction and is defined as

$$\vec{\Omega} = \frac{b^2\sigma}{n_0 k_0}\hat{\mathbf{z}}. \qquad (9.1.15)$$

Equation (9.1.14) governs a ray's trajectory inside a GRIN medium and includes the spin-orbit coupling through the last term. It reduces to the harmonic-oscillator equation obtained in Section 3.1 when the last term is neglected. As discussed there, the resulting periodic solution for a ray's path is responsible for the self-imaging phenomenon discussed in earlier chapters. The spin-orbit coupling term in Eq. (9.1.14) can be interpreted as a Coriolis force resulting from a reference frame rotating around the z axis with the angular velocity Ω [20]. The origin of this rotation can be understood from Figure 9.4. As the direction of the unit vector **u** changes on a ray's path in response to refractive-index variations, the other two orthogonal unit vectors, say **v** and **w**, also change. Such changes appear as a rotation in a local coordinate system when the ray follows a helical path. As seen in Eq. (9.1.14), the spatial frequency of this rotation depends on the SAM of the optical beam through σ. As a result, a circularly polarized beam follows a different trajectory for $\sigma = \pm 1$. This is the photonic analog of the spin-Hall effect [18, 19].

The spin-orbit coupling term in Eq. (9.1.14) is also responsible for a geometrical phase [17]. This can be seen by noting that the unit vector **u** represents the direction of the wave vector (or a photon's momentum $\hbar \mathbf{k}$). As a ray follows a helical path inside a GRIN medium, **k** returns to its original direction after each period, tracing a closed loop in the **k** space. The resulting geometrical phase depends on the SOP of the optical field. The phase shift produced by the Ω term can be found by solving Eq. (9.1.14). It depends on the SOP of an optical beam through σ and has opposite signs for $\sigma = \pm 1$. In essence, the spin-orbit coupling inside a GRIN fiber induces a circular birefringence. It leads to rotation of the plane of polarization of a Gaussian beam that is linearly polarized and is launched off-center into a GRIN fiber because the beam follows a helical path [17]. If we treat the initially linear SOP as a linear combination of the LCP and RCP parts, which acquire different phase shifts because of circular birefringence, we can show that the plane of polarization of the incident beam rotates as it propagates down a GRIN medium [15].

9.2 Polarization-Dependent Phenomena

In this section we consider evolution of the transverse electric field $\mathbf{E}_t$ along a ray's path inside a GRIN medium. In one approach, changes in the curvature and torsion of the trajectory are used for this purpose [17]. Here we follow an alternative method discussed in Ref. [21].

9.2.1 Circular Birefringence and Geometrical Phase

Consider the trajectory shown in Figure 9.4, where the ray's local direction **u** changes along its path. The transverse electric field $\mathbf{E}_t$ along this path is found to satisfy the following equation [21]:

$$\nabla^2 \mathbf{E}_t + n^2 k_0^2 \mathbf{E}_t + 2ink_0 \Lambda (\mathbf{u} \times \mathbf{E}_t) = 0, \qquad (9.2.1)$$

where the last term results from the Coriolis force and depends on Λ, defined as $\Lambda = (d\mathbf{v}/ds) \cdot \mathbf{w}$. This term couples the two components of $\mathbf{E}_t$ along the $\mathbf{v}$ and $\mathbf{w}$ directions (see Figure 9.4).

The coupling between the linearly polarized components in Eq. (9.2.1) disappears when the electric field is written in terms of its circularly polarized components as

$$\mathbf{E}_t = E^+\hat{\mathbf{c}} + E^-\hat{\mathbf{c}}^*, \qquad \hat{\mathbf{c}} = (\mathbf{v} + i\mathbf{w})/\sqrt{2}. \tag{9.2.2}$$

The components E^+ and E^- are found to satisfy

$$\nabla^2 E^\pm + n^2 k_0^2 E^\pm + 2\sigma n k_0 \Lambda E^\pm = 0, \tag{9.2.3}$$

with $\sigma = \pm 1$ for the LCP and RCP states, respectively. As the Λ term is relatively small, we can write Eq. (9.2.3) in the form

$$\nabla^2 E^\pm + k_0^2 (n + \sigma \Lambda/k_0)^2 E^\pm = 0. \tag{9.2.4}$$

The preceding equation shows that the refractive indices become slightly different for the LCP and RCP states, i.e., the GRIN medium exhibits circular birefringence. The polarization-dependent phase shift induced by the new term is the geometrical phase, which accumulates along the ray's path as

$$\phi_g = k_0 \sigma \int \frac{\Lambda}{k_0} ds = \sigma \int \Lambda \, ds. \tag{9.2.5}$$

When the input beam is linearly polarized, the relative phase difference between its LCP and RCP components leads to rotation of the plane of polarization as the beam propagates along a helical path [15].

For a GRIN medium with the parabolic profile given in Eq. (9.1.11), Λ is related to the GRIN parameters $\Lambda = b^2/(n_0 k_0)$. The two refractive indices in this case can be written as

$$n_\pm = n + \frac{\sigma b^2}{n_0 k_0^2} = n_0 \left[1 - \frac{1}{2} b^2 \rho^2 + \frac{\sigma b^2}{(n_0 k_0)^2} \right]. \tag{9.2.6}$$

When the input beam is linearly polarized, the relative phase difference between its LCP and RCP components after a distance z is found to be

$$\delta\phi = k_0 (n_+ - n_-) z = 2b^2 z/(n_0 k_0). \tag{9.2.7}$$

For an estimate, consider a GRIN fiber with $\Delta = 0.05$ and $a = 25$ μm, resulting in $b = 12.6$ mm^{-1}. Using $n_0 = 1.5$ at the input wavelength of 1550 nm, the SOP-dependent index change from Eq. (9.2.6) is found to be about 7×10^{-6}. The resulting phase difference in Eq. (9.2.7) produces rotation of the linear SOP of the beam by about 16°/cm. This estimate indicates that the trajectory-dependent effects should be observable, both in GRIN fibers and GRIN rods. Recent experiments have shown that this is indeed the case [22].

9.2.2 *Photonic Spin-Hall Effect*

As mentioned earlier, the trajectories followed by an optical beam become different for $\sigma = \pm 1$ as a result of spin-orbit coupling inside a GRIN medium. This photonic analog of the spin-Hall effect has been discussed in several different contexts [18–20]. In a 2008 experiment, it was observed when a circularly polarized beam at a wavelength of 633 nm followed a helical path on the surface of a 9.6-cm-long glass rod of 1.6-cm diameter [20]. The shift occurred in the opposite directions for the LCP and RCP beams and its magnitude was relatively small (about 2 μm).

A similar shift occurs in the case of a GRIN medium, where an optical beam also follows a helical path if it is launched such that its center is shifted from the fiber's axis. To find the magnitude of this shift, we must solve Eq. (9.1.14) governing the path taken by the beam's center. When written in terms of the two components of $\mathbf{r}_t = (x, y)$, we obtain two coupled equations

$$\frac{d^2x}{dz^2} + b^2 x = -\Omega \frac{dy}{dz}, \qquad \frac{d^2y}{dz^2} + b^2 y = \Omega \frac{dx}{dz}, \tag{9.2.8}$$

where Ω depends on the parameters of the GRIN medium as indicated in Eq. (9.1.15). The last term is responsible for the shift of the trajectory in a direction normal to the local $\mathbf{r}_t$ vector. This direction is opposite for the LCP and RCP states because Ω depends on the SAM through σ.

Consider a collimated Gaussian beam that is launched into a GRIN fiber such that its center is offset by an amount (x_0, y_0). To find its trajectory, we can use a perturbation approach and first neglect the Ω term in Eq. (9.2.8). The beam's center in this case follows a helical path such that

$$\bar{x}(z) = x_0 \cos(bz) + (x_0'/b) \sin(bz), \qquad \bar{y}(z) = y_0 \cos(bz) + (y_0'/b) \sin(bz), \tag{9.2.9}$$

where x_0' and y_0' are in the initial direction of the wavefront. Replacing x and y with $\bar{x}$ and $\bar{y}$ on the right side in Eq. (9.2.8), we obtain equations that correspond to a driven harmonic oscillator. Their solution indicates how the perturbation provides an external force that shifts the trajectory in opposite directions for the RCP and LCP beams. The relative shift of the beam's trajectory induced by the spin-Hall effect depends on the parameters Ω, b, and x_0 through the combination $\Omega/(b^2 x_0)$. It follows from Eq. (9.1.15) that the shift depends on the input wavelength and on the initial position of the beam (through x_0), but not on the GRIN medium's parameters. Its magnitude is also limited to an amount comparable to the wavelength of incident light.

Much larger shifts are possible when the input beam is tightly focused onto a GRIN rod and contains non-paraxial rays traveling at large angles from the rod's axis. This was the case in a 2019 experiment [22]. It employed a 4-mm-long GRIN rod with 1.8-mm diameter, resulting in a numerical aperture of 0.55. A linearly

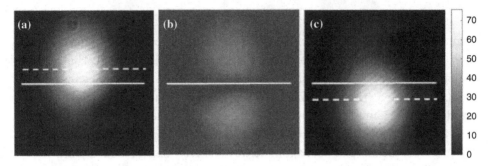

Figure 9.5 Observation of the spin-Hall effect in a GRIN rod. Intensity patterns for three rotations of a polarizer ($-9°$, 0, and 9°) used for the weak-measurement scheme. (After Ref. [22]; ©2019 Optica.)

polarized Gaussian beam was focused tightly onto this rod such that its center was shifted by 0.7 mm from the rod's center. The output intensity exhibited a two-spot pattern because the LCP and RCP components of the beam shifted in the opposite directions, owing to the photonic spin-Hall effect. Figure 9.5 shows the observed intensity patterns for three settings of a polarizer, when the magnitude of the spin-Hall shift was amplified through a weak-measurement scheme [23]. The observed spin-Hall shift (about 95 μm) in this experiment was comparable in magnitude to the beam's width, as the two spots are visibly separated in part (b) of Figure 9.5.

9.2.3 SAM-to-OAM Conversion

The periodic focusing of optical beams, related to the self-imaging phenomenon in a GRIN media, can result in the so-called SAM-to-OAM conversion, in which an LCP beam, without any OAM, is converted into a vortex beam with an OAM of $\ell = 2$ such that its SAM changes from $\sigma = 1$ to -1, and the total angular momentum $J = \ell + \sigma$ remains conserved. The situation is reversed for an RCP beam, which is converted into an LCP beam with $\ell = -2$ The SAM-to-OAM conversion has attracted considerable attention in recent years in several different contexts [24–26]. In the case of a locally isotropic medium, it occurs in the non-paraxial region realized through tight focusing of an optical beam or scattering of light by small particles [27].

A simple way to understand the SAM-to-OAM conversion of non-paraxial light is to consider the angular spectrum of an optical beam focused tightly by a lens with a large numerical aperture:

$$\mathbf{E}_f(\mathbf{r}) = \iint \mathbf{A}(k_x, k_y) \exp(i\mathbf{k} \cdot \mathbf{r}) dk_x \, dk_y, \tag{9.2.10}$$

where E_f is the electric field in the focal region of the lens. The angular spectrum A is a function of $k_x = k \sin\theta \cos\phi$ and $k_y = k \sin\theta \sin\phi$, which are the transverse components of the **k** vector in the spherical coordinates, assuming that the beam is propagating along the z axis. In the non-paraxial region, the maximum value of θ is relatively large ($> 10°$), and $\sin\theta$ cannot be replaced with θ. The input field $\mathbf{E}_0$ is paraxial with a narrow angular spectrum and its SOP is nearly the same for all plane waves. However, the SOP of the focused beam becomes different because of tight focusing. The Jones matrix responsible for this SOP change has a geometric origin and depends on the angles θ and ϕ in the circular-polarization basis as [27]

$$M(\theta,\phi) = \begin{pmatrix} \cos^2(\theta/2) & -\sin^2(\theta/2)e^{-2i\phi} \\ -\sin^2(\theta/2)e^{2i\phi} & \cos^2(\theta/2) \end{pmatrix}. \tag{9.2.11}$$

Consider what happens when the input beam is a circularly polarized vortex beam with both SAM and OAM such that its total angular momentum is $J = \ell_0 + \sigma$. Using an input Jones vector in the form $E_0 = E^\sigma e^{i\ell_0\phi}(1,0)^T$, where T denotes the transpose, the Jones vector in the focal plane is found to be

$$E_f = ME_0 = \begin{pmatrix} \cos^2(\theta/2)E^\sigma e^{i\ell_0\phi} \\ -\sin^2(\theta/2)E^{-\sigma}e^{i(\ell_0+2\sigma)\phi} \end{pmatrix}. \tag{9.2.12}$$

This relation shows that a part of the original beam is converted into a new beam whose OAM is $\ell = \ell_0 + 2\sigma$ and whose SAM is $-\sigma$ such that the total angular momentum J is conserved. The conversion efficiency depends on the angle θ through the factor $\sin^2(\theta/2)$. This conversion process is similar to that of a q-plate made with an anisotropic material (see Section 8.3.2). The underlying physics is the same in both cases.

The SAM-to-OAM conversion also occurs inside a GRIN rod when a tightly focused Gaussian beam is launched into it [22]. The periodic focusing of this beam inside the GRIN rod produces an output beam that also contains a vortex with $l = \pm 2$. We can analyze this case using the LCP and RCP components of the input beam. Each of these components will be transformed by the Jones matrix in Eq. (9.2.11). Using $\ell_0 = 0$ for the input Gaussian beam in Eq. (9.2.12), the input component with $\sigma = +1$ produces a vortex beam with $\ell = 2$ and $\sigma = -1$. In contrast, the input component with $\sigma = -1$ produces a vortex beam with $\ell = -2$ and $\sigma = 1$. The SOP of the output beam is found to vary in a complex fashion when analyzed using a polarizer. The situation becomes even more complex when the input Gaussian beam is offset from the GRIN rod's central axis because of the beam's shift leads to the spin-Hall effect.

In a 2019 experiment, a 4-mm-long GRIN rod of 1.8-mm diameter was used to observe the process of SAM-to-OAM conversion [22]. A suitably polarized Gaussian beam was focused tightly onto the rod to ensure the propagation of light

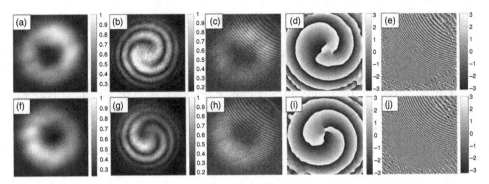

Figure 9.6 Observation of the SAM-to-OAM conversion in a GRIN rod for RCP (top row) and LCP beams (bottom row). Five patterns correspond to (a) output intensity, (b) on-axis and (c) off-axis interference patterns, and extracted phases (d and e) from them. (After Ref. [22]; ©2019 Optica.)

in the non-paraxial regime. Figure 9.6 shows several types of patterns recorded when the RCP (top row) and LCP beams (bottom row) were launched into the GRIN rod. The five patterns in each case correspond to (a) output intensity, (b) on-axis interference and (c) off-axis interference, and the extracted phases (d and e) corresponding to those in (b) and (c). These patterns occur for an SOP state that is orthogonal to the input SOP, i.e, nothing would be observed in the absence of the SAM-to-OAM conversion. They provide clear evidence that the SAM-to-OAM conversion occurs inside a GRIN medium under suitable conditions.

The opposite situation in which the OAM of an optical beam is converted into SAM can also occur. It was found in a recent study that considerable polarization changes occur when a radially polarized vortex beam (see Section 8.2.3) is propagated through a GRIN medium [28]. As expected from the discussion in Section 8.4, such a beam exhibits periodic self-imaging and its size is reduced considerably at the focal planes of the GRIN medium. It turns out that the state of polarization (SOP) of the focused beam becomes spatially nonuniform. Even though the beam maintains cylindrical symmetry, it exhibits LCP in some spatial regions and RCP in other regions. Such OAM-SAM conversion occurs only for vortex beams with some topological charge and disappears when the input beam has no OAM ($\ell = 0$).

9.3 Vector Modes of GRIN Media

In this section we look for modal solutions of Eq. (9.1.7) and consider how a relatively small index-gradient term containing ∇n affects the HG_{mn} or LP_{lm} modes of a GRIN medium obtained in Section 2.2. We first discuss a perturbation approach that can be used to find the vector modes and their propagation constants approximately.

After that, we describe how Eq. (9.1.7) can be solved more accurately to obtain the circularly polarized modes of a GRIN medium.

9.3.1 *Perturbative Approach*

As the last term in Eq. (9.1.7) is small relative to other terms, it can be treated as a perturbation, and its effects on the HG_{mn} or LP_{lm} modes included using the standard perturbation theory [29]. Indeed, such an approach was used in several studies for solving Eq. (9.1.7) approximately [30–32]. To find the modified modes, we write the electric field in the Cartesian coordinates as

$$\tilde{\mathbf{E}} = (\hat{\mathbf{x}}E_x + \hat{\mathbf{y}}E_y + \hat{\mathbf{z}}E_z)e^{i\beta z}, \tag{9.3.1}$$

where β is the mode's propagation constant. As the longitudinal component E_z can be obtained from the transverse components, we focus on the transverse components, which are found to satisfy from Eq. (9.1.7) the following coupled equations:

$$(\nabla_T^2 + n^2 k_0^2)E_x + \frac{\partial}{\partial x}\left[E_x\frac{\partial}{\partial x}(\ln n^2) + E_y\frac{\partial}{\partial y}(\ln n^2)\right] = \beta^2 E_x, \tag{9.3.2}$$

$$(\nabla_T^2 + n^2 k_0^2)E_y + \frac{\partial}{\partial y}\left[E_x\frac{\partial}{\partial x}(\ln n^2) + E_y\frac{\partial}{\partial y}(\ln n^2)\right] = \beta^2 E_y. \tag{9.3.3}$$

These equations show that the index-gradient term couples the transverse components E_x and E_y. We can write them in a matrix form using a Jones vector. Its use allows us to write Eq. (9.1.7) as the following two-dimensional eigenvalue problem:

$$(\hat{H}_0 + \hat{H}_1)|\psi\rangle) = \beta^2|\psi\rangle, \qquad |\psi\rangle = \begin{pmatrix} E_x \\ E_y \end{pmatrix}, \tag{9.3.4}$$

where the operators $\hat{H}_0$ and $\hat{H}_1$ are defined as

$$\hat{H}_0 = (\nabla_T^2 + n^2 k_0^2)\begin{pmatrix} 1 & 0 \\ 0 & 1 \end{pmatrix}, \quad \hat{H}_1|\psi\rangle = \begin{pmatrix} \frac{\partial}{\partial x}\left[E_x\frac{\partial}{\partial x}(\ln n^2)\right] & \frac{\partial}{\partial x}\left[E_y\frac{\partial}{\partial y}(\ln n^2)\right] \\ \frac{\partial}{\partial y}\left[E_x\frac{\partial}{\partial x}(\ln n^2)\right] & \frac{\partial}{\partial y}\left[E_y\frac{\partial}{\partial y}(\ln n^2)\right] \end{pmatrix}. \tag{9.3.5}$$

As is usual in quantum mechanics, we first find the modes by solving the eigenvalue problem in Eq. (9.3.4) with $\hat{H}_1 = 0$, and then expand E_x and E_y in terms of these modes. In the case of a parabolic index profile, for which $n^2 = [1 - b^2(x^2 + y^2)]$, these are just the Hermite–Gauss modes $U_{mn}(x, y)$ obtained in Section 2.2 with the eigenvalues given in Eq. (2.2.11). The perturbation problem is complicated because of the degenerate nature of these modes. For given values of the integers m and n, there exist $m + n + 1$ states with the same eigenvalue. Perturbation removes the degeneracy and provides a new set of modes that are specific linear combinations of the U_{mn} modes. The eigenvalues given in Eq. (2.2.11) are modified as [30]

$$\bar{\beta}^2_{mn} = \beta^2_{mn} - (m + n + 1 - 2j)b^2, \tag{9.3.6}$$

where the integer j takes values from 0 to $(m+n)$. The modes resulting from a linear combination of the unperturbed modes have SOPs that are far from being linear. For example, when $m + n = 1$, the perturbation combines four degenerate modes to produce modes such that two of them have radial and azimuthal SOPs. This case corresponds to the situation discussed in Section 8.2.3.

It is evident from Eq. (9.3.5) that $\hat{H}_1$ is not diagonal in a basis based on linear SOPs. However, as we found in Section 9.2, coupling between the linearly polarized components in Eq. (9.3.3) disappears when the electric field is written in terms of its circularly polarized components as

$$\tilde{\mathbf{E}} = (\hat{\mathbf{e}}_+ E^+ + \hat{\mathbf{e}}_- E^-)U(x, y)e^{i\beta z}, \tag{9.3.7}$$

where $U(x, y)$ governs the spatial distribution of the modes. In this basis, the operator in $\hat{H}_1$ in Eq. (9.3.4) becomes diagonal. We can use perturbation theory to calculate how the eigenvalues β_{lm} of the LP$_{lm}$ modes obtained for $\hat{H}_1 = 0$ are affected by this perturbation. To the first-order, the change in β_{lm} can be calculated using

$$\delta\beta_{lm} = \iint U^*_{lm}(\rho, \phi)H_1 U_{lm}(\rho, \phi)\rho \, d\rho \, d\phi, \tag{9.3.8}$$

where $U_{lm}(\rho, \phi)$ is assumed to be normalized, and we employed the cylindrical coordinates for integration in the xy plane.

Both integrals in Eq. (9.3.8) can be done using the form of U_{lm} given in Eq. (2.2.20). The result is different for the LCP and RCP modes and can be written as [32]

$$\delta\beta^\sigma_{lm} = (1 + \sigma\ell)\kappa, \qquad \kappa = b^2/(n_0 k_0), \tag{9.3.9}$$

where $\sigma = \pm 1$ represents the SAM of the mode. It shows that the perturbation-induced shift in the value of β_{lm} does not depend on the integer m and is the same for all modes with a fixed value of l. The shift occurs even for the radially symmetric LP$_{0m}$ modes with $\ell = 0$, but it does not depend on the mode's SOP. The shift becomes larger for modes with finite values ℓ and depends on the mode's SOP because it has opposite signs for $\sigma = \pm 1$. The case $\ell = \pm 1$ is special because the perturbation-induced change vanishes when ℓ and σ have opposite signs.

The σ dependence of $\delta\beta_{lm}$ in Eq. (9.3.9) is responsible for the circular birefringence induced by the gradient of the refractive index in a GRIN medium. The magnitude of this birefringence for any optical beam depends on the number of modes excited by that beam. It follows from Eq. (9.3.9) that no birefringence occurs when an optical beam is launched on-axis and excites only radially symmetric LP$_{0m}$ modes with $\ell = 0$. As we saw in Section 2.4.1, modes with $\ell \neq 0$ are excited when the input beam's center is offset from the central axis of a GRIN medium. Such beams experience circular birefringence inside this medium. However, its magnitude must be calculated using the modal expansion given in Eq. (2.4.3). Adding $\delta\beta_{lm}^{\sigma}$ to β_{lm} in that equation, we can write the result as

$$\mathbf{E}^{\pm}(\mathbf{r}) = \hat{\mathbf{e}}_{\pm} \sum_{l} \sum_{m} C_{lm} R_{lm}(\rho) e^{il(\phi\pm\kappa z)} \exp(i\beta_{lm}z + i\kappa z). \tag{9.3.10}$$

The ℓ-dependent phase term in Eq. (9.3.10) shows the impact of circular bire-fringence. Its z dependence implies that the spatial pattern of the beam's intensity should exhibit a rotation in the opposite directions for the LCP and RCP beams. Such a rotation was observed in a 1992 experiment using a few-mode step-index fiber [32]. The shift was about $1.4°$ for a 1-m-long fiber. In a related experiment, a linearly polarized beam was launched into a GRIN fiber and changes in its direction of linear SOP induced by the circular birefringence were observed [15].

9.3.2 Analytic Form of Vector Modes

It is possible to solve Eq. (9.1.7) directly using $\tilde{\mathbf{E}}(\mathbf{r}) = \mathbf{E}(x, y) e^{i\beta z}$, where $\mathbf{E}(x, y)$ does not depend on z as required for any mode and β is a propagation constant yet to be determined. The main difference from the modal analysis of Section 2.2 is that $\mathbf{E}$ is not assumed to be polarized linearly with a single component in the transverse plane. The equation for the magnetic field is also solved using $\tilde{\mathbf{H}}(\mathbf{r}) = \mathbf{H}(x, y) e^{i\beta z}$.

At this point, one can follow the technique used to find the vector modes of a step-index fiber [33, 34]. In this approach, equations for the longitudinal components, E_z and H_z, are solved first. Maxwell's equations are then used to express the transverse components of $\mathbf{E}$ and $\mathbf{H}$ in terms of E_z and H_z as

$$E_{\rho} = \frac{i\beta}{p^2}\left[\frac{\partial E_z}{\partial\rho} + \frac{\omega\mu_0}{\beta\rho}\frac{\partial H_z}{\partial\varphi}\right], \qquad E_{\phi} = \frac{i\beta}{p^2}\left[\frac{1}{\rho}\frac{\partial E_z}{\partial\varphi} - \frac{\omega\mu_0}{\beta}\frac{\partial H_z}{\partial\rho}\right], \tag{9.3.11}$$

$$H_{\rho} = \frac{i\beta}{p^2}\left[\frac{\partial H_z}{\partial\rho} - \frac{\omega\epsilon}{\beta\rho}\frac{\partial E_z}{\partial\varphi}\right], \qquad H_{\phi} = \frac{i\beta}{p^2}\left[\frac{1}{\rho}\frac{\partial H_z}{\partial\varphi} + \frac{\omega\epsilon}{\beta}\frac{\partial E_z}{\partial\rho}\right], \tag{9.3.12}$$

where $p^2 = n^2 k_0^2 - \beta^2$ and $\epsilon = n^2\epsilon_0$. The cylindrical coordinates were used for the transverse components because the refractive index n varies in the ρ direction in a GRIN medium.

The E_z and H_z equations that need to be solved are

$$[\nabla_T^2 + (n^2 k_0^2 - \beta^2)]E_z + \frac{\partial}{\partial z}\left[\frac{\partial}{\partial\rho}(\ln n^2)E_{\rho}\right] = 0, \tag{9.3.13}$$

$$[\nabla_T^2 + (n^2 k_0^2 - \beta^2)]H_z + \frac{\partial}{\partial\rho}(\ln n^2)\left[\frac{\partial H_z}{\partial z} - \frac{\partial H_z}{\partial\rho}\right] = 0, \tag{9.3.14}$$

where $n^2(\rho)$ for a GRIN medium has the form

$$n^2(\rho) = n_0^2[1 - h(\rho)], \qquad h(\rho) = 2\Delta f(\rho/a). \tag{9.3.15}$$

The E_z and H_z equations can be solved in the weakly guiding approximation, $|h(\rho)| \ll 1$, that holds when $\Delta \ll 1$. In the cylindrical coordinates, $n(\rho)$ does not depend on ϕ. Using a method based on the separation of variables, the solution can be written as [35]

$$E_z(\rho, \phi) = n_0^{-1/2} U(\rho) e^{i\ell\phi}, \qquad H_z(\rho, \phi) = i(n_0\epsilon_0/\mu_0)^{1/2} V(\rho) e^{i\ell\phi}, \qquad (9.3.16)$$

where the functions $U(\rho)$ and $V(\rho)$ satisfy

$$\hat{M} U(\rho) = V(\rho), \qquad \hat{M} V(\rho) = U(\rho), \qquad (9.3.17)$$

and $\hat{M}$ is the following second-order differential operator:

$$\hat{M} = \left[\frac{d^2}{d\rho^2} + \left(\frac{1}{\rho} + \frac{dh/d\rho}{\chi - h} \right) \frac{d}{d\rho} + \left(n_0^2 k_0^2 (\chi - h) - \frac{\ell^2}{\rho^2} \right) \right] \frac{\rho(h - \chi)}{\ell(dh/d\rho)}. \quad (9.3.18)$$

Here $\chi = 1 - (\beta/n_0 k_0)^2$ has values such that $\chi \ll 1$.

It follows from Eq. (9.3.17) that $\hat{M}^2 U = U$ or $\hat{M}^2 = \hat{I}$, where $\hat{I}$ is an identity operator. Using $\hat{M} = \pm\hat{I}$, U can be a solution of the following two equations:

$$\hat{M} U_+(\rho) = U_+(\rho), \qquad \hat{M} U_-(\rho) = -U_-(\rho). \qquad (9.3.19)$$

Each of these equations has two independent solutions. However, only one of them is finite at $\rho = 0$. As a result, E_z and H_z have two groups of solutions that depend on U_+ and U_-, respectively. We can use them in Eq. (9.3.12) to find the transverse components of the electric field. The result can be written as [35]

$$E_t^\pm(\rho, \phi) = \frac{1}{\sqrt{2}} \left(\hat{\rho} \pm i\hat{\phi} \right) F_\pm(\rho) e^{i\ell\phi}, \qquad (9.3.20)$$

where the functions $F_\pm(\rho)$ satisfy

$$\frac{d^2 F_\pm}{d\rho^2} + \frac{1}{\rho} \frac{dF_\pm}{d\rho} + \left[n_0^2 k_0^2 (\chi - h) - \frac{(\ell \pm 1)^2}{\rho^2} \right] F_\pm = 0. \qquad (9.3.21)$$

In contrast to the modes found in Section 2.2, the vector modes are not linearly polarized. This can be seen by noting

$$\hat{\rho} + i\hat{\phi} = \hat{x} \cos\phi + \hat{y} \sin\phi + i(\hat{y} \cos\phi - \hat{x} \sin\phi)$$

$$= \hat{x}(\cos\phi - i\sin\phi) + i\hat{y}(\cos\phi - i\sin\phi) = \sqrt{2}\hat{e}_+ e^{-i\phi}. \qquad (9.3.22)$$

Similarly, $\hat{\rho} - i\hat{\phi} = \sqrt{2}\hat{e}_- e^{i\phi}$. Thus, $E_t^\pm$ are circularly polarized fields with the $+$ and $-$ signs standing for the RCP and LCP modes. Using Eq. (9.3.22) in Eq. (9.3.20), we can write $E_t^\pm$ in the form

$$E_t^\pm(\rho, \phi) = \hat{e}_\pm F_\pm(\rho) e^{i(\ell \mp 1)\phi}. \qquad (9.3.23)$$

Thus, we have found that a GRIN medium supports circularly polarized vector modes when the last term in Eq. (9.1.7) is taken into account. Notice that $J = \ell \mp 1$

in the phase term represents total angular momentum of the mode because ± 1 has its origin in the SAM associated with the circularly polarized light.

It is remarkable that the preceding analysis holds for a general form of the index profile given in Eq. (9.3.15). In the case of a parabolic index profile, $h(\rho) = b^2 \rho^2$ with $b^2 = 2\Delta/a^2$. In this case, Eq. (9.3.21) can be solved analytically in terms of the Laguerre–Gauss functions. The resulting modes have the same spatial profiles found in Section 2.2 for the linearly polarized scalar modes. However, the polarization degeneracy is removed because the propagation constants become different for the RCP and LCP modes.

9.4 Quantum-Mechanical Approach

A quantum-mechanical approach has also been used for solving Eq. (9.3.4). It allows one to find the vector modes of a GRIN medium in an analytic form [36–38]. Its use allows one to consider the Fock and coherent states that are well known in quantum mechanics. We discuss this technique and its applications in this section.

9.4.1 Approximate Solution

We begin with the Schrödinger equation (9.3.4). As we saw in Section 2.5.1, the H_0 part of the Hamiltonian for a parabolic GRIN medium can be written as

$$\hat{H}_0 = \frac{1}{2}\left(\hat{p}_1^2 + \hat{p}_2^2\right) + \frac{1}{2}\left(\hat{q}_1^2 + \hat{q}_2^2\right), \qquad (9.4.1)$$

where $\hat{q}_1 = x/w_g$, $\hat{q}_2 = y/w_g$ are the position operators with $w_g = (kb)^{-1/2}$ and the corresponding momentum operators are defined as $\hat{p}_j = -i(d/dq_j)$ for $j = 1, 2$. In terms of the annihilation and creation operators, defined as

$$\hat{a}_j = (\hat{q}_j + i\hat{p}_j)/\sqrt{2}, \qquad \hat{a}_j^\dagger = (\hat{q}_j - i\hat{p}_j)/\sqrt{2}, \qquad (9.4.2)$$

the Hamiltonian in Eq. (9.4.1) becomes

$$\hat{H}_0 = (\hat{a}_1^\dagger \hat{a}_1 + \hat{a}_2^\dagger \hat{a}_2 + 1)\hat{I}, \qquad (9.4.3)$$

where $\hat{I}$ is the identity operator. Its form corresponds to a two-dimensional harmonic oscillator.

The operator $\hat{H}_1$ given in Eq. (9.3.5) can also be expressed using the same a_1 and a_2 operators. In terms of the Pauli spin matrices, it takes the form [38]:

$$\hat{H}_1 = \eta[(\hat{B}_0 + 1)\hat{I} + \hat{B}_1\hat{\sigma}_1 + \hat{B}_2\hat{\sigma}_2 + \hat{B}_3\hat{\sigma}_3], \qquad (9.4.4)$$

where $\eta = b/(n_0 k_0)$ is a small parameter and the $\hat{B}_j$ operators depend on $\hat{a}_1$ and $\hat{a}_2$ as

$$\hat{B}_0 = \tfrac{1}{2}[(\hat{a}_1^2 - \hat{a}_1^{\dagger 2}) + (\hat{a}_2^2 - \hat{a}_2^{\dagger 2})], \qquad \hat{B}_1 = (\hat{a}_1 \hat{a}_2 - \hat{a}_1^{\dagger} \hat{a}_2^{\dagger}), \qquad (9.4.5)$$

$$\hat{B}_2 = i(\hat{a}_1^{\dagger} \hat{a}_2 - \hat{a}_1 \hat{a}_2^{\dagger}), \qquad \hat{B}_3 = \tfrac{1}{2}[(\hat{a}_1^2 - \hat{a}_1^{\dagger 2}) - (\hat{a}_2^2 - \hat{a}_2^{\dagger 2})]. \qquad (9.4.6)$$

The presence of the small parameter η in Eq. (9.4.4) allows one to treat $\hat{H}_1$ as a perturbation and employ the tools of quantum mechanics to calculate the eigenfunctions and eigenvalues of Eq. (9.3.4) in an approximate manner. The eigenfunctions of H_0 are found first. They correspond to the Fock (or number) states, whose spatial profiles represent the Hermite–Gauss modes, or the LP$_{lm}$ modes given in Eq. (2.2.20). Their eigenvalues β_{mn} are given in Eq. (2.2.12).

When perturbation theory is applied to the second order in η, both the modes and their eigenvalues are modified and depend on the value of the integer $\sigma = \pm 1$. It is useful to write the electric field of the resulting vector modes in the form [36]

$$\tilde{\mathbf{E}}_{\ell m \sigma}(\rho, \phi) = \left[(\hat{\mathbf{x}} + i\sigma \hat{\mathbf{y}}) R_{\ell m}(\rho) + \hat{\mathbf{z}} E_z \right] \exp(i\beta_{\ell m \sigma} z + i\ell \phi), \qquad (9.4.7)$$

where $R_{\ell m}(\rho)$ is the radial part of the mode. The subscript σ has been added to ℓ and m to indicate that the modes also depend on the SAM of the state. The propagation constants of these modes are found to be [37]

$$\beta_{\ell m \sigma} = n_0 k_0 \left(1 - \eta g - \frac{\eta^2}{32}[11g^2 - J^2 - 2J\sigma] \right), \qquad (9.4.8)$$

where $J = \ell + \sigma$ is the total angular momentum of the mode. Following Section 2.2, we have defined the group integer g as $g = \ell + 2m - 1$.

As we found in Section 9.3.2, the vector modes of a GRIN medium are circularly polarized and exhibit both the SAM and OAM, whose magnitudes are governed by the integers σ and ℓ, respectively. To the first-order in η, $\beta_{\ell m \sigma}$ depends only on $g = \ell + 2m - 1$, and each mode group contains multiple degenerate modes. This degeneracy is removed by $\hat{H}_1$, which produces the η^2 term in Eq. (9.4.8) and makes $\beta_{\ell m \sigma}$ depend on both the OAM and SAM through the total angular momentum $J = \ell + \sigma$. The last term $2J\sigma$ has its origin in the spin-orbit coupling. It shows that for fixed values of ℓ and m, β becomes different for the LCP and RCP modes by a small amount

$$\delta\beta = \eta^2 n_0 k_0 \ell / 4 = b^2 \ell / (4n_0 k_0). \qquad (9.4.9)$$

This difference is a measure of circular birefringence induced by the spin-orbit coupling in a GRIN medium. Different refractive indices also imply that the phase and group velocities of modes are slightly different for the LCP and RCP modes. Notice that these effects disappear for $\ell = 0$. For this reason, their observation requires the use of a vortex beam.

9.4.2 Evolution of Coherent States

The quantum state of optical beams emitted by lasers is well represented by coherent states, which represent the eigenstates of the annihilation operator $\hat{a}$. In the two-dimensional case, the coherent states satisfy the eigenvalue equation:

$$\hat{a}_j|\alpha_1, \alpha_2\rangle = \alpha_j|\alpha_1, \alpha_2\rangle, \quad j = 1, 2. \tag{9.4.10}$$

The wave function associated with a coherent state, $A(q_1, q_2) = \langle q_1, q_2|\alpha_1, \alpha_2\rangle$, takes the following form:

$$A(q_1, q_2) = (\pi)^{-1/2} \exp\left[-\tfrac{1}{2}(q_1^2 + q_2^2) + \sqrt{2}(\alpha_1 q_1 + \alpha_2 q_2). \right.$$
$$\left. -\tfrac{1}{2}(\alpha_1^2 + \alpha_2^2 + |\alpha_1|^2 + |\alpha_2|^2) \right]. \tag{9.4.11}$$

We can think of this wave function as the wave packet associated with an optical beam at the input of a GRIN medium and ask how it evolves inside such a medium. For this purpose, we work in the Heisenberg picture and follow the technique discussed in Section 2.5.2 for calculating the average of any operator [see Eq. (2.5.10)]:

$$\langle O(\xi)\rangle = \int A^*(q_1, q_2)\, \hat{O}(\xi) A(q_1, q_2)\, dq_1\, dq_2. \tag{9.4.12}$$

Such an approach requires the solution of the Heisenberg's equation of motion

$$i\frac{d\hat{O}}{d\xi} = [\hat{O}, \hat{H}], \tag{9.4.13}$$

where we have assumed that the operator $\hat{O}$ does not depend on ξ explicitly. This approach was used as early as 1980 [39] and has also been employed more recently [38].

The beam's trajectory inside a GRIN medium is found by calculating the averages $\langle q_1 \rangle$ and $\langle q_2 \rangle$ that specify the location of the beam's center. The beam's widths in the two directions are related to the variances as

$$w_j^2 = \langle q_j^2 \rangle - \langle q_j \rangle^2, \quad j = 1, 2. \tag{9.4.14}$$

The important question is how $\langle q_j \rangle$ and w_j depend on the SOP of the input field. In the case of a linear SOP, the beam's center follows the same trajectory as predicted by Eq. (2.5.15) of Section 2.5.2. Recall also that the beam's width oscillates in a periodic fashion because of the self-imaging phenomenon. The SOP of the input beam plays a minor role in the case of linear polarization.

The situation becomes different in the case of a circularly polarized beam, launched off-axis such that it follows a helical path inside the GRIN medium. For example, when the beam is launched off-axis with its center at $(x_0, 0)$ and with a slope $(0, s_0)$, the beam's center moves with $\xi = bz$ as [38]

$$\langle x(z) \rangle = x_0 \cos(bz) - \frac{\sigma b s_0 z}{2 n_0 k_0} \sin(bz), \tag{9.4.15}$$

$$\langle y(z) \rangle = \frac{s_0}{b} \sin(bz) + \frac{\sigma b^2 x_0 z}{2 n_0 k_0} \cos(bz), \tag{9.4.16}$$

where $\sigma = \pm 1$ for the LCP and RCP states, respectively. The beam's trajectories are clearly different for the LCP and RCP states. The dependence of the beam's path on the SAM of the beam is the photonic spin-Hall effect discussed earlier in Section 9.2.2.

The same technique can be used to show that the plane of polarization of a linearly polarized beam undergoes rotation because of the circular birefringence introduced by the last term in Eq. (9.1.7). It can also be used to find how the beam's width changes inside a GRIN medium. When the input beam is radially symmetric, the widths in the x and y directions remain the same and evolve as

$$w_x(z) = w_y(z) = w(0)[1 + (b/n_0 k_0)\sin^2(bz)]^{1/2}. \tag{9.4.17}$$

In the quantum description, the beam's width varies with distance because a coherent state is not an eigenstate of a harmonic oscillator. As the input beam excites many modes (or the Fock states), it is not surprising that the beam's width does not remain constant. It is important to stress that changes in the width in Eq. (9.4.17) do not depend on the SOP of the input beam.

References

[1] M. Born and E. Wolf, *Principles of Optics*, 7th ed. (Cambridge University Press, 1999).

[2] A. Papp and H. Harms, *Appl. Opt.* **14**, 2406 (1975); https://doi.org/10.1364/AO.14 .002406.

[3] W. Su and J. A. Gilbert, *Appl. Opt.* **35**, 4772 (1996); https://doi.org/10.1364/AO.35 .004772.

[4] J. L. Rouke and D. T. Moore, *Appl. Opt.* **38**, 6574 (1996); https://doi.org/10.1364/AO .38.006574.

[5] G. P. Agrawal, *Fiber-Optic Communication Systems*, 5th ed. (Wiley, 2021).

[6] C. D. Poole and C. R. Giles, *Opt. Lett.* **13**, 155 (1988); https://doi.org/10.1364/OL.13 .000155.

[7] P. K. A. Wai and C. R. Menyuk, *J. Lightwave Technol.* **14**, 148 (1996); https://doi.org/ 10.1109/50.482256.

[8] G. J. Foschini, R. M. Jopson, L. E. Nelson, and H. Kogelnik, *J. Lightwave Technol.* **17**, 1560 (1999); https://doi.org/10.1109/50.788561.

[9] J. Camacho and D. Tentori, *J. Opt. A: Pure Appl. Opt.* **3**, 89 (2001); https://doi.org/ 10.1088/1464-4258/3/1/315.

[10] C. He, J. Chang, Q. Hu et al., *Nat. Commun.* **10**, 4264 (2019); https://doi.org/10.1038/ s41467-019-12286-3.

[11] S. Pancharatnam, *Proc. Ind. Acad. Sci. A* **44**, 247 (1956); https://doi.org/110.1007/ BF03046050.

[12] M. V. Berry, *Proc. R. Soc. Lond. A*, **392**, 45 (1984); https://doi.org/10.1098/rspa.1984 .0023.

[13] J. N. Ross, *Opt. Quantum Electron.* **16**, 455 (1984); https://doi.org/10.1007/FBF00619638.

[14] A. Tomita and R. Y. Chiao, *Phys. Rev. Lett.* **57**, 937 (1986); https://doi.org/10.1103/PhysRevLett.57.937.

[15] B. Y. Zel'dovich and V. S. Liberman, *Sov. J. Quantum Electron.* **20**, 427 (1990); https://doi.org/10.1070/QE1990v020n04ABEH005947.

[16] V. S. Liberman and B. Y. Zel'dovich, *Phys. Rev. A* **46**, 5199 (1992); https://doi.org/10.1103/PhysRevA.46.5199.

[17] J. Liñares, M. C. Nistal, and D. Baldomir, *Appl. Opt.* **33**, 4293 (1994); https://doi.org/10.1364/AO.33.004293.

[18] M. Onoda, S. Murakami, and N. Nagaosa, *Phys. Rev. Lett.* **93**, 083901 (2004); https://doi.org/10.1103/PhysRevLett.93.083901.

[19] A. Kavokin, G. Malpuech, and M. Glazov, *Phys. Rev. Lett.* **95**, 136601 (2005); https://doi.org/10.1103/PhysRevLett.95.136601.

[20] K. Y. Bliokh, A. Niv, V. Kleiner, and E. Hasman, *Nat. Photon.* **2**, 748 (2008); https://doi.org/10.1038/nphoton.2008.229.

[21] K. Y. Bliokh, *J. Opt. A: Pure Appl. Opt.* **11**, 094009 (2009); https://doi.org/10.1088/1464-4258/11/9/094009.

[22] T. P. Chakravarthy and N. K. Viswanathan, *OSA Continuum* **2**, 1576 (2019); https://doi.org/10.1364/OSAC.2.001576.

[23] O. Hosten and P. Kwiat, *Science* **319**, 787 (2008); https://doi.org/10.1126/science.1152697.

[24] L. Marrucci, C. Manzo, and D. Paparo, *Phys. Rev. Lett.* **96**, 163905 (2006); https://doi.org/10.1103/PhysRevLett.96.163905.

[25] K. Y. Bliokh and Y. P. Bliokh, *Phys. Rev. E* **75**, 066609 (2007); https://doi.org/10.1103/PhysRevE.75.066609.

[26] Y. Zhao, J. S. Edgar, G. D. M. Jeffries, D. McGloin, and D. T. Chiu, *Phys. Rev. Lett.* **99**, 073901 (2007); https://doi.org/10.1103/PhysRevLett.99.073901.

[27] K. Bliokh, E. A. Ostrovskaya, M. A. Alonso et al., *Opt. Express* **19**, 26132 (2011); https://doi.org/10.1364/OE.19.026132.

[28] X. Yin, C. Zhao, C. Yang, and J. Li, *Opt. Express* **30**, 16432 (2022); https://doi.org/10.1364/OE.457375.

[29] J. J. Sakurai and J. Napolitano, *Modern Quantum Mechanics*, 3rd ed. (Cambridge University Press, 2020).

[30] M. Matsuhara, *J. Opt. Soc. Am.* **63**, 135 (1973); https://doi.org/10.1364/JOSA.63.000135.

[31] K. Thyagarajan and A. K. Ghatak, *Opt. Commun.* **11**, 417 (1974); https://doi.org/10.1016/0030-4018(74)90250-8.

[32] A. V. Dooghin, N. D. Kundikova, V. S. Liberman, and B. Y. Zel'dovich, *Phys. Rev. A* **45**, 8204 (1992); https://doi.org/10.1103/PhysRevA.45.8204.

[33] J. A. Buck, *Fundamentals of Optical Fibers*, 2nd ed. (Wiley, 2004).

[34] K. Okamoto, *Fundamentals of Optical Waveguides*, 2nd ed. (Academic Press, 2010).

[35] C. N. Kurtz and W. Streifer, *IEEE Trans. Microw. Theory Techn.* **17**, 11 (1969); https://doi.org/10.1109/TMTT.1969.1126872.

[36] N. I. Petrov, *Phys. Rev. A* **88**, 023815 (2013); https://doi.org/10.1103/PhysRevA.88.023815.

[37] N. I. Petrov, *J. Opt. Soc. Am. A* **33**, 1363 (2016); https://doi.org/10.1364/JOSAA.33.001363.

[38] N. I. Petrov, *Fibers* **9**, 34 (2021); https://doi.org/10.3390/fib9060034.

[39] S. G. Krivoshlykov and I. N. Sissakian, *Soviet J. Quantum Electron.* **10**, 312 (1980); https://doi.org/10.1070/QE1980v010n03ABEH009979.

10

Photonic Crystals and Metamaterials

Starting around the year 2000, considerable effort was directed toward developing new types of artificial materials known now as photonic crystals and metamaterials. Even though the initial focus was not on creating a spatially varying refractive index, it was soon realized that such materials can be fabricated with an index gradient in one or more dimensions. In this chapter, we focus on the novel GRIN devices whose design is based on photonic crystals and metamaterials. Section 10.1 introduces the basic concepts needed to understand the physics behind these two types of materials. Section 10.2 is devoted to GRIN structures based on the concept of photonic crystals. Metamaterials designed with an index gradient are discussed in Section 10.3. The focus of Section 10.4 is on a subgroup of metamaterials, known as metasurfaces, which contain nanoscale objects made with dielectric or metallic materials and are thinner than the wavelength of radiation they are intended for.

10.1 Basic Concepts

Photonic crystals are the optical analog of electronic crystals. Just as a crystal exhibits a periodic arrangement of atoms that affects the movement of electrons inside it, photonic crystals make use of the periodicity of the refractive index in one or more dimensions to control the propagation of light inside them [1–3]. The development of metamaterials, on the other hand, was motivated by the desire to develop devices whose effective refractive index is negative or close to zero in some wavelength range [4–6]. In this section, we review the underlying concepts behind these two types of structures.

10.1.1 Photonic Crystals

Since 1987, when the concept of photonic crystals was introduced [7, 8], such devices have been fabricated and used for a variety of applications [9–12]. Photonic

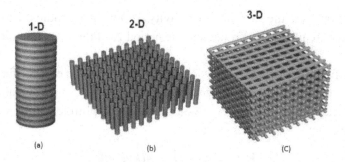

1-D   2-D   3-D

(a)   (b)   (C)

Figure 10.1 Schematic of photonic crystals exhibiting periodicity in one, two, and three dimensions. (After Ref. [12]; ©2013 CC BY.)

crystals combine two materials with different refractive indices on a sub-wavelength scale such that the effective refractive index of the resulting structure is periodic. Figure 10.1 shows schematically three types of photonic crystals with periodicity in one, two, and three dimensions. In the one-dimensional (1-D) case, alternating thin films of two different materials are deposited on a substrate. Periodicity in two dimensions (2-D) can be realized by stacking dielectric rods in a grid-like fashion, or by drilling holes in a periodic fashion on a layer of uniform material such as silicon; air acts as the second material in such a structure. The situation is more complicated in the three-dimensional (3-D) case, but the underlying idea remains the same.

Physics behind the photonic crystals can be understood by considering a one-dimensional structure similar to that seen in Figure 10.1(a). When an optical beam is launched at one end of such a device, the refractive index varies in a periodic fashion along the direction of propagation, as the beam travels from one layer to another, because the two materials are different by design. An example of a one-dimensional photonic crystal is provided by fiber Bragg gratings [13], in which the index difference between the high- and low-index regions is relatively small ($\delta n < 0.001$). Another example is provided by thin-film Bragg mirrors [14], made by depositing alternating thin layers of two dielectric materials with a relatively large index difference ($\delta n > 0.1$).

Propagation of plane waves in such periodic structures is governed by Eq. (2.1.11) for the electric field at a specific frequency ω:

$$\nabla^2 \mathbf{E} + \mu_0 \epsilon(z)\omega^2 \mathbf{E} + \nabla \left(\frac{1}{\epsilon}\nabla\epsilon \cdot \mathbf{E} \right) = 0, \qquad (10.1.1)$$

where $\epsilon(z)$ is a periodic function in the z direction in which light is propagating. For this reason, $\epsilon(z)$ can be expanded in a Fourier series as

$$\epsilon(z) = \epsilon_0 n^2(z) = \epsilon_0 \sum_{m=-\infty}^{\infty} d_m \exp(imGz). \qquad (10.1.2)$$

In this equation, ϵ_0 is the vacuum permittivity, $n(z)$ is the periodic refractive index, and the spatial frequency G is related to the period Λ of the 1-D photonic crystal as $G = 2\pi/\Lambda$.

To simplify the following discussion, we assume that the electric field $\mathbf{E}$ in Eq. (10.1.1) is in the form of a plane wave propagating in the z direction and does not vary in the transverse dimensions. As a result, its amplitude can be written as a Fourier transform,

$$\mathbf{E}(z) = \hat{\mathbf{p}} \frac{1}{2\pi} \int_{-\infty}^{\infty} A(k) e^{ikz} dk, \tag{10.1.3}$$

where $\hat{\mathbf{p}}$ is the polarization unit vector in the plane transverse to the direction of propagation. The last term in Eq. (10.1.1) vanishes in this situation because $\nabla\epsilon$ is orthogonal to $\hat{\mathbf{p}}$. With these simplifications, Eq. (10.1.1) is reduced to the ordinary differential equation,

$$\frac{d^2\mathbf{E}}{dz^2} + \mu_0\epsilon(z)\omega^2\mathbf{E} = 0. \tag{10.1.4}$$

When we substitute Eqs. (10.1.2) and (10.1.3) into Eq. (10.1.4) and equate the terms with the same e^{ikz} factor, we obtain [15]

$$k^2 A(k) = (\omega/c)^2 \sum_{m=-\infty}^{\infty} d_m A(k - mG), \tag{10.1.5}$$

where we used the relation $\mu_0\epsilon_0 = 1/c^2$. By removing the $m = 0$ term from the sum, we can write Eq. (10.1.5) as

$$(k^2 - d_0\omega^2/c^2)A(k) = (\omega/c)^2 \sum_{m \neq 0} d_m A(k - mG). \tag{10.1.6}$$

This equation shows that $A(k)$ is coupled to other Fourier components of the form $A(k - mG)$, resulting in an infinite set of equations. In the absence of periodicity, all terms on the right side of Eq. (10.1.6) vanish, and $A(k)$ is finite only when the dispersion relation $k = \pm\sqrt{d_0}\omega/c$ is satisfied. These two values of k correspond to the forward and backward propagating waves inside a homogeneous medium.

The infinite set of equations obtained from Eq. (10.1.6) can be truncated when $A(k)$ couples strongly to a specific $A(k - mG)$. For the choice $m = 1$, only two components, $A(k)$ and $A(k - G)$, are coupled strongly by the periodic structure, and others can be neglected. The resulting two equations can be written in a matrix form as

$$\begin{pmatrix} k^2 - d_0\omega^2/c^2 & -d_1\omega^2/c^2 \\ -d_{-1}\omega^2/c^2 & (k - G)^2 - d_0\omega^2/c^2 \end{pmatrix} \begin{pmatrix} A(k) \\ A(k - G) \end{pmatrix} = 0. \tag{10.1.7}$$

A nontrivial solution of Eq. (10.1.7) exists only if the determinant of the coefficient matrix vanishes. This condition leads to the dispersion relation

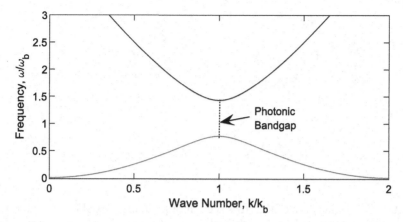

Figure 10.2 Dispersion curves of a one-dimensional photonic crystal obtained from Eq. (10.1.7). The shortest gap between the two branches is called the photonic bandgap.

$$(k^2 - \bar{n}^2\omega^2/c^2)[(k - G)^2 - \bar{n}^2\omega^2/c^2] = |d_1|^2(\omega/c)^4, \qquad (10.1.8)$$

where we used $d_{-1} = d_1^*$ from Eq. (10.1.2), assuming $n(z)$ is real. We also replaced d_0 with $\bar{n}^2$, where $\bar{n}$ is the average refractive index of the periodic structure.

Figure 10.2 shows the dispersion curves by plotting ω/ω_b as a function of k/k_b, where $k_b = \pi/\Lambda$ and $\omega_b = ck_b/\bar{n}$ are used for normalizing k and ω respectively. The ratio $|d_1|/d_0$ was chosen 0.3 for Figure 10.2. The two branches with a gap in between are analogous to the energy bands founds for electrons moving inside a crystal. The minimum frequency gap occurring for $k = k_b$ is known as the photonic bandgap. Two- and three-dimensional photonic crystals exhibit a much more complicated band structure with multiple bandgaps. Even in the one-dimensional case, the band diagram contains multiple bands when more terms are included in the sum on the right side of Eq. (10.1.6).

The shortest gap in Figure 10.2 occurs for $k = k_b$. Using this specific value of k in Eq. (10.1.8), the frequencies that correspond to the band edges are found to be:

$$\omega_\pm = ck\left/\sqrt{\bar{n}^2 \pm |d_1|}\right. \qquad (10.1.9)$$

In the frequency range between ω_- and ω_+, the propagation constant k becomes complex, and its imaginary part leads to the exponential decay of the optical field inside a photonic crystal. This is the reason why incident light at such frequencies cannot propagate through such a medium and is mostly reflected. Outside of this range, two real values of k exist for any frequency. These correspond to the two waves with different propagation constants that become coupled inside a photonic crystal. The Bloch modes of a periodic structure represent specific linear

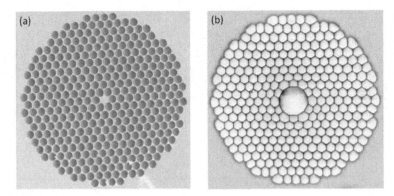

Figure 10.3 Schematic of the cross section of photonic crystal fibers with (a) a solid core and (b) a hollow core. In both cases, cladding is made using a periodic array of hollow silica tubes.

combinations of these two waves. When the period of the photonic crystal is a fraction of the incident wavelength, the two waves may even propagate in opposite directions. This is the situation, for example, in a Bragg grating.

Photonic crystal fibers (PCFs), developed during the 1990s, are examples of two-dimensional photonic crystals [16–18]. Figure 10.3 shows the cross sections of two types of PCFs that have found a wide variety of applications. The cladding of both PCFs is made using a periodic array of hollow silica tubes. However, the central core contains silica in part (a) but air in part (b). As a result, the confinement of light to the core occurs via two different mechanisms. In the case of a solid core, light is confined through total internal reflection because the effective refractive index of the cladding containing ar holes is lower than that of the core. In the case of a hollow core, the confinement occurs through the photonic bandgap, which does not let the light traveling within the core of a PCF to escape through its periodic cladding.

PCFs containing a solid narrow core have found many applications. The main reason behind their usefulness is that the nonlinear effects are enhanced considerably inside such PCFs compared to step-index optical fibers designed with a uniform cladding of slightly lower refractive index compared to that of the core [19]. At the same time, dispersive properties of such PCFs can be tailored by adjusting the size and spacing of air holes in their cladding. These properties have made the solid-core PCFs a material of choice for supercontinuum generation, a nonlinear phenomenon that expands the spectrum of short optical pulses from a few nanometers to hundreds of nanometers [20–22]. Such wideband sources are useful for biomedical imaging, among other things. PCFs with a hollow core offer the possibility of filling their core with another material such as a gas or a liquid. They are often referred to as hybrid PCFs and are useful for a variety of applications [18].

10.1.2 Metamaterials

Similar to photonic crystals, metamaterials contain components with a repeating unit cell that is much smaller than the wavelength of light they are designed for. Each unit cell typically contains both metallic and dielectric materials and is designed such that the effective refractive index of the whole device is quite different from the refractive indices of the materials used to make it. The initial motivation behind developing metamaterials was on fabricating a device capable of exhibiting a negative value of the refractive index in some frequency range [23–27].

Maxwell's equations, given in Section 2.1, relate the refractive index n of a material to its electric permittivity ϵ and magnetic permeability μ as $n = \pm c\sqrt{\epsilon\mu}$. For dielectric materials, both ϵ and μ are positive, resulting in $n > 0$. In the case of metals, μ is positive but ϵ can be negative. Metamaterials are designed to make the effective values of both ϵ and μ negative in some frequency range. As a result, they provide a negative value of n when the frequency of incident light lies in that range. Such artificial materials exhibit remarkable properties and are also known as left-handed or negative-index materials [23]. One such property is that the phase front of an electromagnetic wave moves in a direction opposite to the direction of power flow. Another property is related to the unusual bending of a beam occurring when it enters a negative-index material. This property can be used to make a "perfect" lens that does not suffer from the fundamental limitation set by the wavelength of incident light on the resolution of conventional lenses [28].

Metamaterials were initially designed to exhibit a negative value of the refractive index in the microwave region. They employed arrays of metallic wires for realizing negative ϵ and split-ring resonators for realizing negative μ in certain microwave frequency range [24]. In 2005, a different approach based on pairs of parallel metal nanorods was employed to fabricate a metamaterial capable of exhibiting a negative refractive index at wavelengths near 1.5 μm [25]. In another experiment, two thin gold layers, separated by a 60-nm thick dielectric layer and containing periodic circular holes, were found to exhibit $n < 0$ at wavelengths near 2 μm [26]. By 2007, a fish-net structure was used for a metamaterial to realize $n < 0$ at wavelengths near 780 nm [27]. More recently, the focus has shifted to metamaterials whose refractive index becomes nearly zero in a range of frequencies [29–31].

A variety of techniques can be used used to fabricate metamaterials [32]. As the dimensions of a unit cell are relatively small in the visible and infrared regions, photolithography or electron-beam lithography is often used for this purpose. Such techniques require a layer-by-layer approach for making a 3-D metamaterials. Focused ion-beam milling is useful for making devices with a large aspect ratio. In practice, fabrication of metamaterials becomes less time-consuming as the number of layers is reduced. Metamaterials have found a variety of applications including

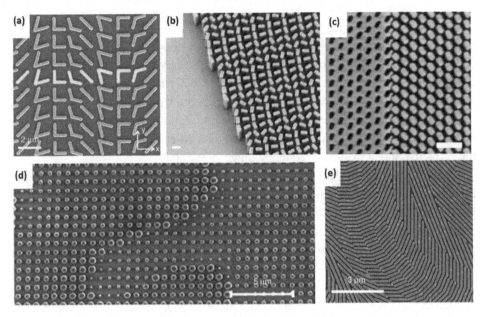

Figure 10.4 Several designs for plasmonic and dielectric metasurfaces. See text
for details. (After Ref. [37]; ©2018 CC BY.)

enhancing an antenna's efficiency, cloaking, improving resolution of lenses, reduced
scattering, and energy harvesting [33].

A new class of devices, the so-called metasurfaces, emerged around 2010. These
constitute a subgroup of metamaterials and are designed such that the entire structure
fits on a surface whose thickness is a fraction of the wavelength they are designed
for [34–38]. In the case of metasurfaces, the objective is not to produce a negative
refractive index. Rather, such devices affect the wavefront of incoming light through
phase shifts that vary spatially or depend on its state of polarization. As a result,
they act as a lens, or a wave plate, and can replace such bulky elements with a thin
and compact device.

Figure 10.4 shows several designs for metasurfaces that have been developed in
recent years [37]. They can be divided into two groups, plasmonic and dielectric
metasurfaces, depending on whether metallic or dielectric materials are used to make
them. Plasmonic metasurfaces (top row) are designed to operate in the reflection
mode because of losses of metals. The metasurface in part (a) of Figure 10.4 contains
V-shaped nano-size antennas that are capable of producing phase shifts in the entire
range of 0 to 2π [39, 40]. A metal-insulator-metal structure is also employed for
this purpose [41]. The metasurface in part (b) is made with TiO_2 nanofins on a glass
substrate and acts as a lens [42]. The metasurface in part (c) contains nanopillars

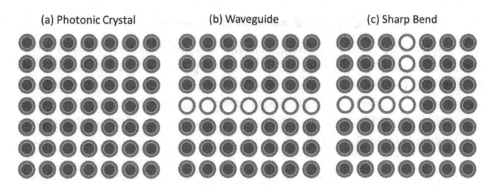

Figure 10.5 Schematic of (a) a photonic crystal, (b) a waveguide formed by a row
of missing holes, and (c) a waveguide with a sharp bend. (After Ref. [51]; ©2016
CC BY.)

and works as an achromatic lens [43]. Dielectric metasurfaces (bottom row) contain
silicon posts in part (d) and silicon nanobeams in part (e) on top of a SiO_2 substrate
[44]. They act as lenses that can be used for imaging. Another type of metasurface
makes use of a hyperbolic material that behaves as a metal in one direction and as a
dielectric in an orthogonal direction [45].

Dielectric metasurfaces are attracting attention because light can be transmitted
through them without suffering excessive losses [46–48]. They are fabricated by
placing nano-size dielectric elements (posts, pillars, or holes) on a flat dielectric surface such as glass. They may also employ aperiodic patterns or multilayer elements
to minimize reflection. Excitation of quadrupoles and high-order multipoles plays a
significant role in the case of dielectric arrays of materials with a large difference in
their refractive indices [34].

10.2 Graded-Index Photonic Crystals

As we discussed in Section 10.1.1, photonic crystals employ a periodic structure in
one, two, or three dimensions to realize a photonic bandgap, resulting in a range
of incident frequencies that cannot be propagated through such devices. A two-
dimensional periodic structure often consists of nano-size holes on the surface of a
dielectric layer, as shown schematically in Figure 10.5(a). However, defects can be
introduced selectively in such a periodic structure to make waveguides that allow
transmission of light through them [49, 50]. If holes are missing along a specific
row or column, that region acts as a waveguide [51]. This situation is shown in
part (b) of Figure 10.5. Light can also be forced to bend around corners along a
two-dimensional pattern of missing holes as shown in part (c) of Figure 10.5.

10.2.1 Graded-Index Planar Waveguides

One way to understand the guiding behavior in a photonic crystal makes use of the effective-medium theory [52–54]. This theory was first put forward in 1904 by Maxwell-Garnett [55]. According to it, when a composite material contains spherical inclusions whose density is not too large and whose size is smaller than the wavelength of incident light, the effective dielectric constant of the whole structure can be written in the form [56]

$$\epsilon_{\text{eff}} = \epsilon_h + \frac{f\epsilon_h\delta\epsilon}{\epsilon_h + (1-f)\delta\epsilon/2}, \tag{10.2.1}$$

where f is the volume fraction of the inclusions, ϵ_h is the dielectric constant of the host material and $\delta\epsilon = \epsilon_i - \epsilon_h$ is the difference between dielectric constants of two materials. This formula works reasonably well for $f \ll 1$.

The effective dielectric constant in Eq. (10.2.1) applies to a three-dimensional structure. In the case of a planar waveguide with a two-dimensional array of inclusions (or holes), it applies only in the plane of this array [56]. In the direction normal to the plane, effective dielectric constant is different and is given by $\epsilon_\perp = (1 - f)\epsilon_h + f\epsilon_i$. In effect, such a two-dimensional photonic crystal behaves as an anisotropic material and exhibits geometrically-induced birefringence.

In the case of a photonic crystal waveguide such as shown in Figure 10.5(b), the refractive index in the central region without holes is just $n_h = \sqrt{\epsilon_h}$. In the cladding region containing holes, the effective refractive index, $n_c = \sqrt{\epsilon_{\text{eff}}}$, is smaller than n_h because $\delta\epsilon = 1 - \epsilon_h$ is negative in Eq. (10.2.1). Moreover, the decrease in n_c depends on the size of holes through the volume fraction f. As long as $n_c < n_h$, light is confined to the central region through total internal reflection. It is important to note that holes in the cladding region do not have to follow a strict periodic pattern because n_c depends only on the volume fraction f. This feature suggests a simple way of creating an index gradient; it consists of changing the size of air holes or their spacing in the vicinity of the core region of the waveguide. Another option is to keep the hole size and spacing constants but change the refractive index in the hole region. However, its practical implementation is more difficult compared to changing the size or spacing of the holes.

Starting in 2005, several experiments showed that GRIN photonic crystals can be made by tailoring the size or spacing of the holes, and used such devices for focusing or bending the light transmitted through them [57–61]. In a 2005 experiment, the two-dimensional photonic crystal contained an array of identical holes of 190-nm radius within an InP layer, but the spacing between holes was varied to produce an index gradient [57]. It was found that gradual changes in the period of a photonic crystal can bend light on a micrometer scale. In a 2006 numerical study, it was shown that a GRIN photonic crystal, made by tailoring the radii of air holes, can

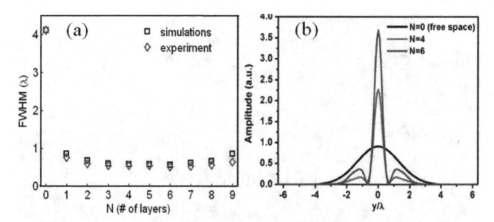

Figure 10.6 (a) Measured FWHM at the focal point of a Gaussian beam as a function of the number of GRIN layers used to make the GRIN device. (b) Corresponding numerically simulated field profiles at the focal point. (After Ref. [61]; ©2008 AIP.)

act as a tiny lens capable of focusing an incoming optical beam [59]. The focusing action was demonstrated in a 2008 experiment, where the GRIN structure was fabricated by modulating the spacing of the dielectric rods in each layer, with air acting as the host material [61]. Several such layers with a constant spacing were employed to make the GRIN device. Figure 10.6(a) shows the measured full width at half-maximum (FWHM) of a beam at the focal point as a function of the number of GRIN layers. A Gaussian beam with a width of ten times the spacing between the layers was launched at the input end. The numerically simulated field profiles are shown in part (b) of Figure 10.6. Clearly, this GRIN device acts as a lens when the number of layers exceeds three.

GRIN devices discussed so far exhibit an index gradient only in the direction in which the size or the spacing of holes (or rods) is modulated. In principle, one can do this type of modulation in the entire plane of the photonic crystal. This was demonstrated in a 2012 experiment in which the silicon-on-insulator technology was adopted to produce GRIN devices [62]. More specifically, a two-dimensional array of holes was created within a 260-nm-thick silicon layer with a constant spacing of $a = 390$ nm. However, radii of holes were modulated according to $r = 0.35a \exp(-\rho^2/2R^2)$ with $R = 62$ μm, where ρ is the radial distance in the plane of the silicon layer. The resulting device produced an index gradient by varying the air-hole filling factor in both directions of a square lattice. Numerical simulations indicated that such a GRIN device should bend an incoming beam by $90°$ at the wavelength $\lambda = 4a$ falling near 1560 nm. The experimental results agreed with this prediction. The wavelength selectivity of the GRIN device was also used to make a two-channel demultiplexer.

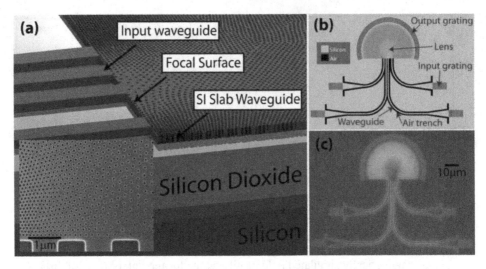

Figure 10.7 (a) The GRIN structure used for making a Luneberg lens. Inset shows an SEM image of the distribution of holes etched on a silicon waveguide. (b) Schematic and (c) SEM image of the fabricated device. (After Ref. [64]; ©2012 Optica.)

The same approach has been used for making a Luneberg lens [63]. The refractive index in such a spherical lens varies in all three dimensions as

$$n(\rho) = \sqrt{2 - (r/R)^2}, \qquad 0 \le r \le R, \tag{10.2.2}$$

where R is the radius of the lens. A Luneburg lens focuses incoming rays from infinity to a point on a spherical surface. In its three-dimensional form, the lens has a curved focal surface that is not compatible with planar detectors. It was shown in 2012 that the transformation optics, developed in the context of metamaterials, can be used to make a planar version of the Luneberg lens [64]. The distribution of sub-wavelength holes within the silicon waveguide followed the pattern dictated by a numerical algorithm. Figure 10.7 shows the fabricated device. It was tested at the 1550-nm wavelength, and the results agreed well with theoretical predictions. A rod-based GRIN structure has also been proposed for making a Luneberg lens [65].

In recent years, GRIN devices based on photonic crystals have been used for a variety of applications ranging from sensing to imaging [66–69]. In a 2022 study, it was found numerically that a GRIN flat lens, designed with a two-dimensional photonic crystal, can exhibit resolution better than the fundamental $\lambda/2$ limit in the wavelength region in which such a device exhibits negative values for its effective refractive index [69]. It should be stressed that the distinction between a metamaterial and a photonic crystal becomes vague for such devices. It is common to think of photonic crystals as a subgroup of metamaterials.

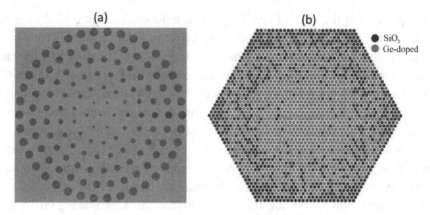

Figure 10.8 (a) GRIN fibers made by (a) varying the radii of air holes and (b) using two types of rods containing materials with different refractive indices. (After Ref. [72]; ©2018 CC BY.)

10.2.2 *Photonic Crystal Fibers*

As we discussed in Section 10.1.1, PCFs are an example of two-dimensional photonic crystals. Externally, a PCF looks like a conventional optical fiber and contains a cylindrical central core that is surrounded with a cladding. However, it differs from such fibers internally because its cladding contains a periodic array of air holes [16–18]. Also, the core is made of pure silica, as no dopants are used to increase its refractive index. Rather, the effective index of the cladding becomes smaller than that of the core because of its holey nature.

It was realized soon after the advent of PCFs that a cladding containing periodic holes is essential only when the core contains air. When the core is made of silica glass, air holes in cladding do not have to follow a periodic pattern. This feature gave rise to a wider class of optical fibers, known as microstructured fibers [70]. Even when a fiber has a hollow core, a periodic array of holes is not required if the core is filled with a fluid. Such fibers are useful for sensing applications [71].

Similar to the planar GRIN devices based on photonic crystals, a GRIN PCF can be made by varying the radii or the spacing of hollow silica tubes that are used during the fabrication of such a fiber. Figure 10.8 shows two approaches that have been used to make GRIN PCFs. Part (a) shows a structure in which the radius of the holes is quite small near the center but increases in all directions moving away from the center [66]. In the structure shown in part (b), two types of rods containing materials with different refractive indices are employed to make a GRIN fiber [72]. Such a fiber does not have a cladding with holes, but an index gradient can be realized by varying the density of two types of rods.

The operation of the GRIN fiber, designed as indicated in Figure 10.8(a), is easy to understand from Eq. (10.2.1) based on the effective medium theory. The volume fraction f in this equation is small in the central region and increases with the radial distance. Using $\epsilon_i = 1$ and $\epsilon_h \approx 2.1$ for silica glass, $\delta\epsilon$ is a negative quantity. It follows that the effective refractive index of the composite material is the largest at the center and decreases moving away from the center. With a proper design, a parabolic or any other power-law index profile can be realized in practice. The propagation characteristics of such fibers have been analyzed [73]. As expected for a GRIN fiber, the width of an optical beam propagating inside such a PCF varies periodically owing to the self-imaging phenomenon discussed in Section 3.2.

The structure shown in Figure 10.8(b) was used in 2018 for making GRIN fibers [72]. The preform was made by stacking more than 2100 glass rods with a diameter of only 0.45 mm. Two types of rods were used, one type containing pure silica and the other Ge-doped silica. The refractive index of the latter rods was larger by about 5% because of germanium doping. All rods were packed together in such a fashion that the density of Ge-doped rods was higher near the center and decreased away from the center such that the effective refractive index of the composite material varied in a parabolic fashion. The fiber was drawn using such a preform until the diameter of each rod was reduced to about 190 nm. Figure 1.9 of Chapter 1 shows images of the fiber's cross section at three magnification levels.

In a 2020 experiment, a 33-cm-long piece of such a nanostructured GRIN fiber was used for supercontinuum generation by launching 1-ns pulses at 1550 nm with up to 4 kW of peak power [74]. At this wavelength, the GRIN fiber exhibited anomalous GVD ($\beta_2 < 0$). However, pump pulses were so wide that soliton effects played little role, and spectral broadening was mostly due to self-phase modulation (SPM) and stimulated Raman scattering (see Section 5.5.1). Nevertheless, the output spectrum was relatively flat and extended from 1.5 to 2.1 μm for pump pulses with 400-nJ energy.

A nanostructured GRIN fiber was also used in a 2022 experiment with the objective to produce a supercontinuum in the mid-infrared spectral region [75]. This was accomplished by using two types of lead-bismuth-gallate glass rods with different refractive indices. A 33-cm-long piece of such a GRIN fiber was pumped with 350-fs pulses at 1700 nm. The resulting supercontinuum extended from 0.7 to 2.8 μm when the energy of pump pulses exceeded 250 nJ (see Figure 5.17). Such a large spectral broadening was possible because of the spatiotemporal modulation instability, facilitated by the self-imaging phenomenon occurring in a GRIN fiber (see Section 5.3). The GRIN fiber also exhibited anomalous GVD at wavelengths beyond 2 μm.

10.3 GRIN Metamaterials

As discussed in Section 10.1.2, metamaterials are artificial structures whose cells (also called meta-atoms) combine metallic and dielectric materials to produce negative values of the effective refractive index in some frequency range. It was realized soon after such a device was fabricated at microwave frequencies [24] that a GRIN metamaterial can be fabricated by grading a parameter internal to individual cells. In a 2005 demonstration of this concept, the thickness of the dielectric material around the metallic split-ring resonator was varied to modify the effective refractive index from one region to another [76]. Since then, GRIN metamaterials have been used in several different contexts, where a negative value of the effective refractive index is not a requirement [77–82].

In a 2007 experiment, focusing of an optical beam inside a GRIN metamaterial was demonstrated at the 1550-nm wavelength region [77]. Such a planar GRIN lens was fabricated by creating a spatially nonuniform structure on a nanoscale within a silicon waveguide. The effective medium theory was used to estimate its index gradient and to show that the device acts as a GRIN lens. A two-dimensional array of spatially nonuniform holes (different radii and spacing) was employed in a 2010 experiment to realize devices that acted as a GRIN lens or were able to steer an optical beam [78]. As the devices were designed to operate at the microwave frequencies, holes were of relatively large size ($\sim$ 1 mm).

The fabrication of an integrated Luneberg lens has attracted attention in recent years. A silicon waveguide was used in 2011 for this purpose [85]. A plasmonic luneburg lens with a diameter of 13 μm was fabricated in another 2011 experiment by depositing a nonuniform circular layer of polymer on top of a thin gold film [79]. The same approach was used to make an Eaton lens [83]. By 2014, a spherical Eaton lens with a large index gradient could be fabricated using only dielectric materials [84]. The concept of digital metamaterials is also useful for making flat GRIN lenses of various types [80]. In this approach, two materials with constant refractive indices act as "bits" that are used to digitally synthesize a metamaterial with a desired spatially varying refractive-index profile.

In a 2020 study, an anisotropic metamaterial was used for making the Luneberg and fisheye lenses, capable of working over a broad range of wavelengths in the microwave region [82]. A Luneberg lens requires the radial GRIN profile given in Eq. (10.2.2). The refractive-index profile for a fish-eye lens takes the form

$$n(\rho) = n_0/[1 + (r/R)^2], \qquad 0 \le r \le R. \tag{10.3.1}$$

Figure 10.9 shows the configuration of anisotropic cells (or meta-atoms) used in 2020 for making the Luneberg and fisheye lenses in a planar form. Each cell

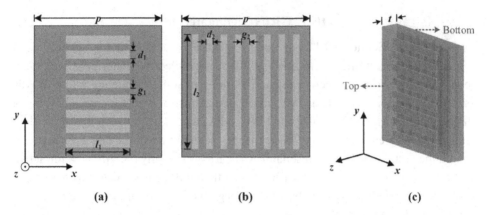

Figure 10.9 Schematic of the anisotropic unit cell for metamaterials used for making the Luneberg and fisheye lenses. (a) Top view; (b) bottom view; (c) perspective view. (After Ref. [82]; ©2020 CC BY.)

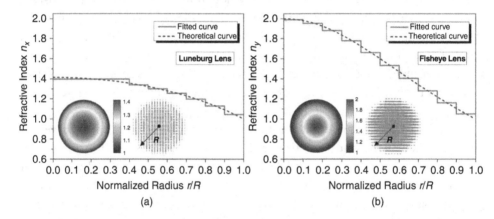

Figure 10.10 Comparison of the actual (solid line) and theoretical (dashed line) radial index profiles for the metasurfaces used for making the Luneberg and fisheye lenses. The two insets show the ideal and synthesized distributions of the refractive index. (After Ref. [82]; ©2020 CC BY.)

contained linear metallic arrays on the top and bottom of a dielectric substrate such that they were oriented in orthogonal directions. The thickness of metallic stripes (d_1 and d_2) and their spacings (g_1 and g_2) were < 1 mm in the range of microwave frequencies (2–8 GHz) used in the experiment, and they were varied from cell to cell for creating an index gradient. Each cell of the metamaterial had a square shape with 6-mm sides. The resulting spatially varying refractive index of the entire structure was calculated with the effective medium theory discussed earlier.

Figure 10.10 compares the actual refractive-index profile with the one required for the Luneberg and fisheye lenses. In each case, the insets show the ideal and the

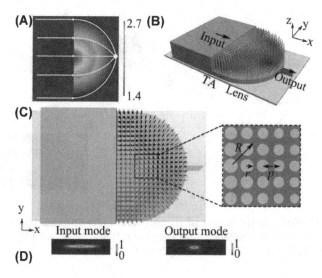

Figure 10.11 Design of a fisheye lens built with nanosize silicon rods of different sizes on a SiO$_2$ layer. (A) focusing by the lens; (B) whole device with the input waveguide; (C) top view of the nanorods with a small portion magnified; (D) input and output spatial profiles. (After Ref. [86]; ©2022 CC BY.)

synthesized distributions of the refractive index. Despite the staircase nature of the index profile over the surface of the metamaterial, experimental data confirmed that the lenses were able to focus microwave radiation at frequencies in the range of 5–7 GHz.

The silicon-on-insulator technology was employed in 2022 for making a planar fisheye lens. Figure 10.11 shows schematically the design of this device [86]. As seen there, a semi-circular structure, built with nanosize silicon rods of different sizes on a SiO$_2$ layer on top of the Si substrate, provided the index gradient required for a fisheye lens. The input waveguide was 8 μm wide and its output was sent to the GRIN lens (containing nanorods) that was only 5.4 μm long. The effective refractive index across the fisheye lens varied from 2 to 2.8 at the 1550-nm wavelength, depending on the filling fraction of the nanorods. Such a tiny fisheye lens was able to focus an 8 μm-wide input beam to a < 1 μm spot size.

Another direction of research on metamaterials involves devices that exhibit a dielectric constant ϵ that nearly vanishes in some frequency range. Such structures are referred to as epsilon-near-zero metamaterials [87–89]. They can also be fabricated such that the refractive index varies spatially, while remaining near zero [90, 91]. In a 2017 experiment, such an epsilon-near-zero metamaterial was used as a GRIN lens designed to operate at terahertz Frequencies [92]. It employed an array of narrow, hollow, rectangular waveguides near their cutoff frequencies.

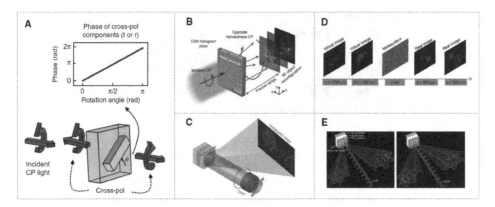

Figure 10.12 Some applications of metasurfaces. (A) Polarization conversion; (B) transmissive and (C) reflective holography; (D) holographic multiplexing; (E) OAM-multiplexing. (After Ref. [98]; ©2019 CC BY.)

10.4 GRIN Metasurfaces

As mentioned in Section 10.1.2, metasurfaces are a subgroup of metamaterials that are designed such that the thickness of the entire structure is a fraction of the wavelength it is used for. For operation in the visible and near-infrared regions, a metasurface contains a two-dimensional array of nanosized cells that may be made of metallic or dielectric materials, depending on the intended application.

The concept of an index gradient can also be applied to a metasurface. As early as 2013, a GRIN metasurface was designed to operate at terahertz Frequencies [93]. It contained split-ring resonators (made of gold) on a 90-nm-thick layer of vanadium dioxide (VO_2), a material that allowed for memory-enabled tuning of the device. The effective refractive index of the device was varied spatially through an applied current. It was possible to change the refractive index by as much as 50% over a distance of few wavelengths. Such a reconfigurable device can be used for focusing or steering an incoming beam.

Many other GRIN metasurfaces have been developed in recent years for a variety of applications [94–99]. Figure 10.12 shows several ways a GRIN metasurface can be used in practice. These include: (A) SAM-to-OAM conversion, (B) transmissive and (C) reflective holography, (D) holographic multiplexing, and (E) OAM-multiplexing. All such applications require phase shifts that vary spatially on the surface of a metasurface.

SAM-to-OAM conversion of circularly polarized light by a metasurface has been discussed in Section 8.3.2 in the context of vortex beams. A birefringent metasurface was used in 2014 for creating a vortex beam in the same way as a q-plate does [100]. It contained nanosize antennas (made of gold), whose dimensions and separations

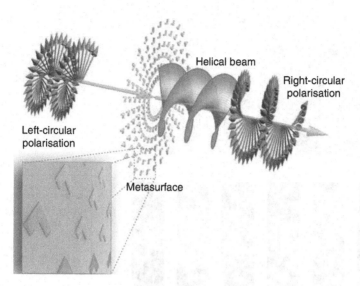

Figure 10.13 Schematic showing how a suitably designed metasurface can convert an LCP Gaussian beam into an RCP vortex beam that has an OAM of $\ell = 2$. (After Ref. [100]; ©2014 CC BY.)

were chosen to introduce a π phase shift between the orthogonally polarized components of an incoming optical beam. When a left-circularly polarized (LCP) Gaussian beam is reflected from such a metasurface, it becomes right-circularly polarized (RCP) and is converted into a vortex beam with an OAM of $\ell = 2$. The opposite happens when the incident beam is RCP, resulting in a vortex beam with $\ell = -2$. Figure 10.13 shows the conversion process schematically in the case of an LCP input beam.

The holographic applications of metasurfaces make use of unit cells whose response to incident light depends on its state of polarization. This can be realized by using metallic structures that are not radially symmetric. Two examples are provided by a cylindrical post with an elliptical cross section and a cross-like shape with different lengths in the orthogonal directions [98]. Wavelength multiplexing of holograms can also be realized with the help of a suitably designed metasurface. If the angle of incident beam is varied with wavelength during recording of a computer-generated hologram, multiple images can be stored in the same hologram. A specific image is retrieved by choosing the correct incident angle for that specific wavelength.

When the unit cell of a metasurface is designed using a material whose refractive index changes with external bias, the same hologram can be used for storing a large number of images that are retrieved by choosing the correct voltage. This feature is referred to as the tunable metasurface holography [98]. In the optical

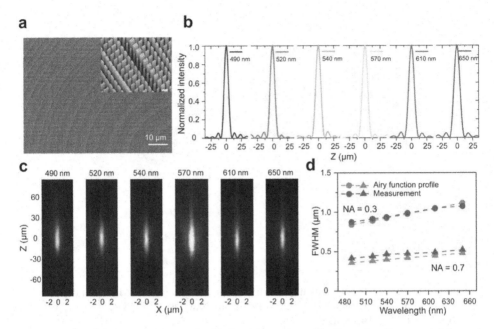

Figure 10.14 (a) SEM image of a fabricated metalens. The inset shows a zoomed-in view by a factor of 25. (b) Simulated focal intensity distribution at six wavelengths. (c) Measured focal intensity distribution and (d) widths at the same six wavelengths. Widths are also compared to the ideal values. (After Ref. [99]; ©2022 CC BY.)

domain, materials such as liquid crystals, vanadium oxide, and indium tin dioxide can provide tuning of the refractive index with an applied voltage. Another material that has been used in recent years is the alloy GeSbTe, whose nature changes from amorphous to crystalline in response to an electrical bias or temperature change.

A shortcoming of many metasurface designs is that the physical size of the resulting device is often limited to an area of 1 mm^2 or less. This issue has recently been addressed with an inverse-design approach that allows fabrication of complex GRIN metasurfaces of relatively large sizes [99]. Such a technique is powerful because it can be used to realize any prescribed refractive-index distribution over the metasurface at a much reduced fabrication cost. With its help, it has become possible to design and fabricate aberration-corrected lenses (called metalenses) with large numerical apertures and with large diameters (up to a few centimeters).

As an example, Figure 10.14(a) shows the SEM image of a metalens fabricated with the inverse-design approach [99]. It has a numerical aperture of 0.3 and is designed to focus an incoming beam to a small spot size over a wide range of input wavelengths. Part (b) shows the simulated focal intensity distribution at six wavelengths covering a wide wavelength range from 490 to 650 nm. Part (c) shows the measured focal intensity distribution at these six wavelengths. Part (d) compares the measured full-widths at half-maximum (FWHMs) of the focal spots to those expected to occur for an Airy profile under ideal conditions. The performance of the planar metalens is remarkable. It should be stressed that metamaterials and metasurfaces constitute an active area of research. As a result, further advances can be expected in the near future.

References

[1] J. D. Joannopoulos, S. G. Johnson, J. N. Winn, and R. D. Meade, *Photonic Crystals: Molding the Flow of Light*, 2nd ed. (Princeton University Press, 2008).

[2] D. W. Prather, A. Sharkawy, S. Shi, J. Murakowski, and G. Schneider, *Photonic Crystals, Theory, Applications and Fabrication* (Wiley, 2009).

[3] A. R. McGurn, *Introduction to Photonic and Phononic Crystals and Metamaterials* (Morgan and Claypool, 2020).

[4] W. Cai and V. M. Shalaev, *Optical Metamaterials: Fundamentals and Applications* (Springer, 2009).

[5] B. Banerjee, *An Introduction to Metamaterials and Waves in Composites* (CRC Press, 2011).

[6] N. Shankhwar and R. K. Sinha, *Zero Index Metamaterials: Trends and Applications* (Springer, 2021).

[7] E. Yablonovitch, *Phys. Rev. Lett.* **58**, 2059 (1987); https://doi.org/10.1103/PhysRevLett.58.2059.

[8] S. John, *Phys. Rev. Lett.* **58**, 2486 (1987); https://doi.org/10.1103/PhysRevLett.58.2486.

[9] C. M. Soukoulis, *Nanotechnology* **13**, 420 (2002); https://doi.org/10.1088/0957-4484/13/3/335.

[10] P. Russell, *Science* **299**, 358 (2003); https://doi.org/10.1126/science.1079280.

[11] R. V. Nair and R. Vijaya, *Prog. Quantum Electron.* **34**, 89 (2010); https://doi.org/10.1016/j.pquantelec.2010.01.001.

[12] S. Robinson and R. Nakkeeran, Photonic Crystal Ring Resonator-Based Optical Filters in *Advances in Photonic Crystals*, Ed. V. M. N. Passaro (IntechOpen, 2013), 1; https://doi.org/10.5772/54533.

[13] R. Kashyap, *Fiber Bragg Gratings*, 2nd ed. (Academic Press, 2009).

[14] W. Heiss, T. Schwarzl, J. Roither et al., *Prog. Quantum Electron.* **25**, 193 (2001); https://doi.org/10.1016/S0079-6727(01)00011-8.

[15] A. Yariv and P. Yeh, *Optical Waves in Crystals* (Wiley, 2002), 155.

[16] P. Russell, *J. Lightwave Technol.* **12**, 4729 (2006); https://doi.org/10.1109/JLT.2006.885258.

[17] J. C. Knight, *J. Opt. Soc. Am. B* **14**, 1661 (2007); https://doi.org/10.1364/JOSAB.24.001661.

[18] C. Markos, J. C. Travers, A. Abdolvand, B. J. Eggleton, and O. Bang, *Rev. Mod. Phys.* **89**, 045003 (2017); https://doi.org/10.1103/RevModPhys.89.045003.

[19] G. P. Agrawal, *Nonlinear Fiber Optics*, 6th ed. (Academic Press, 2019).

[20] J. M. Dudley, G. Genty, and S. Coen, *Rev. Mod. Phys.* **78**, 1135 (2006); https://doi.org/10.1103/RevModPhys.78.1135.

[21] A. Dubietis, A. Couairon, and G. Genty, *J. Opt. Soc. Am. B* **36**, SG1 (2019); https://doi.org/10.1364/JOSAB.36.000SG1.

[22] R. R. Alfano, Ed., *The Supercontinuum Laser Source*, 4th ed. (Springer, 2022).

[23] V. G. Veselago, *Sov. Phys. Uspekhi* **10**, 509 (1968); https://doi.org/10.1070/PU1968v010n04ABEH003699.

[24] R. A. Shelby, D. R. Smith, and S. Schultz, *Science* **292**, 77 (2001); https://doi.org/10.1126/science.1058847.

[25] V. M. Shalaev, W. Cai, U. K. Chettiar et al., *Opt. Lett.* **30**, 3356 (2005); https://doi.org/10.1364/OL.30.003356.

[26] S. Zhang, W. Fan, N. C. Panoiu et al., *Phys. Rev. Lett.* **95**, 137404 (2005); https://doi.org/10.1103/PhysRevLett.95.137404.

[27] G. Dolling, M. Wegener, C. M. Soukoulis, and S. Linden, *Opt. Lett.* **32**, 53 (2007); https://doi.org/10.1364/OL.32.000053.

[28] J. B. Pendry, *Phys. Rev. Lett.* **85**, 3966 (2000); https://doi.org/10.1103/PhysRevLett.85.3966.

[29] R. Maas, J. Parsons, N. Engheta, and A. Polman, *Nat. Photon.* **7**, 907 (2013); https://doi.org/10.1038/nphoton.2013.256.

[30] X. Niu, X. Hu, S. Chu, and Q. Gong, *Adv. Opt. Mater.* **6**, 1701292 (2018); https://doi.org/10.1002/adom.201701292.

[31] Y. Yang, J. Lu, and A. Manjavacas et al., *Nat. Phys.* **15**, 1022 (2019); https://doi.org/10.1038/s41567-019-0584-7.

[32] A. Ali, A. Mitra, and B. Aïssa, *Nanomaterials* **12**, 1027 (2022); https://doi.org/10.3390/nano12061027.

[33] J. Jung, H. Park, J. Park, T. Chang, and J. Shin, *Nanophotonics* **9**, 3165 (2020); https://doi.org/10.1515/nanoph-2020-0111.

[34] S. B. Glybovski, S. A. Tretyakov, P. A. Belov, Y. S. Kivshar, and C. R. Simovski, *Phys. Rep.* **634**, 1 (2016); https://doi.org/10.1016/j.physrep.2016.04.004.

[35] H.-T. Chen, A. J. Taylor, and N. Yu, *Rep. Prog. Phys.* **79**, 076401 (2016); https://doi.org/10.1088/0034-4885/79/7/076401.

[36] V.-C. Su, C. H. Chu, G. Sun, and D. P. Tsai, *Opt. Express* **26**, 13148 (2018); https://doi.org/10.1364//OE.26.013148.

[37] D. Neshev and I. Aharonovich, *Light Sci. Appl.* **7**, 58 (2018); https://doi.org/10.1038/s41377-018-0058-1.

[38] S. Sun, Q. He, J. Hao, S. Xiao, and L. Zhou, *Adv. Opt. Photon.* **11**, 380 (2019); https://doi.org/10.1364//AOP.11.000380.

[39] N. Yu, P. Genevet, M. A. Kats et al., *Science* **334**, 333 (2011); https://doi.org/10.1126/science.1210713.

[40] A. V. Kildishev, A. Boltasseva and V. M. Shalaev, *Science* **338**, 1232009 (2013); https://doi.org/10.1126/science.1232009.

[41] A. Pors and S. I. Bozhevolnyi, *Opt. Express* **21**, 27438 (2013); https://doi.org/10.1364//OE.21.027438.

[42] M. Khorasaninejad, W. T. Chen, R. C. Devlin, A. Zhu, and F. Capasso, *Science* **352**, 1190 (2016); https://doi.org/10.1126/science.aaf6644.

[43] S. M. Wang, P. C. Wu, V.-C. Su et al., *Nat. Nanotechnol.* **13**, 227 (2018); https://doi.org/10.1038/s41565-017-0052-4.

[44] D. M. Lin, P. Y. Fan, E. Hasman, and M. L. Brongersma, *Science* **345**, 298 (2014); https://doi.org/10.1126/science.1253213.

[45] J. S. T. Smalley, F. Vallini, S. A. Montoya et al., *Nat. Commun.* **8**, 13793 (2017); https://doi.org/10.1038/ncomms13793.

[46] Y. F. Yu, A. Y. Zhu, R. Paniagua-Domínguez et al., *Laser Photon. Rev.* **9**, 412 (2015); https://doi.org/10.1002/lpor.201500041.

[47] S. Jahani and Z. Jacob, *Nat. Nanotechnol.* **11**, 23 (2016); https://doi.org/10.1038/nnano.2015.304.

[48] S. Kruk, B. Hopkins, I. I. Kravchenko et al., *APL Photon.* **1**, 030801 (2016); https://doi.org/10.1063/1.4949007.

[49] A. Mekis, J. C. Chen, I. Kurand et al., *Phys. Rev. Lett.* **77**, 3787 (1996); https://doi.org/10.1103/PhysRevLett.77.3787.

[50] M. Loncar, T. Doll, J. Vuckovic, and A. Scherer, *J. Lightwave Technol.* **18**, 1402 (2000); https://doi.org/10.1109/50.887192.

[51] B. Wang and M. A. Cappelli, *APL Adv.* **1**, 030801 (2016); https://doi.org/10.1063/1.4954668.

[52] P. Lalanne, *Appl. Opt.* **35**, 5369 (1996); https://doi.org/10.1364/AO.35.005369.

[53] G. Boedecker and C. Henkel, *Opt. Express* **11**, 1590 (2003); https://doi.org/10.1364/OE.11.001590.

[54] Y. Ono, *Appl. Opt.* **45**, 131 (2006); https://doi.org/10.1364/AO.45.000131.

[55] J. C. M. Garnett, *Phil. Trans. R. Soc. A* **203**, 384 (1904); https://doi.org/10.1098/rsta.1904.0024.

[56] P. N. Dyachenko, V. S. Pavelyev, and V. A. Soifer, *Opt. Lett.* **37**, 2178 (2012); https://doi.org/10.1364/OL.37.002178.

[57] E. Centeno and D. Cassagne, *Opt. Lett.* **74**, 2278 (2005); https://doi.org/10.1364/OL.30.002278.

[58] E. Centeno, D. Cassagne, and J. P. Albert, *Phys. Rev. B* **73**, 235119 (2006); https://doi.org/0.1103/PhysRevB.73.235119.

[59] H.-T. Chien and C.-C. Chen, *Opt. Express* **14**, 10759 (2006); https://doi.org/10.1364/OE.14.010759.

[60] H. Kurt and D. S. Citrin, *Opt. Express* **15**, 1240 (2007); https://doi.org/10.1364/OE.15.001240.

[61] H. Kurt, E. Colak, O. Cakmak, H. Caglayan, and E. Ozbay, *Appl. Phys. Lett.* **93**, 171108 (2008); https://doi.org/10.1063/1.3009965.

[62] K.-V. Do, X. Le Roux, D. Marris-Morini, L. Vivien, and E. Cassan, *Opt. Express* **20**, 4776 (2012); https://doi.org/10.1364/OE.20.004776.

[63] R. K. Lüneburg, *Mathematical Theory of Optics* (Brown University, 1944).

[64] J. Hunt, T. Tyler, S. Dhar et al., *Opt. Express* **20**, 1706 (2012); https://doi.org/10.1364/OE.20.001706.

[65] H. Gąo, B. Zhang, S. G. Johnson, and G. Barbastathis, *Opt. Express* **20**, 1617 (2012); https://doi.org/10.1364/OE.20.001617.

[66] Q. Zhu, L. Jin, and Y. Fu, *Ann. Phys.* **527**, 205 (2015); https://doi.org/10.1002/andp.201400195.

[67] S. H. Badri and M. M. Gilarlue, *J. Opt. Soc. Am. B* **36**, 1288 (2019); https://doi.org/10.1364/JOSAB.36.001288.

[68] S. Liang, J. Xie, P. Tang, and J. Liu, *Opt. Express* **27**, 9601 (2019); https://doi.org/10.1364/OE.27.009601.

[69] B. Liang, X. Huang, and J. Zheng, *Mat. Res. Express.* **9**, 016201 (2022); https://doi.org/10.1088/2053-1591/ac4731.

[70] T. M. Monro and H. Ebendorff-Heidepriem, *Ann. Rev. Mat. Res.* **36**, 467 (2006); https://doi.org//10.1146/annurev.matsci.36.111904.135316.

[71] T. Ermatov, J. S. Skibina, V. V. Tuchin, and D. A. Gorin, *Materials* **13**, 921 (2020); https://doi.org/10.3390/ma13040921.

[72] A. Anuszkiewicz, R. Kasztelanic, A. Filipkowski et al., *Sci. Rep.* **8**, 1 (2018); https://doi.org/10.1038/s41598-018-30284-1.

[73] H. Yokota, K. Higuchi, and Y. Imai, *Jpn. J. Appl. Phys.* **55**, 08RE10 (2016); https://doi.org/10.7567/JJAP.55.08RE10.

[74] X. Forestier, T. Karpate, G. Huss et al., *Appl. Nanosci.* **10**, 1997 (2020); https://doi.org/10.1007/s13204-020-01319-9.

[75] Z. Eslami, L. Salmela, A. Filipkowski et al., *Nat. Commun.* **13**, 2126 (2022); https://doi.org/10.1038/s41467-022-29776-6.

[76] D. R. Smith, J. J. Mock, A. F. Starr, and D. Schurig, *Phys. Rev. E* **71**, 036609 (2005); https://doi.org/10.1103/PhysRevE.71.036609.

[77] U. Levy, M. Abashin, K. Ikeda et al., *Phys. Rev. Lett.* **98**, 243901 (2007); https://doi.org/10.1103/PhysRevLett.98.243901.

[78] Z. L. Mei, J. Bai, and T. J. Cui, *J. Phys. D* **43**, 055404 (2010); https://doi.org/10.1088/0022-3727/43/5/055404.

[79] T. Zentgraf, Y. Liu, M. H. Mikkelsen, J. Valentine, and X. Zhang, *Nat. Nanotechnol.* **6**, 151 (2011); https://doi.org/10.1364/OE.19.020122.

[80] C. D. Giovampaola and N. Engheta, *Nat. Mater.* **13**, 1115 (2014); https://doi.org/10.1038/nmat4082.

[81] Y. Jin, B. Djafari-Rouhani, and D. Torrent, *Nanophotonics* **8**, 685 (2018); https://doi.org/10.1515/nanoph-2018-0227.

[82] J. Chen, Y. Zhao, L. Xing, Z. He, and L. Sun, *Sci. Rep.* **10**, 20381 (2020); https://doi.org/10.1038/s41598-020-77270-0.

[83] J. E. Eaton, *Trans. IRE Antennas Propag.* **4**, 66 (1952); https://doi.org/10.1109/TPGAP.1952.237341.

[84] M. Yin, X. Y. Tian, L. L. Wu, and D. C. Li, *Appl. Phys. Lett.* **104**, 094101 (2014); https://doi.org/10.1063/1.4867704.

[85] L. H. Gabrielli and M. Lipson, *Opt. Express* **19**, 20122 (2011); https://doi.org/10.1364/OE.19.020122.

[86] J. Shen, Y. Zhang, Y. Dong et al., *Nanophotonics* **11**, 3603 (2022); https://doi.org/10.1515/nanoph-2022-0242.

[87] A. Alú and N. Engheta, *Phys. Rev. B* **78**, 045102 (2008); https://doi.org/10.1103/PhysRevB.78.045102.

[88] E. J. R. Vesseur, T. Coenen, H. Caglayan, N. Engheta, and A. Polman, *Phys. Rev. Lett.* **110**, 013902 (2013); https://doi.org/10.1103/PhysRevLett.110.013902.

[89] P. Moitra, Y. Yang, Z. Anderson et al., *Nat. Photon.* **7**, 791 (2013); https://doi.org/10.1038/nphoton.2013.214.

[90] M. Navarro-Cía, M. Beruete, M. Sorolla, and N. Engheta, *Phys. Rev. B* **86**, 165130 (2012); https://doi.org/10.1103/PhysRevB.86.165130.

[91] V. Torres, V. Pacheco-Peña, P. Rodríguez-Ulibarri et al., *Opt. Express* **21**, 9156 (2013); https://doi.org/10.1364/OE.21.009156.

[92] V. Pacheco-Peña, N. Engheta, S. Kuznetsov, A. Gentselev, and M. Beruete, *Phys. Rev. Appl.* **8**, 034036 (2017); https://doi.org/10.1103/PhysRevApplied.8.034036.

[93] M. D. Goldflam, T. Driscoll, D. Barnas et al., *Appl. Phys. Lett.* **102**, 224103 (2013); https://doi.org/10.1063/1.4809534.

[94] D. Ohana, B. Desiatov, N. Mazurski, and U. Levy, *Nano Lett.* **16**, 7956 (2016); https://doi.org/10.1021/acs.nanolett.6b04264.

[95] Y. Fan, X. Le Roux, A. Korovin, A. Lupu, and A. de Lustrac, *ACS Nano* **11**, 4599 (2017); https://doi.org/110.1021/acsnano.7b00150.

[96] A. She, S. Zhang, S. Shian, D. R. Clarke, and F. Capasso, *Opt. Express* **26**, 1573 (2018); https://doi.org/10.1364/OE.26.001573.

[97] Y. Jin, R. Kumar, O. Poncelet, O. Mondain-Monval, and T. Brunet, *Nat. Commun.* **10**, 143 (2019); https://doi.org/10.1038/s41467-018-07990-5.

[98] J. Sung, G.-Y. Lee, and B. Lee, *Nanophotonics* **8**, 1701 (2019); https://doi.org/10.1515/nanoph-2019-0203.

[99] Z. Li, R. Pestourie, J. S. Park et al., *Nat. Commun.* **12**, 2409 (2022); https://doi.org/10.1038/s41467-022-29973-3.

[100] E. Karimi, S. A Schulz, I. De Leon et al., *Light Sci. Appl.* **3**, e167 (2019); https://doi.org/10.1038/lsa.2014.48.

11

Impact of Partial Coherence

So far in this book, the optical field incident on a GRIN medium is assumed to be free of noise, without any fluctuations in its amplitude or phase. In practice, any incoming beam is likely to exhibit both amplitude and phase fluctuations that render it partially coherent. In this chapter, we focus on the impact of partial coherence on the propagation of optical beams inside a GRIN medium. Section 11.1 introduces the basic coherence-related concepts needed to understand the later material. Section 11.2 uses the evolution of cross-spectral density to study whether periodic self-imaging, an intrinsic property of a GRIN medium, is affected by partial coherence of an incoming beam. Section 11.3 employs the Gaussian–Schell model to discuss how the optical spectrum, the spectral intensity, and the degree of coherence associated with a Gaussian beam change with the beam's propagation inside a GRIN medium. The focus of Section 11.4 is on Gaussian beams that are only partially polarized. The concept of the polarization matrix is used to study how the degree of polarization evolves when such a partially coherent Gaussian beam is transmitted through a GRIN medium.

11.1 Basic Concepts

The primary concept in the theory of partial coherence is related to the correlation between the electric fields at two spatially separated points at two different times [1–3]. This correlation is governed by the coherence function defined as

$$\Gamma_{ij}(\mathbf{r}_1, \mathbf{r}_2, \tau) = \langle E_i^*(\mathbf{r}_1, t)E_j(\mathbf{r}_2, t+\tau)\rangle, \qquad (11.1.1)$$

where $E_i(\mathbf{r}, t)$ is a component of the electric field vector ($i, j = x, y, z$) and the angular brackets denote an ensemble average over random fluctuations in the optical field. The coherence function depends only on the time difference τ for continuous-wave (CW) beams because fluctuations of $E_i(\mathbf{r}, t)$ are governed by a stationary random process [2]. Equation (11.1.1) can be used to understand the origins of temporal coherence, spatial coherence, and polarization coherence.

11.1.1 Degree of Coherence

Let us simplify Eq. (11.1.1) to the scalar case in which light is polarized linearly and has only one component. In that situation, we can drop the subscripts i and j and write the equation as

$$\Gamma(\mathbf{r}_1, \mathbf{r}_2, \tau) = \langle E^*(\mathbf{r}_1, t)E(\mathbf{r}_2, t + \tau)\rangle. \tag{11.1.2}$$

It follows that $\Gamma(\mathbf{r}, \mathbf{r}, 0)$ corresponds to the average intensity at the point $\mathbf{r}$ defined as $I(\mathbf{r}) = \langle E^*(\mathbf{r}, t)E(\mathbf{r}, t)\rangle$. For CW beams, the average does not depend on time t for a stationary random process. It is useful to normalize the coherence function as

$$\gamma(\mathbf{r}_1, \mathbf{r}_2, \tau) = \frac{\Gamma(\mathbf{r}_1, \mathbf{r}_2, \tau)}{\sqrt{I(\mathbf{r}_1)I(\mathbf{r}_2)}}. \tag{11.1.3}$$

In general, γ is a complex quantity. However, one can show that the values of $|\gamma|$ lie in the range $0 \le |\gamma| \le 1$. For this reason, $|\gamma|$ is a useful measure of the partial coherence of an optical field and is known as the *degree of coherence*.

If we consider the correlation at the same spatial point and set $\mathbf{r}_1 = \mathbf{r}_2 = \mathbf{r}$, Eq. (11.1.3) describes temporal coherence at that point. For a beam with uniform intensity, the degree of coherence $|\gamma(\tau)|$ depends only on a single variable τ such that $|\gamma| = 1$ when $\tau = 0$. The value of $|\gamma(\tau)|$ decreases as τ increases. The time delay $\tau = \tau_c$ for which $|\gamma|$ becomes negligible is called the coherence time of the optical source emitting the beam under consideration. The coherence time of any optical field depends inversely on the bandwidth of its spectrum. This bandwidth is so large for most thermal sources that their coherence time is typically shorter than 1 ps. In contrast, it can exceed 1 μs for semiconductor lasers designed to operate in a single longitudinal mode. The distance traveled by a beam during its coherence time is known as the coherence length. A Michelson interferometer produces a fringe pattern only when the path difference in its two arms is shorter than the coherence length.

To discuss spatial coherence of an optical source, we set $\tau = 0$ in Eq. (11.1.2). In this situation, $\gamma(\mathbf{r}_1, \mathbf{r}_2, 0)$ provides us with a measure of the spatial coherence. In the case of an optical beam with uniform intensity, this degree of coherence depends only on the distance, $\delta r = |\mathbf{r}_2 - \mathbf{r}_1|$, between the two spatial points. Clearly, $\gamma = 1$ when $\delta r = 0$ because fluctuations at the same point are fully correlated. However, its magnitude decreases as δr increases. The distance δr_c for which γ becomes negligible is called the coherence radius of the optical source emitting the beam. This radius is typically small (< 0.1 mm) for thermal sources but becomes larger for laser beams.

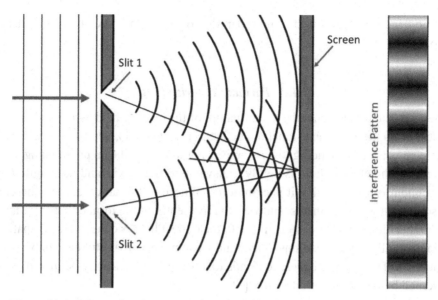

Figure 11.1 Schematic of a Young's interferometer in which two slits are used to create an interference pattern on a screen.

The importance of spatial coherence can be understood through the Young's interferometer shown schematically in Figure 11.1. As seen there, when light passes through its two slits, the two transmitted waves interfere and create a fringe pattern on a screen located at some distance away. This fringe pattern is produced only when two slits are within a coherent radius of the incoming beam. Its creation can be understood by noting that the light at any point on the screen comes from both slits, and the average intensity at that point can be written as

$$I(\mathbf{r}, t) = \langle |K_1 E(\mathbf{r}_1, t - t_1) + K_2 E(\mathbf{r}_2, t - t_2)|^2 \rangle, \tag{11.1.4}$$

where K_j is a constant ($j = 1, 2$) and t_j is the time taken by the light at slit j to arrive at the screen. The preceding equation can be written as

$$I(\mathbf{r}, t) = |K_1|^2 I(\mathbf{r}_1, t - t_1) + |K_1|^2 I(\mathbf{r}_2, t - t_2) + 2\mathrm{Re}[K_1^* K_2 \Gamma(\mathbf{r}_1, \mathbf{r}_2, \tau)], \tag{11.1.5}$$

where $\tau = t_2 - t_1$. The last term in this relation depends on the degree of coherence of the beam used to produce the interference pattern. For a plane wave of constant intensity, the visibility of the fringe pattern depends on the spatial degree of coherence as [2]

$$V_f = \frac{I_{\max} - I_{\min}}{I_{\max} + I_{\min}} = |\gamma(d)|, \tag{11.1.6}$$

where $I_{\max}$ and $I_{\min}$ are the maximum and minimum values of the intensity given in Eq. (11.1.5) and d is the distance between the two slits. It follows from Eq. (11.1.6) that no fringes are produced if the separation of two slits is such that $|\gamma|$ is nearly zero.

11.1.2 Partially Polarized Light

When we take into account the vector nature of the electric field associated with an optical beam, the coherence function, defined in Eq. (11.1.1), becomes a tensor with nine elements corresponding to $i,j = x,y,z$. For a paraxial beam propagating along the z axis, the electric field is nearly transverse to the z axis, and its longitudinal component is so small that it is common to ignore it. In that case, the coherence tensor has only four elements, and we can represent it in the form of a 2×2 matrix. In the specific case of $\mathbf{r}_1 = \mathbf{r}_2$ and $\tau = 0$, the optical field is coherent, both spatially and temporally, but fluctuations in its x and y components may still have only partial coherence. This type of coherence leads to a beam being partially polarized and is governed by the polarization matrix [1]

$$J(\mathbf{r}) = \begin{pmatrix} \Gamma_{xx}(\mathbf{r},\mathbf{r},0) & \Gamma_{xy}(\mathbf{r},\ \mathbf{r},0) \\ \Gamma_{yx}(\mathbf{r},\mathbf{r},0) & \Gamma_{yy}(\ \mathbf{r},\mathbf{r},0) \end{pmatrix}, \tag{11.1.7}$$

where the $\mathbf{r}$ dependence shows that J can still vary over a beam's cross section. The matrix J is Hermitian as $\Gamma_{yx} = \Gamma_{xy}^*$, from Eq. (11.1.1).

It is useful to define a normalized cross-correlation coefficient between the x and y polarized components of an optical beam as

$$j_{xy} = \Gamma_{xy}/\sqrt{\Gamma_{xx}\Gamma_{yy}}. \tag{11.1.8}$$

In general, j_{xy} is a complex quantity. However, one can show that its magnitude satisfies the relation $0 \le |j_{xy}| \le 1$. For a fully polarized beam, $j_{xy} = 1$. In contrast, light is said to be unpolarized when $j_{xy} = 0$. For intermediate values of j_{xy}, light is only partially polarized.

An important question is how the matrix J is related to the state of polarization (SOP) of an optical beam. To see this connection, we consider the three components of the Stokes vector defined in Eq. (8.1.2) of Section 8.1. For a noisy electric field, we must average this equation over fluctuations. It is easy to conclude that the averaged components of the Stokes vector can be written as

$$S_1 = J_{xx} - J_{yy}, \qquad S_2 = J_{xy} + J_{yx}, \qquad S_3 = i(J_{yx} - J_{xy}). \tag{11.1.9}$$

Also, the magnitude of the Stokes vector can be expressed as $S_0 = J_{xx} + J_{yy}$. Recall that S_0 is the total intensity at any point in the beam in the case of fully coherent light.

It turns out that the polarization matrix of any optical beam can be split as $J = J_p + J_u$, where J_p and J_u are the matrices for the polarized and unpolarized parts of the beam. For an unpolarized beam, $J_{xy} = 0$ and $J_{xx} = J_{yy}$. As a result, the matrices J_u and J_p take the form

$$J_u = \begin{pmatrix} A & 0 \\ 0 & A \end{pmatrix}, \qquad J_p = \begin{pmatrix} J_{xx} - A & J_{xy} \\ J_{yx} & J_{yy} - A \end{pmatrix}, \qquad (11.1.10)$$

where A is a constant. To find its value, we note from Eq. (11.1.8) that the determinant of the matrix J_p must vanish to ensure $|j_{xy}| = 1$ for the polarized part of the beam, that is,

$$\text{Det}\, J_p = (J_{xx} - A)(J_{yy} - A) - J_{xy}J_{yx} = 0. \qquad (11.1.11)$$

This quadratic equation has two solutions for A, and the physically valid solution is given by

$$A = \frac{1}{2}\left[\text{Tr}\, J - \sqrt{(\text{Tr}\, J)^2 - 4\text{Det}\, J}\,\right], \qquad (11.1.12)$$

where Tr and Det stand for the trace and the determinant of a matrix.

Noting that $\text{Tr}\, J$ provides the beam's total intensity, the intensity of the polarized and unpolarized parts are given by

$$I_p = \text{Tr}\, J_p = \text{Tr}\, J - 2A, \qquad I_u = \text{Tr}\, J_u = 2A. \qquad (11.1.13)$$

The degree of polarization of a beam is defined as the fraction of the total intensity that is polarized [1]. Using Eqs. (11.1.12) and (11.1.13), it can be written as

$$\mathscr{P} = \frac{I_p}{I_p + I_u} = \sqrt{1 - \frac{4\text{Det}\, J}{(\text{Tr}\, J)^2}}. \qquad (11.1.14)$$

It is easy to deduce that $0 \leq \mathscr{P} \leq 1$, and the limiting values of 0 and 1 correspond to unpolarized and polarized beams, respectively.

We have seen in Section 8.1 that the normalized Stokes vector represents the SOP of a fully polarized beam on the surface of the Poincaré sphere of unit radius. In the case of partially polarized light, the tip of the Stokes vector lies inside the Poincaré sphere at a distance $\mathscr{P} < 1$. The SOP of the beam is still governed by its location on the surface of the smaller sphere. In the case of an unpolarized beam, the length of the Stokes vector shrinks to zero, and the SOP is represented by the center of the Poincaré sphere.

11.1.3 Frequency-Domain Description

As we have seen in Section 2.1, it is often easier to solve the propagation problem in the frequency domain by making use of the Fourier transform of the electric field, as indicated in Eq. (2.1.3). Thus, we should consider correlations at two different spatial points in the frequency domain as well. Such a correlation function is called the cross-spectral density and is defined as [1]

$$W_{ij}(\mathbf{r}_1, \mathbf{r}_2, \omega) = \langle \tilde{E}_i^*(\mathbf{r}_1, \omega)\tilde{E}_j(\mathbf{r}_2, \omega) \rangle, \tag{11.1.15}$$

where the average is over an ensemble in the frequency domain. According to the Wiener–Khintchine theorem, W_{ij} is just the Fourier transform of the coherence function, that is,

$$W_{ij}(\mathbf{r}_1, \mathbf{r}_2, \omega) = \int_{-\infty}^{\infty} \Gamma_{ij}(\mathbf{r}_1, \mathbf{r}_2, \tau) e^{i\omega t} d\tau. \tag{11.1.16}$$

Let us focus first on the scalar case and drop the subscripts i and j. The cross-spectral density in this case takes the form

$$W(\mathbf{r}_1, \mathbf{r}_2, \omega) = \langle \tilde{E}^*(\mathbf{r}_1, \omega)\tilde{E}(\mathbf{r}_2, \omega) \rangle. \tag{11.1.17}$$

It represents the correlation at two different spatial points of a single spectral component of the optical field at frequency ω. If the two points coincide, $W(\mathbf{r}, \mathbf{r}, \omega)$ becomes the spectral density of the field at that point, that is,

$$S(\mathbf{r}, \omega) = W(\mathbf{r}, \mathbf{r}, \omega) = \langle |\tilde{E}(\mathbf{r}, \omega)|^2 \rangle. \tag{11.1.18}$$

The spectral degree of coherence is introduced by considering the Young's interferometer in the frequency domain (see Figure 11.1). Focusing on a specific frequency component, the spectral density at any point on the screen can be written as

$$S(\mathbf{r}, \omega) = \langle |K_1\tilde{E}(\mathbf{r}_1, \omega)e^{ikR_1} + K_2\tilde{E}(\mathbf{r}_2, \omega)e^{ikR_2}|^2 \rangle, \tag{11.1.19}$$

where R_1 and R_2 are the distances from the two slits. This equation can be written as

$$I(\mathbf{r}, t) = |K_1|^2 S(\mathbf{r}_1, \omega) + |K_1|^2 S(\mathbf{r}_2, \omega) + 2\text{Re}[K_1^* K_2 W(\mathbf{r}_1, \mathbf{r}_2, \omega)e^{ik(R_1 - R_2)}]. \tag{11.1.20}$$

The last term depends on the cross-spectral density of the beam used to produce the interference pattern. We use it to define the spectral degree of coherence as

$$\mu(\mathbf{r}_1, \mathbf{r}_2, \omega) = \frac{W(\mathbf{r}_1, \mathbf{r}_2, \omega)}{[S(\mathbf{r}_1, \omega)S(\mathbf{r}_2, \omega)]^{1/2}}. \tag{11.1.21}$$

Similar to the time-domain case, the spectral degree of coherence is a complex quantity, but its magnitude $|\mu|$ satisfies the relation $0 \le |\mu| \le 1$.

11.2 Propagation-Induced Changes in Coherence

It is well known that the coherence of light emitted by a source can change during its propagation in any medium, including vacuum [1–3]. This feature was discovered during the 1930s and is known as the van Cittert–Zernike theorem [4, 5]. It implies that the degree of coherence and the degree of polarization [6–8] of a partially coherent optical beam change when the beam propagates through a medium. Indeed, such changes attracted considerable attention during the 1970s in the context of GRIN fibers [9–11].

11.2.1 Evolution of Cross-Spectral Density

Let us consider how the cross-spectral density changes during propagation of an optical beam inside a linear medium. We have seen in Section 3.2 that any component of the electric field at a given frequency evolves as

$$\tilde{E}(\mathbf{r}, \omega) = \int_{-\infty}^{\infty} K(\mathbf{r}, \mathbf{s}) \tilde{E}(\mathbf{s}, \omega) \, d\mathbf{s}, \tag{11.2.1}$$

where $\mathbf{s}$ represents a point at the input plane $z = 0$ and the integral is over the surface of this plane. As discussed in Section 3.2, the propagation kernel can be written in terms of the ABCD matrix elements of a linear medium as

$$K(\mathbf{r}, \mathbf{s}) = \left(\frac{k e^{ikz}}{2\pi i B} \right) \exp \left(\frac{ik}{2B} [D\mathbf{u} \cdot \mathbf{u} - 2\mathbf{u} \cdot \mathbf{s} + A\mathbf{s} \cdot \mathbf{s}] \right), \tag{11.2.2}$$

where $\mathbf{u} = (x, y)$ is the transverse part of the vector $\mathbf{r}$ at the location z. In the case of a homogeneous medium, $A = D = 1$ and $B = z$. In the case of a GRIN medium with a parabolic index profile $n(\rho) = n_0(1 - b^2\rho^2/2)$, these matrix elements become $A = D = \cos(bz)$ and $B = \sin(bz)/b$.

Using Eq. (11.2.1) in Eq. (11.1.17) at both spatial points, the cross-spectral density at a distance z is related to its input value at $z = 0$ as [9]

$$W(\mathbf{r}_1, \mathbf{r}_2, \omega) = \iint_{-\infty}^{\infty} K^*(\mathbf{r}_1, \mathbf{s}_1) K(\mathbf{r}_2, \mathbf{s}_2) W(\mathbf{s}_1, \mathbf{s}_2, \omega) \, d\mathbf{s}_1 \, d\mathbf{s}_2. \tag{11.2.3}$$

This form requires a four-dimensional integration over the input plane variables. Even though the algebra can be tedious, it provides a straightforward approach to study how the spectral degree of coherence changes on propagation inside a medium for which the ABCD matrix is known. Changes in the degree of polarization can also be calculated by replacing W with W_{ij} in Eq. (11.2.3) for $i, j = x$ or y. The reason for this simple change is that Eq. (11.2.3) applies to each element of the polarization matrix. The coherence function in the time domain is obtained by taking the inverse Fourier transform of $W(\mathbf{r}_1, \mathbf{r}_2, \omega)$. From Eq. (11.1.16), it is given by

$$\Gamma(\mathbf{r}_1, \mathbf{r}_2, \tau) = \frac{1}{2\pi} \int_{-\infty}^{\infty} W(\mathbf{r}_1, \mathbf{r}_2, \omega) e^{-i\omega\tau} \, d\omega. \tag{11.2.4}$$

Before using Eq. (11.2.3), we need to specify the input cross-spectral density $W(\mathbf{s}_1, \mathbf{s}_2, \omega)$ at $z = 0$. The spectral degree of coherence in Eq. (11.1.21) is unity when the two points coincide and depends only on the distance between them. Thus, $W(\mathbf{s}_1, \mathbf{s}_2, \omega)$ can be written in the form

$$W(\mathbf{s}_1, \mathbf{s}_2, \omega) = [S(\mathbf{s}_1, \omega)S(\mathbf{s}_2, \omega)]^{1/2} \mu(\mathbf{s}_2 - \mathbf{s}_1, \omega), \qquad (11.2.5)$$

where $S(\mathbf{s}, \omega)$ is the spectral intensity at the input plane at the point $\mathbf{s}$. If the spectrum is the same at all spatial points, we can write the spectral intensity in the form $S(\mathbf{s}, \omega) = S_0(\omega)U(\mathbf{s})$, where $S_0(\omega)$ is the input spectrum and $U(\mathbf{s})$ is the dimensionless spatial profile of the beam's intensity. Combining Eqs. (11.2.3) and (11.2.5), we obtain

$$W(\mathbf{r}_1, \mathbf{r}_2, \omega) = S_0(\omega) \iint_{-\infty}^{\infty} K^*(\mathbf{r}_1, \mathbf{s}_1)K(\mathbf{r}_2, \mathbf{s}_2)\sqrt{U(\mathbf{s}_1)U(\mathbf{s}_2)}\mu(\mathbf{s}_2 - \mathbf{s}_1, \omega)\,d\mathbf{s}_1\,d\mathbf{s}_2.$$
$$(11.2.6)$$

The spectral intensity of the beam at a distance z is obtained by setting $\mathbf{r}_1 = \mathbf{r}_2 = \mathbf{r}$ in Eq. (11.2.6), that is, $S(\mathbf{r}, \omega) = W(\mathbf{r}, \mathbf{r}, \omega)$. The resulting expression implies that the spectrum of a partially coherent beam can change during its propagation inside a linear medium. This phenomenon is sometimes called the Wolf effect [12]. Spectral changes occurring inside a GRIN medium are discussed later in Section 11.3.2.

Before we can use Eq. (11.2.6) to calculate propagation-induces changes in the spectrum or the degree of coherence of an optical beam, we need to specify its spatial profile $U(\mathbf{s})$ as well as its degree of coherence $\mu(\mathbf{s}_2 - \mathbf{s}_1, \omega)$ at $z = 0$. In one model, known as the Gaussian–Schell model [1], both are assumed to have a Gaussian shape, and the two functions appearing in Eq. (11.2.6) are written as

$$U(\mathbf{s}) = \exp(-|\mathbf{s}|^2/w_0^2), \qquad \mu(\rho, \omega) = \exp(-\rho^2/2\sigma_c^2), \qquad (11.2.7)$$

where w_0 is the spot size of the input beam and σ_c is its coherence radius. Another simplification is sometimes made in Eq. (11.2.6) when the condition $w_0 \gg \sigma_c$ is satisfied. In that situation, the intensity varies much more slowly compared to variations in the degree of coherence, and it becomes possible to use the approximation

$$\sqrt{U(\mathbf{s}_1)U(\mathbf{s}_2)} \approx U[(\mathbf{s}_1 + \mathbf{s}_2)/2]. \qquad (11.2.8)$$

This is often referred to as the quasi-homogeneous approximation [1].

11.2.2 Self-Imaging of Partially Coherent Beams

We have seen in Section 3.2 that a fully coherent beam (no fluctuations) exhibits periodic self-imaging inside a GRIN medium such that the input situation at $z = 0$ is reproduced at distances separated by the period $L_p = 2\pi/b$. One should ask whether such self-imaging can occur even for partially coherent beams. In a 2015 study, an approach based on the coherent states associated with a harmonic oscillator was used to show that self-imaging occurs for partially coherent beams [13]. We follow here a simpler approach based on the properties of the kernel appearing in Eq. (11.2.3).

Following the analysis in Section 3.2.2, if we use $\mathbf{u}_j = (x_j, y_j)$ and $\mathbf{s}_j = (x'_j, y'_j)$ with $j = 1, 2$ for the components of four vectors appearing in Eq. (11.2.2), we can factor the kernel as $K(\mathbf{r}_j, \mathbf{s}_j) = F(x_j, x'_j)F(y_j, y'_j)$, where the function $F(x, x')$ is defined as

$$F(x, x') = \left(\frac{ke^{ikz}}{2\pi iB}\right)^{1/2} \exp\left(\frac{ik}{2B}[Dx^2 - 2xx' + Ax'^2]\right), \tag{11.2.9}$$

with $A = D = \cos(bz)$ and $B = \sin(bz)/b$ for a GRIN medium. At distances such that bz is a multiple of 2π, $A = D = 1$ but $B \to 0$. It was found in Section 3.2.2 that the limit $B \to 0$, $F(x, x')$ reduces to a delta function such that $F(x, x') = \delta(x-x')e^{ikz/2}$. It follows immediately that the four integrals appearing in Eq. (11.2.3) can easily be done to obtain the simple result

$$W(\mathbf{u}_1, \mathbf{u}_2, \omega, mL_p) = W(\mathbf{u}_1, \mathbf{u}_2, \omega, 0), \tag{11.2.10}$$

where m is an integer. We have added the distance argument to W to show that the entire structure of the cross-spectral density at $z = 0$ is reproduced precisely at the self-imaging planes separated by the period L_p.

The expression in Eq. (11.2.10) is remarkable for several reasons. It shows that the cross-spectral density of any input beam at the input plane $z = 0$ is self-imaged inside a GRIN medium at distances $z = mL_p$, where m is an integer and $L_p = 2\pi/b$ is the self-imaging period of the GRIN medium. Because the cross-spectral density is related to many properties of the beam, such as its spatial shape, optical spectrum, and the spatial and temporal degrees of coherence, all these properties of the input beam are reproduced precisely every time the beam is self-imaged. The degree of polarization is also reproduced because Eq. (11.2.10) applies to each component to the polarization matrix given in Eq. (11.1.7). It should be stressed that all of these properties change considerably within each period, as will become clear in later sections.

One more general feature is noteworthy. As we saw in Section 3.2.2, the middle point of each period also has a special property. At distances such that $z = (m - \frac{1}{2})L_p$, where m is an integer, the function $F(x, x')$ in Eq. (11.2.9) is reduced o a delta function such that $F(x, x') = \delta(x + x')e^{ikz/2}$. In this case, Eq. (11.2.10) becomes

$$W[\mathbf{u}_1, \mathbf{u}_2, \omega, (m - \tfrac{1}{2})L_p] = W(-\mathbf{u}_1, -\mathbf{u}_2, \omega, 0). \tag{11.2.11}$$

This relation implies that the cross-spectral density and all coherence properties of the input beam are inverted around the central axis in the transverse plane at distances that correspond to the middle of each self-imaging period. In the case of a radially symmetric beam, this inversion has no impact, and one can conclude that the beam is self-imaged at these distances as well.

It is interesting to consider the limiting case of a totally incoherent beam. This case was considered in a 1974 study devoted to the propagation of partially coherent beams through a GRIN fiber [9]. The cross-spectral density of an incoherent beam can be written from Eq. (11.2.5) as

$$W(\mathbf{s}_1, \mathbf{s}_2, \omega) = S(\mathbf{s}_1, \omega)\delta(\mathbf{s}_1 - \mathbf{s}_2), \tag{11.2.12}$$

where the delta function ensures that no correlation exists between two nearby points. Using this form in Eq. (11.2.3), the cross-spectral density inside the GRIN medium takes the form

$$W(\mathbf{r}_1, \mathbf{r}_2, \omega) = \int_{-\infty}^{\infty} K^*(\mathbf{r}_1, \mathbf{s})K(\mathbf{r}_2, \mathbf{s})S(\mathbf{s}, \omega)\, d\mathbf{s}. \tag{11.2.13}$$

This equation shows that the beam becomes partially coherent inside the GRIN medium because $W(\mathbf{r}_1, \mathbf{r}_2, \omega)$ does not always vanish when $\mathbf{r}_1 \neq \mathbf{r}_2$.

Using the form of the kernel given in Eq. (11.2.3), Eq. (11.2.13) becomes

$$W(\mathbf{r}_1, \mathbf{r}_2, \omega) = \left(\frac{k}{2\pi B}\right)^2 \exp\left(\frac{ikD}{2B}(|\mathbf{u}_2|^2 - |\mathbf{u}_1|^2)\right)$$
$$\int_{-\infty}^{\infty} S(\mathbf{s}, \omega)\exp\left(\frac{ikA}{B}[(\mathbf{u}_1 - \mathbf{u}_2)\cdot\mathbf{s}]\right) d\mathbf{s}. \tag{11.2.14}$$

The integral in this equation represents a two-dimensional Fourier transform of the spectral intensity at the plane $z = 0$. This result is related to the van Cittert–Zernike theorem, which states that an incoherent beam becomes partially coherent on propagation inside a homogeneous medium. The difference in the case of a GRIN medium is that the beam's coherence properties vary periodically inside such a medium because of the self-imaging phenomenon, which also implies that the beam must become incoherent periodically at distances that are a multiple of L_p.

11.3 Partially Coherent Beams in GRIN Media

As we saw in Section 11.2, both the degree of coherence and the degree of polarization can change when a partially coherent beam is transmitted through a GRIN medium. Moreover, the optical spectrum of such a beam can also change during its propagation. Changes in the coherence properties of a beam inside optical fibers were studied during the 1970s [9–11] and have remained a topic of interest since then [14–21].

11.3.1 Evolution of a Partially Coherent Gaussian Beam

The Gaussian–Schell model in Eq. (11.2.7) applies to a Gaussian beam whose spatial coherence decreases with distance in a Gaussian fashion. The cross-spectral density inside a GRIN medium at a distance z is found from Eq. (11.2.6) using the kernel in Eq. (11.2.2). The four-dimensional integration can be done using the cartesian coordinates with $\mathbf{u}_j = (x_j, y_j)$ and $\mathbf{s}_j = (x'_j, y'_j)$ for $j = 1, 2$. Their use allows us to factor $W(\mathbf{r}_1, \mathbf{r}_2, \omega)$ as

$$W(\mathbf{r}_1, \mathbf{r}_2, \omega) = S_0(\omega) H(x_1, x_2, z) H(y_1, y_2, z), \qquad (11.3.1)$$

where the function $H(x_1, x_2, z)$ is defined as

$$H(x_1, x_2, z) = \iint_{-\infty}^{\infty} \exp\left[-\frac{(x'^2_1 + x'^2_2)}{2w_0^2} - \frac{(x'_1 - x'_2)^2}{2\sigma_c^2}\right] K^*(x_1, x'_1) K(x_2, x'_2) \, dx'_1 \, dx'_2,$$

$$(11.3.2)$$

and the one-dimensional kernel $K(x, x')$ is obtained from Eq. (11.2.2):

$$K(x, x') = \left(\frac{ke^{ikz}}{2\pi iB}\right)^{1/2} \exp\left(\frac{ik}{2B}(Ax^2 - 2xx' + Ax'^2)\right). \qquad (11.3.3)$$

Here, $A = \cos(bz)$, $B = \sin(bz)/b$, w_0 is the spot size of the input beam, and σ_c is its coherence radius at $z = 0$.

The two-fold integration in Eq. (11.3.2) can be simplified by introducing two new variables representing the average and the difference of old variables [15]:

$$\xi = \tfrac{1}{2}(x'_2 + x'_1), \qquad \xi' = x'_2 - x'_1. \qquad (11.3.4)$$

In terms of the new variables, Eq. (11.3.2) can be written as

$$H(x_1, x_2, z) = \left(\frac{k}{2\pi B}\right) \int_{-\infty}^{\infty} \exp\left[-\frac{\xi^2}{w_0^2} + \frac{ik}{B}(x_1 - x_2)\xi\right] G(x_1, x_2, \xi) d\xi, \qquad (11.3.5)$$

where $G(x_1, x_2, \xi)$ involves an integration over ξ':

$$G(x_1, x_2, \xi) = \exp\left[\frac{ikA}{2B}(x_2^2 - x_1^2)\right] \int_{-\infty}^{\infty} \exp\left[-p^2\xi'^2 + \frac{ik}{2B}[2A\xi\xi' - (x_1 + x_2)\xi']\right] d\xi',$$

$$(11.3.6)$$

and p^2 is defined as

$$p^2 = \frac{1}{4w_0^2} + \frac{1}{2\sigma_c^2}. \qquad (11.3.7)$$

The integral in Eq. (11.3.6) can be done using the formula in Eq. (3.2.25), and the result is given by

$$G(x_1, x_2, \xi) = \exp\left[\frac{ikA}{2B}(x_2^2 - x_1^2)\right] \frac{\sqrt{\pi}}{p} \exp\left[-\frac{k^2}{4p^2B^2}[A\xi - (x_1 + x_2)/2]^2\right].$$

$$(11.3.8)$$

This equation indicates that the analysis would be simpler if we introduce the average and the difference of the variables x_1 and x_2 in a way similar to Eq. (11.3.4):

$$\bar{x} = \tfrac{1}{2}(x_2 + x_1), \qquad \delta x = x_2 - x_1. \tag{11.3.9}$$

With the expression for G in Eq. (11.3.8), we use again Eq. (3.2.25) for the integration in Eq. (11.3.5) and obtain the following result:

$$H(x_1, x_2, z) = \left(\frac{k}{2\pi B}\right) \exp\left[\frac{ikA}{B}\bar{x}\delta x - \frac{k^2\bar{x}^2}{4p^2B^2}\right]\frac{\pi}{pq}\exp\left(-\frac{h^2}{4q^2}\right), \tag{11.3.10}$$

where we have used the definitions

$$q^2 = \left(\frac{kA}{2pB}\right)^2 + \frac{1}{w_0^2}, \qquad h = \frac{k^2A}{2p^2B^2}\bar{x} + \frac{ik}{B}\delta x. \tag{11.3.11}$$

If we use the expression for H from Eq. (11.3.10) into Eq. (11.3.1), we obtain an analytical expression for $W(\mathbf{r}_1, \mathbf{r}_2, \omega)$. Although somewhat complex algebraically, we use it to analyze changes in both the spectral intensity and the degree of coherence occurring when a partially coherent Gaussian beam propagates through a GRIN medium.

11.3.2 Changes in Spectral Intensity

The spectral intensity at any point $\mathbf{r}$ is found by setting $\mathbf{r}_1 = \mathbf{r}_2 = \mathbf{r}$ in the expression for the cross-spectral density and can be written as

$$S(\mathbf{r}, \omega) = W(\mathbf{r}, \mathbf{r}, \omega) = S_0(\omega)H(x, x, z)H(y, y, z). \tag{11.3.12}$$

The function $H(x, x, z)$ is obtained from Eq. (11.3.10) by setting $\bar{x} = x$ and $\delta x = 0$ and takes the form

$$H(x, x, z) = \left(\frac{k}{2Bpq}\right) \exp\left[-\frac{k^2x^2}{4p^2B^2}\left(1 - \frac{k^2A^2}{4p^2q^2B^2}\right)\right]. \tag{11.3.13}$$

It can be simplified considerably using q^2 from Eq. (11.3.11):

$$H(x, x, z) = \left(\frac{k}{2Bpq}\right) \exp\left[-\frac{k^2x^2}{(2pqBw_0)^2}\right]. \tag{11.3.14}$$

Using this form of H in Eq. (11.3.12), the spectral intensity is found to be

$$S(\mathbf{r}, \omega) = S_0(\omega)\left(\frac{k}{2pqB}\right)^2 \exp\left[-\frac{k^2(x^2+y^2)}{(2pqw_0B)^2}\right]. \tag{11.3.15}$$

The preceding equation can be simplified further by using q^2 from Eq. (11.3.11) to obtain an effective width of the Gaussian beam in the form

$$w_{\text{eff}}(z) = 2pqw_0B/k = [(Aw_0)^2 + (2pB/k)^2]^{1/2}. \tag{11.3.16}$$

Using $A = \cos(bz)$ and $B = \sin(bz)/b$ for a GRIN medium, we can write w_{eff} as

$$w_{\text{eff}}(z) = w_0\sqrt{f(z)}, \qquad f(z) = \cos^2(bz) + C_p^2 \sin^2(bz), \tag{11.3.17}$$

where we defined the parameter C_p as $C_p = 2p/(kbw_0)$. This result allows us to write Eq. (11.3.15) in the simple form [17],

$$S(\mathbf{r}, \omega) = \frac{S_0(\omega)}{f(z)} \exp\left[-\frac{(x^2 + y^2)}{w_0^2 f(z)}\right]. \tag{11.3.18}$$

The form of Eq. (11.3.18) is identical to that obtained for a coherent Gaussian beam in Section 2.5.3. The only difference lies in the definition of $f(z)$, where C_p appears in place of C_f. We can relate C_p to C_f noting that the constant p, defined in Eq. (11.3.7) depends on both w_0 and the coherence radius σ_c. For a fully coherent beam, $\sigma_c \to \infty$ and p becomes $p = 1/(2w_0)$. In this limit, C_p is reduced to the parameter C_f, defined as $C_f = w_g^2/w_0^2$ in Section 2.5.3, where $w_g = 1/\sqrt{kb}$ is the spot size of the fundamental mode of the GRIN medium. In the case of a partially coherent beam, C_p can be written as

$$C_p = C_f\sqrt{1 + 2w_0^2/\sigma_c^2}. \tag{11.3.19}$$

Equation (11.3.18) is remarkable because it shows that even a partially coherent Gaussian beam evolves in a self-similar fashion and maintains its Gaussian nature inside a GRIN medium. More specifically, the intensity of each spectral component of a partially coherent beam follows a periodic pattern that is similar to that of a coherent beam with one major difference: the parameter C_f is replaced with C_p for partially coherent beams. Whereas a coherent beam undergoes a focusing phase and acquires a small spot size in the focal plane ($C_f < 1$ in practice), a partially coherent beam does not compress much and can even become broader than the input beam at the focal distance, depending on its spatial coherence governed by the ratio σ_c/w_0. The factor C_p in Eq. (11.3.19) exceeds 1 when σ_c is smaller than w_0. It is possible to relate the square-root term in Eq. (11.3.19) to the M^2 factor (indicative of the beam's focusing ability by a lens) associated with such a partially coherent beam as [19]

$$M^2 = \sqrt{1 + 2w_0^2/\sigma_c^2} = C_p/C_f. \tag{11.3.20}$$

As an example, Figure 11.2 shows the width ratio w_{eff}/w_0 as a function of bz for three values of σ_c/w_0 using $C_f = 0.1$. A coherent beam's width would be smaller by a factor of 10 in the focal plane located at $bz = \pi/2$. As seen in Figure 11.2, almost no compression occurs for $\sigma_c/w_0 = 0.2$, and the beam expands by a factor of 2.8 when $\sigma_c/w_0 = 0.05$. It is easy to conclude that the spatial degree of coherence of an input beam affects considerably the evolution of spectral intensity inside a GRIN fiber.

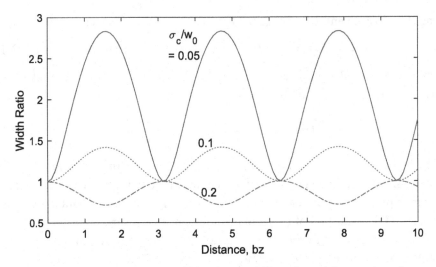

Figure 11.2 Variations in the width of a partially coherent beam plotted as a function of bz for three values of the ratio σ_c/w_0.

11.3.3 Changes in the Optical Spectrum

Spectral changes occurring in GRIN fibers for partially coherent beams were first studied in 1990, and we follow the approach taken in that work [15]. The origin of spectral changes lies in the frequency dependence of the function $f(z)$ defined in Eq. (11.3.17). This function depends on the parameter C_p given in Eq. (11.3.19) and the index-gradient parameter b, both of which can vary with the frequency ω.

How much these two parameters change with frequency depends on the bandwidth of the spectrum of the incident partially coherent beam. If this beam has a narrow spectrum, centered at a specific frequency ω_0 such that its bandwidth satisfies the condition $\Delta\omega \ll \omega_0$, it is a quasi-monochromatic beam with a high temporal coherence. In this case, the parameters C_p and b do not change much over the beam's bandwidth, and its spectrum also does not change much as the beam propagates down a GRIN medium.

The situation becomes quite different for an input beam that is temporally incoherent and has a spectrum wide enough that its bandwidth $\Delta\omega$ is comparable to ω_0. In this case, the spectrum of the beam changes with propagation inside a GRIN medium [15]. The reason can be understood by noting that the parameter C_p depends on ω inversely through $C_f = 1/(kbw_0^2$, where $k = n_0\omega/c$. Moreover, as most GRIN media employ a dispersive medium such as silica glass, the parameters n_0 and b also change with frequency.

To analyze the extent of spectral modification, we choose a fixed distance that corresponds to the first focal plane of the GRIN medium [$L_f = \pi/(2b)$]. Focusing

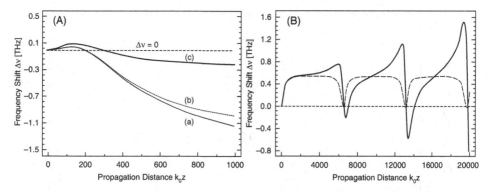

Figure 11.3 (A) Spectral shift of a partially coherent beam plotted as a function of distance in three cases using $k_0\sigma_c = 10$ and $k_0w_0 = 30$: (a) GRIN fiber, (b) step-index fiber, (c) free space. (B) Aperiodic spectral shifts occur when all dispersive effects are included. The dashed curve shows the periodic case. (After Ref. [15]; ©1990 Optica.)

on an on-axis point in this plane and setting $x = y = 0$ in Eq. (11.3.18), we can introduce a spectral modification factor as

$$M(\omega) = \frac{S(L_f, \omega)}{S_0(\omega)} = \frac{1}{C_p^2} = \frac{(kbw_0^2)^2}{1 + 2w_0^2/\sigma_c^2}. \tag{11.3.21}$$

Let us assume that the initial spot size w_0 is chosen such that $M_0 = M(\omega_0) = 1$ at the central frequency ω_0 of the beam's spectrum. If we ignore the chromatic dispersion for the moment, we find that the ratio $M(\omega)/M_0$ scales with the frequency as $(\omega/\omega_0)^2$. Thus, $M(\omega)$ will be larger at frequencies $\omega > \omega_0$ and smaller at frequencies lower than ω_0. As a result, the spectral peak would shift toward the blue side at the focal point compared to the input spectrum at $z = 0$. At other distances, spectrum can also shift toward the red side.

The magnitude of such shifts was calculated in the 1990 study [15] for a typical GRIN fiber using realistic parameter values and taking into account frequency dependence of $n_0(\omega)$. The input spectrum was centered at 532 THz and had a Lorentzian shape with the 34-THz bandwidth. Figure 11.3(A) shows the calculated spectral shifts over short distances (< 1 mm) in three cases using $k_0\sigma_c = 10$ and $k_0w_0 = 30$, where $k_0 = \omega_0/c$. As seen there, the spectral shift remains < 1 THz but is enhanced compared to the free-space case. Part (B) shows that spectral shifts occurring over longer distances exceed the self-imaging period. The spectrum evolves in a periodic fashion when the parameters n_0 and b are treated as constants, as it should from our discussion in Section 11.2.2. However, it does not retain this periodic pattern when the frequency dependence of n_0 and b is included. The spectrum at any distance also becomes asymmetric with more energy on the red or blue side of the central frequency, depending on the distance.

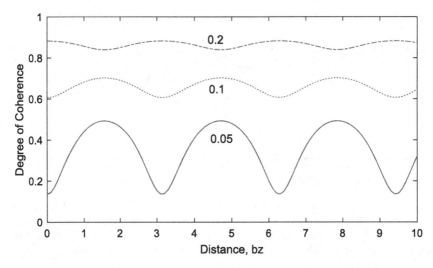

Figure 11.4 Variations in the spectral degree of coherence of a partially coherent Gaussian beam propagating inside a GRIN fiber for three values of the ratio σ_c/w_0.

11.3.4 Spectral Degree of Coherence

The cross-spectral density found in Section 11.3.1 can be used to calculate the spectral degree of coherence using its definition Eq. (11.1.21). The resulting expression is quite complex, but it is simplified considerably when we choose the two points to be on opposite sides of the central axis of the GRIN medium such that $\mathbf{r}_1 = (x, y, z)$ and $\mathbf{r}_2 = (-x, -y, z)$. In this case, the final result can be written as [17]

$$\mu(d, z, \omega) = \exp\left[-\left(p^2 - \frac{1}{4w_0^2}\right)\frac{d^2}{f(z)}\right] = \exp\left[-\frac{d^2}{2\sigma_c^2 f(z)}\right], \quad (11.3.22)$$

where $d = |\mathbf{r}_1 - \mathbf{r}_2| = 2(x^2 + y^2)^{1/2}$ and the function $f(z)$ is defined as in Eq. (11.3.17). The preceding equation shows that the degree of coherence maintains its Gaussian form, but the coherence radius of the beam changes inside the GRIN medium as

$$\sigma(z) = \sigma_c[\cos^2(bz) + C_p^2 \sin^2(bz)]^{1/2}, \qquad C_p = C_f\sqrt{1 + 2w_0^2/\sigma_c^2}. \quad (11.3.23)$$

Notice that $\sigma(z)$ changes with z inside a GRIN medium in the same fashion as the width of a partially coherent Gaussian beam does. The same periodic function that governs the self-imaging phenomenon ensures that the degree of coherence takes its input value at distances such that $bz = m\pi$ for all integer values of $m > 0$.

As an example, Figure 11.4 shows that the degree of coherence varies in a periodic fashion inside a GRIN fiber by plotting μ as a function of bz for three values of the ratio σ_c/w_0 for $d = 0.1w_0$ and $C_f = 0.1$. The input beam is a partially coherent Gaussian beam with a spot size w_0 and coherence radius σ_c. As expected, μ changes in a periodic fashion with the self-imaging period ($L_p = 2\pi/b$). Within each period, beam's coherence is enhanced or reduced, depending on the value of σ_c/w_0, as the Gaussian beam goes through the focal points during each cycle. The beam recovers its original value of μ in the middle and at the end of each self-imaging cycle. Considerable enhancement of coherence at the focal points occurs when the coherence radius of the input beam is relatively small.

11.4 Partially Polarized Gaussian Beams

As we discussed in Section 11.1.2, a partially coherent beam can also be partially polarized. The nature of partial polarization is quantified through the degree of polarization, defined in Eq. (11.1.14) in terms of the elements of a polarization matrix. It was found as early as 1973 that the degree of polarization of an optical beam can change during its propagation in a homogeneous medium, including vacuum [6]. Recently, the degree of polarization was also found to change when an optical beam is focused by a lens [22]. As a GRIN medium acts as a lens, one can expect the degree of polarization to evolve in a periodic fashion inside a GRIN medium. This was indeed the case in a 2006 study devoted to GRIN fibers [17].

11.4.1 Polarization Matrix of an Optical Beam

Before we can calculate the degree of polarization, we need to find the four elements of the polarization matrix associated with a partially coherent vector beam. Rather than using the time-domain polarization matrix given in Eq. (11.1.7), we use its frequency-domain version in the form

$$J(\mathbf{r}) = \begin{pmatrix} W_{xx}(\mathbf{r}, \mathbf{r}, \omega) & W_{xy}(\mathbf{r}, \mathbf{r}, \omega) \\ W_{yx}(\mathbf{r}, \mathbf{r}, \omega) & W_{yy}(\mathbf{r}, \mathbf{r}, \omega) \end{pmatrix}, \tag{11.4.1}$$

where the cross-spectral density W_{ij} for $i, j = x, y$ is related to the components of the electric field as indicated in Eq. (11.1.15).

The evolution of W_{ij} inside a GRIN medium can be studied by noting that Eq. (11.2.3) applies to and element W_{ij} of the cross-spectral density, i.e,

$$W_{ij}(\mathbf{r}, \mathbf{r}, \omega) = \iint_{-\infty}^{\infty} K^*(\mathbf{r}, \mathbf{s}_1) K(\mathbf{r}, \mathbf{s}_2) W_{ij}(\mathbf{s}_1, \mathbf{s}_2, \omega) \, d\mathbf{s}_1 \, d\mathbf{s}_2, \tag{11.4.2}$$

where the propagation kernel is given in Eq. (11.2.2). Thus, if we know W_{ij} at the input plane at $z = 0$, we can find it at any distance z inside a GRIN medium. Following Section 11.2.1, we employ the Gaussian–Schell model to specify $W_{ij}(\mathbf{s}_1, \mathbf{s}_2, \omega)$ and use

$$W_{ij}(\mathbf{s}_1, \mathbf{s}_2, \omega) = [S_i(\mathbf{s}_1, \omega) S_j(\mathbf{s}_2, \omega)]^{1/2} \mu_{ij}(\mathbf{s}_2 - \mathbf{s}_1, \omega), \tag{11.4.3}$$

together with the Gaussian functions $(i, j = x, y)$

$$S_i(\mathbf{s}, \omega) = A_i^2(\omega) \exp(-s^2/w_0^2), \qquad \mu_{ij}(\mathbf{s}, \omega) = B_{ij} \exp(-s^2/2\sigma_{ij}^2), \qquad (11.4.4)$$

where the parameters A_i, w_0, B_{ij}, and σ_{ij} may depend on frequency.

Physically, A_x and A_y in Eq. (11.4.4) represent the amplitudes of the two polarization components of an input Gaussian beam with the spot size w_0. The parameters $B_{xx} = B_{yy} = 1$, but B_{xy} is a complex number whose magnitude $|B_{xy}|$ is less than 1 and governs the degree of polarization of the input beam. The phase of B_{xy} is related to the state of the polarization (SOP) of the input beam. Also, σ_{ij} for $i, j = x, y$ are the coherence radii that govern spatial coherence of the input beam. In general, σ_{xy} is different from the two diagonal components, σ_{xx} and σ_{yy}, which may be equal to each other. It should be stressed that not all parameters can be chosen arbitrarily because the cross-spectral density has requirements that impose constraints on some of these parameters.

The four-fold integral in Eq. (11.4.2) can be carried out following the details given in Section 11.3 in the scalar case. The final result can be written in the form [17]

$$W_{ij}(\mathbf{r}, \mathbf{r}, \omega) = A_i A_j \frac{B_{ij}}{f_{ij}} \exp\left[-\frac{\rho^2}{w_0^2 f_{ij}}\right], \qquad (11.4.5)$$

where ρ is the radial distance and the function $f_{ij}(z)$ is defined as

$$f_{ij}(z) = \cos^2(bz) + C_f^2(1 + 2w_0^2/\sigma_{ij}^2) \sin^2(bz). \qquad (11.4.6)$$

The four elements of the polarization matrix are obtained using the relation

$$J_{ij} = W_{ij}(\mathbf{r}, \mathbf{r}, \omega), \quad i, j = x, y. \qquad (11.4.7)$$

11.4.2 Changes in the Degree of Polarization

We use Eq. (11.1.14) to calculate the degree of polarization. Using $\mathrm{Tr}\, J = J_{xx} + J_{yy}$ and $\mathrm{Det}\, J = J_{xx}J_{yy} - |J_{xy}|^2$, we can write this equation as

$$\mathscr{P}(\rho, z) = \frac{[(J_{xx} - J_{yy})^2 + 4|J_{xy}|^2]^{1/2}}{J_{xx} + J_{yy}}. \qquad (11.4.8)$$

Equations (11.4.5) through (11.4.8) are used to calculate the degree of polarization, $\mathscr{P}$, of a partially coherent Gaussian beam inside a GRIN medium. The resulting expression is complicated but is simplified for an on-axis point where $\rho = 0$:

$$\mathscr{P}(0, z) = \frac{[(A_x^2/f_{xx} - A_y^2/f_{yy})^2 + (2A_xA_y|B_{xy}|/f_{xy})^2]^{1/2}}{(A_x^2/f_{xx} + A_y^2/f_{yy})}. \qquad (11.4.9)$$

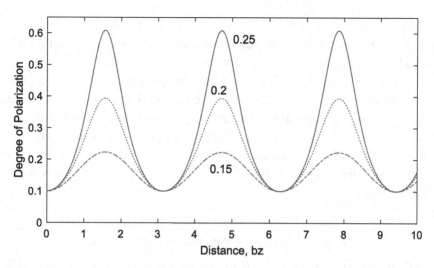

Figure 11.5 Variations in the degree of polarization of a partially coherent Gaussian beam propagating inside a GRIN fiber for three values of the ratio σ_{xy}/w_0 at a fixed value of $\sigma_c/w_0 = 0.1$.

Equations (11.4.9) becomes simpler for an input beam that is either circularly polarized or linearly polarized at a 45° angle so that $A_x = A_y$. If the coherence radii are also the same for the two polarization components ($\sigma_{xx} = \sigma_{yy} = \sigma_c$), Eq. (11.4.9) is reduced to

$$\mathscr{P}(0, z) = \mathscr{P}_0 \frac{f_{xx}(z)}{f_{xy}(z)} = \mathscr{P}_0 \frac{\cos^2(bz) + C_f^2(1 + 2w_0^2/\sigma_c^2)\sin^2(bz)}{\cos^2(bz) + C_f^2(1 + 2w_0^2/\sigma_{xy}^2)\sin^2(bz)}, \qquad (11.4.10)$$

where $\mathscr{P}_0 = |B_{xy}|$ is the degree of polarization of the input beam and σ_{xy} is the degree of coherence for the off-diagonal component of the polarization matrix. As expected, the degree of polarization evolves inside a GRIN medium in a periodic fashion with the self-imaging period $L_p = 2\pi/b$. Notice that no change in the degree of polarization occurs when $\sigma_{xy} = \sigma_c$. At the focal distance $z = L_p/4$, Eq. (11.4.10) provides the expression

$$\mathscr{P}(0, L_p/4) = \mathscr{P}_0 \left(\frac{2w_0^2 + \sigma_c^2}{2w_0^2 + \sigma_{xy}^2} \right) \frac{\sigma_{xy}^2}{\sigma_c^2}. \qquad (11.4.11)$$

As an example, Figure 11.5 shows how the degree of polarization varies in a periodic fashion inside a GRIN fiber by plotting $\mathscr{P}(0, z)$ as a function of bz for three values of the ratio σ_{xy}/w_0 using $P_0 = 0.1$, $\sigma_c/w_0 = 0.1$, and $C_f = 0.1$. The input beam is a partially coherent Gaussian beam with the spot size w_0 and is linearly polarized with equal amplitudes along the x and y directions ($A_x = A_y$). It is evident that the degree of polarization increases considerably as the beam goes through the first focal point located at $bz = \pi/2$, but it recovers its original value at a distance of $bz = \pi = L_p/2$, where L_p is the self-imaging period. This process repeats in the second half of each period. As seen from Eq. (11.4.11), the cross-correlation between the beam's orthogonally polarized components, governed by the magnitude of the coherence radius σ_{xy}, plays an important role in the enhancement of the degree of polarization at each focal point. Its magnitude should be larger than the value of σ_c, which represents the coherence radius of each polarization component of the Gaussian beam.

11.4.3 Changes in the State of Polarization

The SOP of a partially coherent beam also changes with propagation inside a GRIN medium, in addition to its degree of polarization [18]. We can study SOP changes using the average Stokes parameters because they are related to the elements of the polarization matrix as [see Eq. (11.1.9)]

$$S_1 = J_{xx} - J_{yy}, \qquad S_2 = J_{xy} + J_{yx}, \qquad S_3 = i(J_{yx} - J_{xy}). \qquad (11.4.12)$$

Noticing that $J_{ij} = W_{ij}$, we can use W_{ij} given in Eq. (11.4.5) to find the three components of the Stokes vector. It follows that these parameters evolve in a periodic fashion inside a GRIN medium because of the self-imaging phenomenon. As a result, the SOP of the incident Gaussian beam also changes in a periodic fashion such that it recovers its original SOP at distances such that $z = m(L_p/2)$, where m is an integer.

As an example, Figure 11.6 shows how the three Stokes parameter, normalized with respect to $S_0 = J_{xx} + J_{yy}$, vary along the axis of a GRIN fiber ($\rho = 0$) by plotting them as a function of bz. The partially polarized Gaussian beam is initially elliptically polarized and has the following values for its relevant parameters: $A_1 = A_2 = 1$, $B_{xy} = 0.3e^{i\pi/3}$, $\sigma_{xx} = 0.12$, $\sigma_{yy} = 0.1$, $\sigma_{xy} = 0.2$, and $C_f = 0.2$. For these parameters, the two polarization components of the input beam have equal amplitudes but differ in phase by 60°. Moreover, these components are correlated such that the initial degree of polarization is 0.3. The three Stokes parameters change considerably as the beam goes through its first focal point located at $bz = \pi/2$, but they recover their original value at a distance of $bz = \pi = L_p/2$, where L_p is the self-imaging period. At longer distances, this pattern repeats in a periodic fashion.

It is possible to relate the elements J_{ij} of the polarization matrix to the ellipticity and the orientation of the ellipse describing the SOP of the beam at any distance inside the GRIN medium [19]. The results show that both of these quantities also evolve in a periodic fashion. The main conclusion one can draw is that all properties

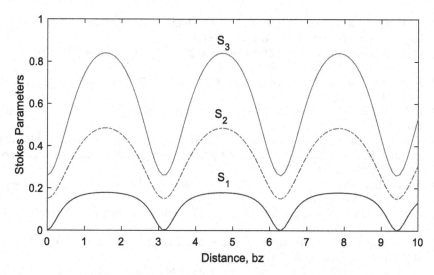

Figure 11.6 Changes in the on-axis values of the three Stokes parameter, S_1, S_2, and S_3 (normalized with respect to S_0), as a function of distance for a partially polarized Gaussian beam propagating inside a GRIN fiber. The parameter values are specified in the text.

of a partially coherent beam, including the SOP and the degree of polarization, follow a periodic evolution pattern dictated by the self-imaging phenomenon that is a unique property of any GRIN medium with a parabolic index profile. Recent work has shown that this conclusion applies even to other types of partially coherent beams, including vortex beams [23–25].

References

[1] E. Wolf, *Introduction to the Theory of Coherence and Polarization of Light* (Cambridge University Press, 2007).

[2] L. Mandel and E. Wolf, *Optical Coherence and Quantum Optics* (Cambridge University Press, 1995).

[3] M. Born and E. Wolf, *Principles of Optics*, 7th ed. (Cambridge University Press, 2020).

[4] P. H. van Cittert, *Physica* **1**, 201 (1934); https://doi.org/10.1016/S0031-8914(34) 90026-4.

[5] F. Zernike, *Physica* **5**, 785 (1938); https://doi.org/10.1016/S0031-8914(38)80203-2.

[6] A. K. Jaiswal, G. P. Agrawal, and C. L. Mehta, *Nuovo Cimento* **15**, 295 (1973); https://doi.org/10.1007/BF02894787.

[7] D. F. V. James, *J. Opt. Soc. Am. A* **11**, 1641 (1994); https://doi.org/10.1364/JOSAA.11 .001641.

[8] G. P. Agrawal and E. Wolf, *J. Opt. Soc. Am. A* **17**, 2019 (1994); https://doi.org/10 .1364/JOSAA.17.002019.

[9] G. P. Agrawal, A. K. Ghatak, and C. L. Mehta, *Opt. Commun.* **12**, 333 (1974); https://doi.org/10.1016/0030-4018(74)90028-5.

[10] D. J. Carpenter and C. Pask, *Opt. Commun.* **22**, 993 (1977); https://doi.org/10.1016/0030-4018(77)90256-5.

[11] S. Piazzolla and G. De Marchis, *Opt. Commun.* **32**, 380 (1980); https://doi.org/10.1016/0030-4018(80)90264-3.

[12] E. Wolf, *Phys. Rev. Lett.* **56**, 1370 (1986); https://doi.org/10.1103/PhysRevLett.56.001370.

[13] S. A. Ponomarenko, *Opt. Lett.* **40**, 566 (2015); https://doi.org/10.1364/OL.40.000566.

[14] M. Imai, S. Satoh, and Y. Ohtsuka, *J. Opt. Soc. Am. A* **3**, 86 (1986); https://doi.org/10.1364/JOSAA.3.000086.

[15] A. Gamliel and G. P. Agrawal, *J. Opt. Soc. Am. A* **7**, 2184 (1990); https://doi.org/10.1364/JOSAA.7.002184.

[16] P. Hlubina, *J. Mod. Opt.* **40**, 1893 (1993); https://doi.org/10.1080/09500349314551921.

[17] H. Roychowdhury, G. P. Agrawal, and E. Wolf, *J. Opt. Soc. Am. A* **23**, 940 (2006); https://doi.org/10.1364/JOSAA.23.000940.

[18] Y. Zhu and D. Zhao, *J. Opt. Soc. Am. A* **25**, 1944 (2008); https://doi.org/10.1364/JOSAA.25.001944.

[19] S. Zhu, L. Liu, Y. Chen, and Y. Cai, *J. Opt. Soc. Am. A* **30**, 2306 (2013); https://doi.org/10.1364/JOSAA.30.002306.

[20] A. Efimov, *Opt. Express* **22**, 15577 (2014); https://doi.org/10.1364/OE.22.015577.

[21] S. A. Wadood, K. Liang, G. P. Agrawal et al., *Opt. Express* **29**, 21240 (2021); https://doi.org/10.1364/OE.425921.

[22] X. Zhao, T. D. Visser, and G. P. Agrawal, *Opt. Lett.* **43**, 2344 (2018); https://doi.org/10.1364/OL.43.002344.

[23] J. Wang, S. Yang, M. Guo, Z. Feng, and J. Li, *Opt. Express* **28**, 4661 (2020); https://doi.org/10.1364/OE.386167.

[24] M. Dong, Yi. Bai, J. Yao, Q. Zhao, and Y. Yang, *Appl. Opt.* **59**, 8023 (2020); https://doi.org/10.1364/AO.396329.

[25] Y. Gao, X. Gao, R. Lin et al., *Optik* **228**, 165755 (2021); https://doi.org/10.1016/j.ijleo.2020.165755.

Appendix A

Quantum Harmonic Oscillator

The Schrödinger equation for a one-dimensional harmonic oscillator takes the form [1–3]:

$$i\hbar \frac{\partial \Psi}{\partial t} = -\frac{\hbar^2}{2m_p} \frac{\partial^2 \Psi}{\partial x^2} + \frac{1}{2} m_p \omega^2 x^2 \Psi, \tag{A.1}$$

where m_p is the mass and ω is the frequency of the harmonic oscillator. Also, $\Psi(x, t)$ is the wave function and $\hbar$ is related to Planck's constant h as $\hbar = h/(2\pi)$. It may appear surprising at first sight that this quantum problem is related to the classical problem of light propagation inside a GRIN medium, for which Planck's constant plays no role.

To reveal the connection between a harmonic oscillator and a GRIN medium, it is useful to transform the variables x and t in Eq. (A.1) into dimensionless forms using

$$q = (m_p \omega/\hbar)^{1/2} x, \qquad \tau = \omega t. \tag{A.2}$$

In terms of these variables, Eq. (A.1) takes the form

$$2i \frac{\partial \Psi}{\partial \tau} + \frac{\partial^2 \Psi}{\partial q^2} - q^2 \Psi = 0. \tag{A.3}$$

Note that $\hbar$ appears in the definition of q but not in Eq. (A.3). A comparison of Eq. (A.3) with Eq. (2.5.3) in Section 2.5 shows why the GRIN-medium problem is analogous to a quantum harmonic oscillator. The only difference is that optical fields evolve along the length of a GRIN medium, rather than evolving in time.

The Schrödinger equation (A.3) is solved in quantum mechanics by writing its solution in the form $\Psi(q, \tau) = \psi(q) e^{-iE\tau}$, where the eigenvalue E plays the role of total energy and $\psi(q)$ satisfies the eigenvalue equation:

$$\hat{H} |\psi\rangle = E |\psi\rangle, \qquad \hat{H} = \frac{1}{2}(\hat{p}^2 + \hat{q}^2), \tag{A.4}$$

where we adopted the Dirac's notation for the wave function, identified the Hamiltonian $\hat{H}$, and defined the momentum operator as $\hat{p} = -i(d/dq)$. If we now introduce the usual annihilation and creation operators as [1]

$$\hat{a} = (\hat{q} + i\hat{p})/\sqrt{2}, \qquad \hat{a}^\dagger = (\hat{q} - i\hat{p})/\sqrt{2}, \tag{A.5}$$

the Hamiltonian in Eq. (A.4) takes the form

$$\hat{H} = \hat{a}^\dagger \hat{a} + \tfrac{1}{2} = \hat{N} + \tfrac{1}{2}, \tag{A.6}$$

where $\hat{N} = \hat{a}^\dagger \hat{a}$ is the so-called number operator.

The number operator satisfies the eigenvalue equation, $\hat{N}|n\rangle = n|n\rangle$, where the integer n starts at 0. The states $|n\rangle$ are known as the number or Fock states [1]. They are also the eigenstates of the harmonic oscillator with the eigenvalues $E_n = n + \tfrac{1}{2}$.

The spatial representation of a Fock state provides the corresponding wave function $\psi_n(q) = \langle q|n\rangle$. To find its explicit form, we consider the action of the annihilation operator on the ground state and note that $\langle q|\hat{a}|0\rangle = 0$ because a lower state does not exist. Using $\hat{a}$ from Eq. (A.5), we obtain a simple differential equation

$$\frac{d\psi_0}{dq} + q\psi_0 = 0. \tag{A.7}$$

Its solution provides us the ground-state wave function in the form

$$\psi_0(q) = \langle q|0\rangle = \pi^{-1/4} \exp(-q^2/2), \tag{A.8}$$

where the prefactor stems from normalization of the state.

The expression for $\psi_0(q)$ can be used to obtain the wave function for all other Fock states by applying the creation operator $\hat{a}^\dagger$ successively to the ground state. The wave function for the nth excited state of a harmonic oscillator is found to be

$$\psi_n(q) = \langle q|n\rangle = N_n H_n(q) \exp(-q^2/2), \tag{A.9}$$

where H_n is the nth-order Hermite polynomial and the normalization constant N_n is given as $N_n = (\sqrt{\pi}\, 2^n n!)^{-1/2}$. In terms of the original variables, the Hermite-Gauss functions in Eq. (A.9) represent the energy eigenstates of a harmonic oscillator with the eigenvalues $E_n = (n + \tfrac{1}{2})\hbar\omega$.

Another set of states that are relevant for harmonic oscillators are known as the coherent states. They are denoted as $|\alpha\rangle$ and are the eigenstates of the annihilation operator, that is,

$$\hat{a}|\alpha\rangle = \alpha|\alpha\rangle, \tag{A.10}$$

where the eigenvalue α is a complex number because the operator $\hat{a}$ is not hermitian. A coherent state can be expanded in terms of the Fock states as [1]

$$|\alpha\rangle = \sum_n c_n|n\rangle, \qquad c_n = \frac{\alpha^n}{(n!)^{1/2}} \exp(-|\alpha|^2/2). \tag{A.11}$$

Coherent states can be generated by applying the displacement operator $D(\alpha)$ to the ground state of a harmonic oscillator such that

$$|\alpha\rangle = \hat{D}(\alpha)|0\rangle, \qquad D(\alpha) = \exp(\alpha\hat{a}^\dagger - \alpha^*\hat{a}). \tag{A.12}$$

They play an important role in optics because the quantum state of a laser operating far above its threshold is well represented by a coherent state $|\alpha\rangle$ such that $\alpha = Ae^{i\phi}$, where both the amplitude A and phase ϕ fluctuate with time.

Coherent states are useful for describing the quantum properties of coherent light emitted by a laser. This can be understood from the wave function $\psi_\alpha(q) = \langle q|\alpha\rangle$ associated with them. It follows from Eq. (A.10) that $\psi_\alpha(q)$ satisfies

$$\frac{1}{\sqrt{2}}\left(q + \frac{d}{dq}\right)\psi_\alpha(q) = \alpha\psi_\alpha(q), \tag{A.13}$$

Using $\alpha = (q_0 + ip_0)/\sqrt{2}$, where $q_0 = \langle\alpha|\hat{q}|\alpha\rangle$ and $p_0 = \langle\alpha|\hat{p}|\alpha\rangle$ are the average values of $\hat{q}$ and $\hat{p}$, the solution of Eq. (A.13) is found to be

$$\psi_\alpha(q) = \pi^{-1/4}\exp\left[-\tfrac{1}{2}(q - q_0)^2 + i(q - q_0)p_0\right]. \tag{A.14}$$

This coherent state represents a Gaussian wave packet centered at q_0. When one considers both transverse dimensions, such a coherent state can be associated with a Gaussian beam with the minimum uncertainty product allowed by quantum mechanics.

Harmonic oscillators constitute a well-studied topic in quantum mechanics. In addition to a standard harmonic oscillator, its perturbations have been analyzed using the tools such as perturbation theory, the WKB method, and the adiabatic approximation [1–3]. In this context, a harmonic oscillator whose frequency varies with time has attracted considerable attention and several techniques have been developed for its analysis. The optical analog of such an oscillator is a tapered GRIN fiber whose core's diameter changes along its length. See Section 7.3 for a discussion of how the techniques developed for quantum harmonic oscillators can be applied to tapered GRIN fibers.

References

[1] J. J. Sakurai and J. Napolitano, *Modern Quantum Mechanics*, 3rd ed. (Cambridge University Press, 2020).
[2] P. Blaise and O. Henri-Rousseau, *Quantum Oscillators* (Wiley, 2011).
[3] D. J. Griffiths and D. F. Schroeter, *Introduction to Quantum Mechanics*, 3rd ed. (Cambridge University Press, 2018).

Appendix B

Fractional Fourier Transform

The fractional Fourier Transform was introduced in 1980 as a generalization of the standard Fourier-transform operation [1]. Since then, this transform has been found useful in many different contexts [2–4]. Particularly noteworthy are its connections to the "propagator" of a quantum harmonic oscillator and to the propagation kernel of a GRIN medium.

Following the notation in Ref. [1], the standard Fourier transform and its inverse operation can be written in an operator form as

$$\hat{F}f(x) = \frac{1}{\sqrt{2\pi}} \int_{-\infty}^{\infty} f(x') e^{ixx'} \, dx', \tag{B.1}$$

$$\hat{F}^{-1}f(x) = \frac{1}{\sqrt{2\pi}} \int_{-\infty}^{\infty} f(x') e^{-ixx'} \, dx', \tag{B.2}$$

where $\hat{F}$ is the Fourier-transform operator and the prefactor has been added to symmetrize the two operations. The eigenfunctions of the operator $\hat{F}$, found by solving the eigenvalue equation [1]

$$\hat{F}\psi_n(x) = \lambda_n \psi_n(x), \tag{B.3}$$

are the Hermite-Gaussian function with the eigenvalues λ_n such that

$$\psi_n(x) = H_n(x) \exp(-x^2/2), \qquad \lambda_n = i^n = \exp(in\pi/2). \tag{B.4}$$

The fractional Fourier Transform of order α is defined [1] by replacing the $\pi/2$ factor in Eq. (B.4) with α and using the eigenvalue equation

$$\hat{F}_\alpha \psi_n(x) = e^{in\alpha} \psi_n(x). \tag{B.5}$$

By definition, Eq. (B.5) reduces to the standard Fourier transform when $\alpha = \pi/2$. To determine how F_α transforms an arbitrary function $f(x)$, we expand $f(x)$ in terms of the Hermite–Gauss functions as

$$f(x) = \sum_n c_n \psi_n(x), \qquad c_n = \int_{-\infty}^{\infty} \psi_n(x) f(x)\, dx, \tag{B.6}$$

where the expansion coefficients c_n were found using the orthogonality relation for the Hermite–Gauss functions. Applying $\hat{F}_\alpha$ operator to Eq. (B.6) leads to

$$\hat{F}_\alpha f(x) = \sum_n c_n e^{in\alpha} \psi_n(x). \tag{B.7}$$

When we substitute c_n from Eq. (B.6) and use the Mehler's formula given in Eq. (3.2.9) to perform the sum, we obtain

$$\hat{F}_\alpha f(x) = \sqrt{\frac{ie^{-i\alpha}}{2\pi \sin \alpha}} \int_{-\infty}^{\infty} \exp\left(-\frac{i}{2\sin\alpha}[\cos\alpha(x^2 + x'^2) - 2xx']f(x')\, dx'\right). \tag{B.8}$$

The inverse operation consists of simply replacing i with $-i$ and takes the form

$$\hat{F}_\alpha^{-1} f(x) = \sqrt{\frac{-ie^{i\alpha}}{2\pi \sin \alpha}} \int_{-\infty}^{\infty} \exp\left(\frac{i}{2\sin\alpha}[\cos\alpha(x^2 + x'^2) - 2xx']f(x')\, dx'\right). \tag{B.9}$$

It is easy to verify that, when $\alpha = \pi/2$, both operations reduces to the standard Fourier-transform relations given in Eqs. (B.1) and (B.2).

The connection of the fractional Fourier Transform to the "propagator" of a quantum harmonic oscillator was already noted in Ref. [1]. As we saw in Appendix A, the propagation of optical beams inside a GRIN medium is closely related to quantum harmonic oscillator. Indeed, the propagation kernel of a GRIN medium found in Section 3.2.1 is related to the fractional Fourier Transform of the input spatial distribution in two transverse dimensions. In this context, the parameter that plays the role of α is the dimensional distance bz in Eq. (3.2.10). The standard Fourier transform corresponds to the choice $bz = \pi/2$ in that equation, a distance that corresponds to the focal plane of a GRIN lens. At other distances, a GRIN lens performs a fractional Fourier Transform on the optical field incident on it.

References

[1] V. Namias, *J. Inst. Maths Applics.* **25**, 241 (1980); https://doi.org/10.1093/imamat/25.3.241.

[2] D. Mendlovic and H. M. Ozaktas, *J. Opt. Soc. Am. A* **10**, 1875 (1993); https://doi.org/10.1364/JOSAA.10.001875.

[3] A. W. Lohmann, *J. Opt. Soc. Am. A* **10**, 2181 (1993); https://doi.org/10.1364/JOSAA.10.002181.

[4] H. M. Ozaktas, Z. Zalevsky, and M. A. Kutay, *The Fractional Fourier Transform: With Applications in Optics and Signal Processing* (Wiley, 2001).

Index

Printed in the United States
by Baker & Taylor Publisher Services